A Leverhulme Public Lecture Series

GHOSTLY MUSCLES, WRINKLED BRAINS, HERESIES AND HOBBITS

A Leverhulme Public Lecture Series

GHOSTLY MUSCLES, WRINKLED BRAINS, HERESIES AND HOBBITS

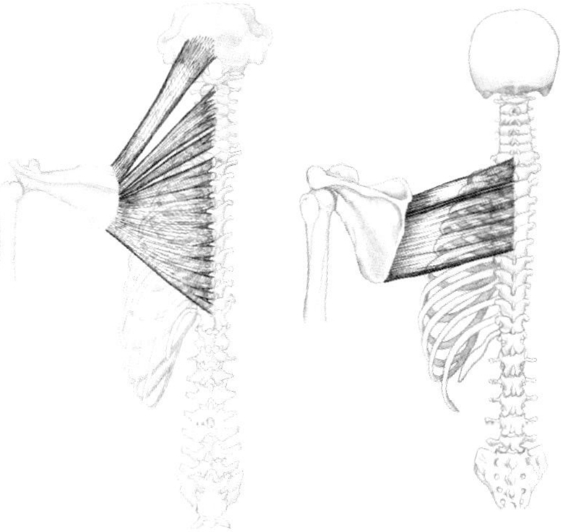

CHARLES OXNARD

The University of Western Australia

World Scientific

NEW JERSEY · LONDON · SINGAPORE · BEIJING · SHANGHAI · HONG KONG · TAIPEI · CHENNAI

Published by

World Scientific Publishing Co. Pte. Ltd.

5 Toh Tuck Link, Singapore 596224

USA office: 27 Warren Street, Suite 401-402, Hackensack, NJ 07601

UK office: 57 Shelton Street, Covent Garden, London WC2H 9HE

Library of Congress Cataloging-in-Publication Data

Oxnard, Charles E., 1933-

Ghostly muscles, wrinkled brains, heresies, and Hobbits : a Leverhulme public lecture series / Charles Oxnard.

p. ; cm.

ISBN-13: 978-981-279-742-1 (hardcover)

ISBN-10: 981-279-742-4 (hardcover)

ISBN-13: 978-981-279-743-8 (softcover)

ISBN-10: 981-279-743-2 (softcover)

1. Anatomy. 2. Human evolution. 3. Physical anthropology. I. Title.

[DNLM: 1. Anatomy. 2. Evolution. 3. Adaptation, Physiological. 4. Anthropology, Physical. 5. Research--methods. GN 281 O98g 2008]

QM23.2.O96 2008

611--dc22

2008033972

British Library Cataloguing-in-Publication Data

A catalogue record for this book is available from the British Library.

Typeset by Stallion Press

Email: enquiries@stallionpress.com

Printed by FuIsland Offset Printing (S) Pte Ltd, Singapore

To my wife Eleanor who has
been in my work and my life for 50 years
and
To the memory of Peter Lisowski,
my colleague and friend for even longer.

Preface

Why This Particular Book?

After I retired I received a Leverhulme Trust Professorship to be held for 1 year in the Department of Anatomy and Developmental Biology, University College, London, and the Department of Human Anatomy and Cell Biology, University of Liverpool. Because I was unable to take a continuous year due to research and teaching commitments in Australia, the Trust graciously allowed me to take up the appointment spread part-time over 3 years: 2001–2004. This permitted me to work with Professors Robin Crompton in Liverpool and Paul O'Higgins then at University College (but now Foundation Professor of Anatomy at the University of York and the Hull York Medical School). The Trust required me to collaborate in research and give research seminars.

However, the Trust also required a series of Leverhulme Lectures. These lectures were to be available to the general public as well as to the academic colleagues and students, but still to explicate my latest research ideas. A tall order! Yet it seemed worth trying.

My work in retirement is also supported by the University of Western Australia through my appointment in 1998 as Emeritus Professor and Senior Honorary Research Fellow in the School of Anatomy and Human Biology, and, in 2006, as Adjunct Professor in the Centre for Forensic Science. In these positions I have obtained continuous Australian Research Council Large and Discovery Grants, the most recent of which goes to 2009. In addition, with Professors O'Higgins and Crompton, and also Dr Michael Fagan (University of Hull) I am a partner in a Leverhulme Trust Research Grant 2004–2006, two Marie Curie Research and Research Training Grants 2005–2008 and a BBSRC, UK, Research Grant 2007–2009. These grants have coincided with my appointments as Honorary Professor of Anatomy in the Hull/York Medical School, and Honorary Professor of Bio-Engineering in the University of Hull.

My work has also been recognised by the Charles Darwin Lifetime Award of the American Association of Physical Anthropology in 2001, a Cambridge University Press Volume 'Shaping Primate Evolution: Form, Function and Behavior', edited by Fred Anapol, Rebecca German and Nina Jablonski, 2004, and the Chancellor's Medal of the University of Western Australia in 2008.

The Leverhulme Trust has agreed that I can include the words 'A Leverhulme Public Lecture Series' as part of the book title.

I have discovered over my lifetime that both students and the public are fascinated by the inside stories of how discoveries are made. Peter Medawar once asked the question: 'Is the scientific paper a fraud?' By this he did not mean that scientific papers were actually fraudulent (though scientific frauds have been perpetrated); rather he meant that the way scientific papers describe research is rarely the way the research actually happened. In this book, in each chapter I have attempted to do what is rarely done, that is, to describe how that chapter's problem actually arose, how the methods used in its examination came about, who and how were collaborators involved, what were the twists and turns of thought involved in the story, what errors were perpetrated (for there are always errors!), what partial solutions have so far appeared, and what new ideas or changed directions or even reversals(!) stem from the work.

I believe it most important never to be wed so firmly to currently popular ideas that the mind is closed to new facts, new interpretations and new possibilities. In particular, I believe I should never be afraid to try new techniques of analysis, though more and more, as I get older and older, I am dependent upon younger colleagues for help in this regard.

Yet, dependent upon them as I now am, I recognise that the younger doctoral student may have a very difficult row to hoe. He, and, nowadays increasingly she, has to work on a single problem for two, three, even sometimes more years. Inevitably, periods occur when the work seems boring, when the muse fails, when the black dog sets in. Of course, these negative periods regress and students do finish. The beauty of being an older academic is, however, that one can have several problems going at the same time. When I run into a block, I merely move over to another problem until the block resolves. I have always thought of this through food. It is important to have several 'bread-and-butter' problems going at once. These are the problems that are highly likely to yield answers. It is also important to have some 'cake' problems (if they have no bread, let them eat cake!). These are somewhat more problematical; the answers are not so obvious, and may indeed be quite surprising. They are, however, still likely, if less so, to be successful. It is further important to have a few 'pie-in-the-sky' problems. These are highly unlikely to be successful; they are, conversely, rather likely to antagonise more conservative colleagues; but if they work out: jack-pot!

Further, the researches of a younger academic are usually relatively linear following a specific line of thought. But as I have become older and worked on problems for longer and longer, the research pathways have become evermore mazelike, evermore complex, with greater interactions among the parts, with unexpected twists and turns along the way, and with a gradual knitting together of questions that initially seemed very distant from one another. I take great delight in finding both the great complexity in what is generally seen as simple, and the simplicity that can often be found in what seems to be hopelessly complex. I feel that I must never be satisfied with the popular version of what is true or false, and, what is also most important, I must never be satisfied that my own efforts have revealed the truth. In this regard I am an unrepentant Popperian; I truly wish to falsify other people's ideas; but I also truly wish to falsify my own ideas. Such an approach should not, however, fall into the trap of negativism. In particular, one should not believe that there is just one recipe (often called nowadays 'world's best practice!') for tackling particular problems. Rather, there is no one world's best practice; there are many different paths that lead towards the acquisition of new evidence; sometimes less rigorous paths, creatively applied, may actually supersede the so-called 'gold standards' applied in a less 'thinking' manner.

One might think that all these are my plea to accept my interpretations. Do not be misled. They are not. They are a plea to do science, and in the doing to enjoy it. As one does science, the part played by serendipity becomes evermore evident. Research is very often a kind of Alice-in-Wonderland task. It is the excitement and complexity and surprise of my little bit of science that I am trying to present in this book.

Acknowledgements

I am grateful to many colleagues and graduate students who, over the years, have participated in my investigations or allowed me to participate in theirs. In addition, I am indebted to several individuals for permitting me to describe their own investigations. All these individuals are cited in the text.

Most of these investigations could not have been carried out without using materials from a number of institutions on four continents. These include: the Powell Cotton Museum, Birchington, UK; the British Museum, Natural History, London, UK; the Field Museum, Chicago, USA; the Los Angeles County Museum, Los Angeles, USA; the Western Australian Museum, Perth, Australia; the Royal College of Surgeons of Edinburgh; the Royal College of Surgeons of England; and the Anatomy Departments of the Universities of Birmingham, Chicago, Southern California, Hong Kong and Western Australia.

I am especially indebted to the late Professor F. Peter Lisowski, Dr Len Freedman, Dr Paul O'Higgins, Professor Robin Crompton, Professor Robert Kidd, Dr Daniel Franklin, Dr Ruliang Pan, Dr Jens Hirschberg and Dr Ken Wessen for the discussions on problems of primate morphology and evolution. In addition, Professor F. Peter Lisowski, Dr Len Freedman and Dr Daniel Franklin are thanked for reading and criticising the text.

I especially wish to document that these ideas have also arisen through discussions with various scientists in the broader area of morphology, especially Professors Brian Hall, Rebecca German, Nina Jablonski and Fred Bookstein. Some of the initial ideas came from discussions years ago with Professor Jacobson during a visit to the University of Utah. They were extended by an invitation to contribute to the Australian Academy of Science Discussion meeting, commemorating the late Professor N. G. W. Macintosh, held in Sydney in 1992. They were further explored in my preparation for keynote lectures for the Wenner Gren Primate Locomotion Conference at the University of California at Davis in 1995, the Linnean Society Symposium on Development Growth and Evolution in 1998 (published 2000), the Beijing meeting of the International Primate Congress in 2002, the Workshop for the 80th birthday of Luis Cavalli–Sforza, at the Institute for Pure Applied Mathematics, UCLA, Los Angeles, also 2002, and the Morphofest at the University of Vienna, 2006. They especially owe much to the information garnered during my attendance at the Tinkering with Evolution Symposium of the Novartis Foundation in 2006 (published in 2007), and through many meetings of the School of Anatomy and Human Biology at the University of Western Australia.

I am especially indebted to a number of scientific illustrators, artists and artistic graduate students who have among them prepared pictures and graphs, and introduced me to modern image processing. These include Bill Pardoe, Birmingham; Joan Hives, Chicago; Erika Oller, Los Angeles; and Martin Thompson, Rebecca Davies and Sue Hayes, Perth.

The investigations have been supported by many funds from the United Kingdom, the United States and Australia during my academic appointments there. Since my retirement, funds have especially come from Australian sources including the Australian Research Council, the Medical and Health Research Funds of Western Australia, and United Kingdom sources including a Leverhulme Research Project Grant, a current BBSRC grant (UK), and current Marie Curie Research and Research Training funds (all courtesy of the Universities of York and Hull). Some of these funds continue to 2009.

Most of all, however, I am indebted to the Leverhulme Trust, for the Leverhulme Professorship that launched the Lecture Series on which this book is based.

Contents

Introduction

As did many investigators in the last century, I started research as a solitary worker in anatomical science. Indeed, my earliest investigations in the 1950s involved dissections in comparative anatomy. Dissecting is a solitary activity. Today, much research involves work at the computer keyboard and this too is often a solitary activity. Today's research lab can degenerate into several desks, each one with a computer, and several research students who may *never*, well hardly *ever*, talk to one another. One of my aims has always been to breach such isolation.

I have, therefore, always been involved in collaboration. In the early days, whether with colleagues or research students, this was generally with individuals, who, like me, had degrees of anatomical expertise, who were anatomists.

However, for a better understanding of the human and animal anatomies, functions, and, later, behaviours, later still, development and evolution, into which anatomical dissection has lead me, I have found that collaborations and consultations with workers in other disciplines (statisticians and applied mathematicians, computer scientists and programmers, applied physicists and engineers) and the use of new concepts and technologies (from mathematics, physics and engineering) have been extremely important. This was not usually available to anatomists in my youth. Hence arose the title of my first book in 1973: *Form and Pattern in Human Evolution: Some Mathematical, Physical and Engineering Approaches*, University of Chicago Press.

Thus, as soon as it was clear that dissection of muscles and observations of bones lead into measurements and their analysis, it became obvious that simple statistics, means and variances, Chi square and

Student '*t*' tests, were not the whole answer. My mentor, Lord Zuckerman (though he would have been horrified at the modern term mentor) insisted that I should visit R. A. Fisher at Rothamstead Experimental Station. Fisher, of course a most senior statistician, was very nice to me but quickly said:

'You should talk to my young man, Yates'.
Yates was no longer a 'young man', and was even then very senior, was equally nice to me but quickly said:
'You must talk to my young man, Healy'.

Michael Healy really was young, about my age. Thus was I introduced to matrix algebra, multivariate statistical methods, programming in binary code, later, Mercury Autocode, and the main frame computers of those days, huge 'tube' or 'valve' machines, first the old Elliot at Rothamstead and later the English Electric KDF 9 in Birmingham.

Years later when I went to the University of Chicago I was worried that I might have lost the Fisherian kind of consultation and collaboration. But at Chicago I immediately discovered (of course) Paul Meier, David Wallace and Bill Kruskal and was amazed to find the same excellent relationships as the Rothamstead group had given me. Later still, when sitting next to, the, by then, venerable Tukey as joint guests of honour at a dinner of the Mathematical Geologists of America in Kansas, I told him about the serendipity of my meetings with these British and US statisticians. He laughed. He knew, as I did not, that the Chicago statisticians were mostly Tukey's students, that there was a continual exchange of students between his lab and Fisher's, and that all these people had developed the same unusual abilities to talk with and advise biological scientists who had not too much knowledge of statistics. Further help in the United States was also given by others. Joel Cohen (who almost immediately left for Harvard) also helped me with multivariate statistical programmes. Peter Neeley (who almost as quickly left for Kansas) helped me with the neighbourhood-limited classification. D. F. Andrews of Bell Telephone Labs (as it was in those days) provided the method of high-dimensional analysis which he had developed using some of our own data. L. A. Zadeh's work (though I never met him) provided early thoughts about fuzzy sets in understanding biological forms. This has even continued with my move to Australia where Norm Campbell (CSIRO) mostly through his later edition of Reyment and Blackith's book, and Adrian Baddeley (University of Western Australia) have been so extremely helpful. I have always been blessed with mathematically gifted colleagues with real abilities to understand my biological problems.

Those first dissections also lead to my taking an experimental stress analysis course in the early sixties in the Department of Mechanical Engineering at the Royal College of Advanced Technology at Salford (now the University of Salford). Other students came to the practical classes with cold steel beams and oily crankshafts; I produced wet, warm and bloody butcher's bones. My mechanical engineering teachers were somewhat horrified, but *very* intrigued.

Similar relationships with many other scientists in the mathematical, physical and engineering worlds, and the 'borrowing' of many of the technologies in their laboratories, also proved essential. Often these took me 'outside' university and 'into' industry. One of the first of these links was with the late Ken Sharples of Sharples Stress Analysis, UK, and an introduction to direct photoelastic analysis. Others were colleagues using reflection photoelasticity at Westland's Aircraft (as it was then) in the Isle of Wight. (This aircraft link reminded me of my father's involvement with Mitchell and the Spitfire at Supermarine Works, Southampton, before the war). These industrial colleagues acted totally altruistically; this was long before the miserable constraints of 'user pays' that have so damaged universities. Industrial collaborations continued when I moved to the United States, where I was involved with Gregory Pincus (Image Analysis at the University of Wisconsin), John Davis (Geology, Kansas State Geological Survey) and Berry and Marbles (Geography, also in Wisconsin). They were all using different kinds of industrial image analysis and optically generated Fourier transforms to examine materials such as aerial pictures of jungle tree tops, sections of oil-bearing rocks, and tree-like patterns of river branches. But I saw in those techniques methods for examining radiographs and sections of bones. Dr Harry Yang in Chicago was my first MD/PhD student who greatly benefited from this technology. Today, of course, Fourier analysis is carried out on desktop computers using Fast Fourier Transforms. But I did not do this until my move to Australia and it required help from my graduate student (now Dr) Alanah Buck (currently a Forensic Anthropologist), programmer, as he was then, Iain Sweetman, and medical physicist, now, Professor Roger Price.

I graduated from the earlier photoelastic stress analysis to computer-derived finite element analyses (FEA) of stress and strain and worked first with another graduate student (later Professor) Artyan Hsu in Southern California. Even in Australia, such relationships have continued with Engineers Chris Windsor and Wayne Robertson in geomechanics at the CSIRO and the University of Western Australia. They introduced me to the Fast Lagrangian Analysis of Continua approach to FEA. As with the statisticians, this relationship with engineers continues to this day with my honorary appointment as Professor of Bioengineering at the University of Hull and new collaborations with Biomedical Engineer Professor Michael Fagan and his staff and research students.

Integration with other disciplines has taken me into areas that questioned the (then) standard ways of thinking about evolution. They involve new integrations of structure, function, genetics and development in evolution, an approach to evolutionary biology that is sometimes called evo-devo. A Chicago undergraduate research student (now Professor at Johns Hopkins), Rebecca German, who has maintained contact with me and in recent years visited Australia many times, has been seminal in this regard. So too, has been another frequent visitor to Western Australia, Professor Brian Hall, of Dalhousie University, of whom it can be said he is one of the 'fathers' of 'evo-devo'.

Most recently of all, my collaborations have involved computer mimicry of evolutionary processes. This has depended upon Dr Ken Wessen, originally my doctoral student in theoretical evolutionary biology but who already held a PhD in theoretical physics. His first draft of a Research Master's Thesis in Human Biology easily transmuted into a Doctoral Thesis worth a Distinction, and then into a Cambridge University Press book: 'Simulating Human Origins and Evolution' (Wessen, 2005). Now with a job in industry, he earns far more than any professor; he continues this modelling research; it is nothing to do with industry; it is how he gets his intellectual jollies!

Alongside this complex set of scientific relationships, and probably because my early education included a medical degree, I have also always kept a weather eye open for medical problems. As a result I have become involved in such topics as vitamin B12 deficiency syndromes, mechanical efficiency of bone prostheses, bone structure in relation to gravitational physiology, early diagnosis and late fracture implications of osteoporosis, anatomy, function and disease of the human incus, and bone growth impairments of iodine deficiency, hypothyroidism and especially the complications of cretinism. Much of this work has been underpinned by extensions of my interests in the mechanical efficiency and adaptation of spongy bone. Some, the vitamin B12 studies and the iodine deficiency effects (with Peter Obendorf and Ben Kefford), were totally serendipitous.

It sometimes frightens me that so many contacts like the above were made serendipitously. How many even better contacts have I missed because of the individuals that I did NOT happen to meet? Indeed, my 'major professor' Lord Zuckerman once hinted to me that I was way behind, that there were far better techniques that he knew about; but that, as Chief Science Advisor to the government, he could not tell me about them; they were classified!

Always these collaborations require that I thank the nimble minds, nimbler than mine anyway, of a series of research students, post-docs and other colleagues over the years, and, happily, this continues into the present. Of course, one usually thinks first of the well-known seniors who were responsible for one's career. Yet my initial stimulus came from the enthusiasm of an unknown headmaster of a tiny primary school in a small Scottish village during the second World War. There was no science in the classical curriculum in Scotland then. But he 'knew', somehow, that this small boy was interested in science. He introduced me to the ideas of

Wegener, Goethe, D'Arcy Thompson and Solly Zuckerman when I was 9 years old!

As a result, by 1952, I may have been the only person in the world who knew about the movements of the continents but who did not know that Wegener's ideas were not accepted for almost half a century. By 1952 the ideas of plate tectonics had become centre-stage. I could not understand what all the excitement was about. I had always known it was so. Likewise, I understood very well that the skull was simply a series of fused vertebrae. It made sense to me. I did not know that Goethe had it wrong until, in 1955, I came to read Gavin de Beer's tome on the vertebrate skull. The new developmental studies in this last couple of decades show that Goethe was a little bit more right than most of us thought, though, of course, for the wrong reasons. I certainly did not understand the mathematical formulae and Greek quotations in Darcy Thompson's 'On Growth and Form'. But there is much in that book to touch a nine-year-old. His pictures of the struts in the interior of a bird's wing were so like the struts between the wings of the biplanes of my childhood. The relationships between stress and architecture of the bridges that he presented were so obvious in the bridge over the Forth that was just down the Firth from my Scottish childhood home. His Cartesian coordinate transformation diagrams of biological forms were fascinating, even to a small boy. His pictures have been with me all my life. They have figured in my investigations even recently through thin plate splines, morphometrics, biomechanics, Fourier transforms and tensegrity towers.

And finally, of course, how could I have known that I would later actually work with Dr Solly Zuckerman, author of that fourth text *Functional Affinities of Man, Monkeys and Apes* (1934) that my headmaster gave me all those years ago?

I thus also owe much to famous people, especially to Lord Zuckerman himself; I think I was his last full-time student. Eric Ashton and Tom Spence with whom I worked in Zuckerman's Department of Anatomy at the University of Birmingham were constant exemplars, colleagues and friends. Indeed, I owe much to that entire department in Birmingham, even to the wider range of academics with whom Zuckerman was associated in those seminal years. All were critical to my own scientific development. When Zuckerman retired (from the chair in Birmingham, he never retired from problem-solving until his death in 1995) his efforts at the University of Birmingham had grown Anatomy from about six academics in 1945 to, at the time of his retirement in 1966, as many as 53. In those few years, he had increased the department's funding from the widow's mite of a typical anatomy department to one that was third largest in the university, being exceeded only by physics and chemistry. He had moved the department from being almost solely occupied with teaching medical anatomy to medical students to one encompassing many other new 'anatomies'. These spanned not only the anatomy of humans, but also the anatomies of apes, monkeys, mammals, vertebrates and even many non-vertebrates. They included the anatomies of reproduction, development, growth and contraception, the anatomies of cells, sub-cellular organelles, membranes and molecules, and the anatomies of brains, behaviours, cognition and psychology. What other anatomy department of those days (or of any day) had biochemists, physicists, engineers, psychologists, schoolmasters, several FRS's, and several peers of the realm on its staff. And all this without the need to change the name!

Other academics outside the Birmingham department also influenced my career: Professor J. Z. Young who, together with Solly, examined me for my Bachelors degree in 1955. Professor Young:

'What do you see down that microscope my boy?'
'The substantia nigra, sir', I said.
Covering the microscope's eyepiece with his hand:
'Is the black stuff inside or outside the cells?' he asked.
Of course, I had no idea, I was not a brain person — but I gave an answer based on standard biological principles. I have never dared go back to see if my answer was right, though I suspect it was.
Solly and JZ roared with laughter. They *knew* I had guessed.

And there were others: A. J. E. Cave, Professor of Anatomy at Barts (St Bartholomew's Hospital Medical College of the old days).

He of Neanderthal fame! Alec Cave knew that the hunched posture and the supposed shambling gait, on its way to modern human walking, as it was supposed, of the brutish Neanderthal was just severe osteoarthritis in a skeleton of an old man. He examined me for my PhD.

'Did you write that thesis?' — indicating my doctoral volume on the table.
'Yes.'
'Then you pass' he said and immediately signed the examination form and appended:
'Highly Commended'.

For the next three hours we had a fascinating discussion about the great ideas and great people of anatomy in earlier days (that was the real examination!) And that discussion with him continued over the years. I had the feeling that he and I were those last anatomists to dissect the really large animals of this world — we had both dissected whales — I was not correct! He was a wonderful scientist who, though he did not know one end of a statistic from the other, understood intuitively just why we were using multivariate statistical methods to characterise animal form and pattern. He had his telegram from the Queen on his 100th birthday but died soon after. It is an especial privilege for me to be able to recognise his part in my life.

Zuckerman's departure from Birmingham in 1966 spawned a diaspora from Birmingham that continued for many years. I was the first emigrant myself in that same year. I then enjoyed further seminal decades at the universities of Chicago, Southern California and Western Australia. I have found, everywhere in the world, members of that original 'Zuckerman Mafia'. Each time that I have moved, I have, while keeping the old: old colleagues, old students, old problems and old techniques, also extended my grasp to the new; and all are part of my acknowledgement.

Perhaps, however, the people to whom I owe most, the people who have contributed most to my researches over the years, indeed, the people who keep me mentally alive even now, are none of the 'old greats' or the 'accidental colleagues'. They are the 'new greats': the students and colleagues, mostly much younger than me, many of them now professors in their turn, in all parts of the world, some of them, however, still students, of mine, or of my students, or even my students' students. Most of them have shown that many of my original 'great' ideas could be 'improved upon'; were, in fact, 'not quite right'; were, actually, even, 'totally wrong'. Yet, a few of them have found that some of my earlier scientific heresies were not so far from reality. It is through these people that I am still able to be involved in the new developments in our discipline. They keep me alive. They are really the ones to be thanked. The current crop are graduate students in Anatomy and Human Biology and in Forensic Science at the University of Western Australia, and in Anatomy and Bioengineering at the Universities of York and Hull, and the Hull York Medical School, UK. They all give me enormous pleasure and help. All are present in spirit as I look at my first 50 years in science, and as I contemplate the many research lines with which I will continue to be involved (hopefully) in the next decades as an honorary investigator.

It is worth recording what has happened to some new younger colleagues: honours students, graduate students and post-doctoral colleagues, in these last few years. Seven of them (all strong students with funding) have opted to leave academia because of the perceived lack of value that our society, our government, our country, even academia itself, places nowadays upon the research and teaching career. You would think this would be a worry; yet no one seems willing to do anything about it.

Chapter 1

The Shape of Bones: Tension and Compression

How It Started

I have been a solitary investigator in anatomical science for much of my life. Indeed my earliest investigations in 1952 involved dissections in the comparative anatomy of mammals. Dissection is a solitary activity.

Yet, for a better understanding of the human and animal morphologies into which dissection has lead me, I have found that collaborations or consultations with workers in other disciplines (mathematicians, physicists and engineers) and the use of physical, mathematical and engineering concepts and technologies not in those days available in anatomical laboratories are extremely important. Indeed, this idea gave me the title for my first book in 1973: *Form and Pattern in Human Evolution: Some Mathematical, Physical and Engineering Approaches*, University of Chicago Press.

Those first studies, extending what can be learnt from dissection, were much aided by my taking an experimental stress analysis course (using Timoshenko, 1955 as one of the texts) in 1960 in the Department of Mechanical Engineering at the Royal College of Advanced Technology at Salford (now the University of Salford). This was actually long before I ever heard the word biomechanics. Other students came to the practical classes with cold steel; I produced wet, warm and bloody bones from the local butcher. My mechanical engineering teachers were somewhat horrified at such material, but *very* intrigued.

Since those early days, it has become apparent to me that consultations and/or collaborations with scientists in the mathematical, physical and engineering worlds, and usage of the many technologies in their laboratories are essential. These collaborations, moreover, involved not only scientists in universities but also those other scientists who were 'out there' in industry. Such industrial colleagues were usually acting altruistically — this was long before today's miserable constraints of 'user pays'.

Finding out about mechanics at first involved talking with individuals such as the late Ken Sharples, at Sharples Stress Analysis Inc., and using equipment from Westland's Aircraft Ltd (in both cases, using photoelastic benches of different types that I employed for experimental stress analysis of bone forms). Photoelastic analysis is very attractive to biologists, providing, as it does, a pictorial representation of the mechanical efficiency of bone form. It has usually been aimed at studying the rather simple external shapes of bones and this is a matter for this chapter. It soon became clear that I could also use it for understanding more of the complex internal texture of bone (especially spongy bone, most evident in sections and radiographs of bones). That is a matter for the next chapter.

Finding out how bones, when loaded, bear stress and are strained was thus the problem. The photoelastic methods though useful initially were only able to give 'ball-park' answers. This was in part because they are mainly two-dimensional.

I did indeed attempt three-dimensional photoelasticity years ago. This involves what was called 'frozen stress analysis'; the word 'frozen' was used because the fringes indicating the stresses were made permanent in a model of a whole bone, although not by freezing but by heating the model during loading. This fixes (freezes) the stress patterns in the model in three dimensions. These are then analysed, however, by carefully cutting the model into two-dimensional sections and looking at the stresses in each section. This turned out to be incredibly long-winded (as the reader can imagine) and I never managed to use the method to solve a real biological problem.

The usefulness of the photoelastic methods was also somewhat less because they assumed isotropic models of the bones (i.e. models of bones as though they were made of a uniform material). This too, could be allowed for by using models with plastic inserts having different optical responses to mechanical loads. Likewise, however, though I was able to do this through examples, it was too complex for me to be able to tackle real biological problems.

Today, however, many of these difficulties and complexities have been overcome through the use of computational methods. At first this was with standard finite element analysis (FEA) on main-line computers (in collaboration with engineer and physical therapy doctoral student, Artyan Hsu). Though initially he did not speak English very well, Hsu spoke Mathematics extremely well, and carried out the first FEA of a section of bone in which I was very interested: the calcaneus (Hsu, 1989). Before that time many workers had assumed that the major stresses acting on the human calcaneus in walking and running were during heel strike. The very words 'heel strike' and the picture in the minds eye that they evoke seem to indicate powerful and violent loads as one rams the heel down. In contrast, Hsu's studies quickly made it clear that by far the largest stresses in walking and running in humans were at the other end of the gait cycle: at 'foot and then toe off'. It was to the stresses generated in this specific loading regime that we found the cancellous structure of the human calcaneus was most correlated (see next chapter).

Later studies came to involve applications of FEA that could be carried out on desktop computers. For me this involved fast Lagrangian analysis of continua (FLAC), thanks to colleagues (David Windsor and Wayne Robertson) in Geomechanics in Western Australia (Runnion, 1991). Their work was on how earthquakes shake buildings, but our application was on how muscles 'shake' bones. This collaboration was funded by the smallest research grant I have ever had ($17,000) but it is the grant of which I am, perhaps, most proud because it was awarded by the Australian Commonwealth Scientific and Industrial Research Organisation to an old-fashioned medical anatomist!

This collaboration also allowed me to work with an ear, nose and throat surgeon (now Professor Francis Lannigan) together with another anatomist (also now Professor Paul O'Higgins) on the form of an ear-bone, the incus, how it works during hearing, and why it resorbs in some older men. Initially this work started with my arm-chair pencil and paper guesses about stresses in the incus. Because that thinking raised a paradoxical possibility, we followed it by simplistic modelling of strain using a two-dimensional rubber sheet shaped like the incus. Because that, in turn, continued the paradoxical idea, the work was extended using FEA through a FLAC study of incudial form and function (I was to learn later that incudes contains the root and is the plural of incus!). The paradox was confirmed (see later in this chapter).

Today I am involved with bioengineering colleagues (at the Universities of Hull and York, and the Hull/York Medical School, UK) Michael Fagan, Catherine Dobson, Kornelius Kupczik and Paul O'Higgins, and with then graduate student Jens Hirschberg at the University of Western Australia (now senior lecturer at the university of Notre Dame, Fremantle). Their abilities in two-dimensional (Hirschberg *et al.*, 2000 and Hirschberg, 2005) and three-dimensional analyses (Kupczik *et al.*, 2007) promise to take us far beyond my initial ideas. As a result of all this I also hope to complete, with joint authorship of course, yet another book (but certainly not my last) to follow that first book written so many years ago. This one is tentatively entitled *The Bone–Joint–Fascia–Muscle Complex: Functional Anatomy and Mechanics.*

The Nub of the Problem

A very early mention of the mechanical significance of bone form was by Galileo (1638) who understood that bones in large animals were not simply scaled up versions of bones in small animals, but had also to be a different shape. Galileo explained it by the upper diagram in Fig. 1. But the differently scaled lower diagram makes the matter more obvious.

For me, however, the beauty of the significance of bone form was thrust upon me when, as a primary school student I saw D'Arcy Thompson's (1917) picture of the bracing struts inside the bone of a bird's wing. It was so similar to the struts and ties between the wings of the biplanes that I saw as a child before the Second World War (Fig. 2).

It was further enhanced when, as a medical student, I heard the story of the German engineer who, upon happening across a sagittal

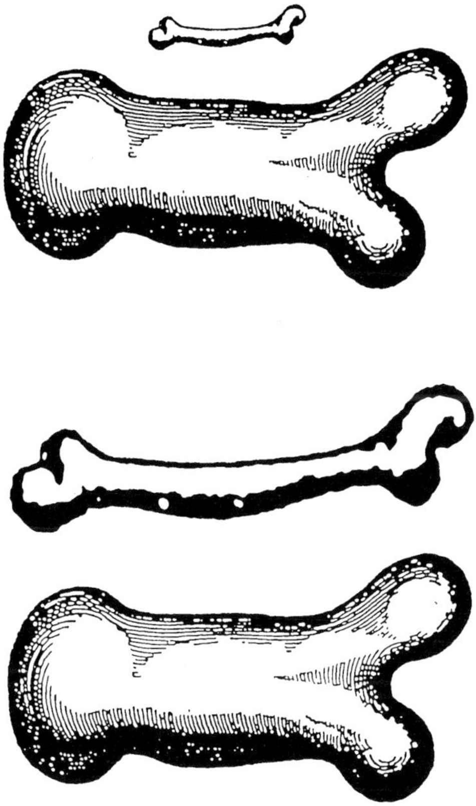

Fig. 1. Top two bones: Galileo's (1638) comparison. Bottom two bones: the same comparison with the bones scaled to the same length.

Fig. 2. D'Arcy Thompson's (1917) figure of the interior of a vulture's wing bone.

section of the upper end of the femur in the laboratory of a biologist colleague, is said to have uttered:

> "Das ist mein crane!"

The relationship that he saw is pictured in Fig. 3 (Murray, 1934) where the pattern of principal stresses that he had calculated in a Fairbairn crane seemed so similar to the pattern of architectural bony spicules inside the head and neck of the human femur. This latter association between mechanics and anatomy, leading to the trajectorial theory of bone architecture as it has been called, so impressed itself upon me that I have taught it to generations of medical students as gospel truth. This, too, however, is a matter for the next chapter.

This all leads to a major problem in attempting to make evolutionary judgements on the basis of bony features, especially the features of fossils, that is, the question of their biological significance. One of the ways of dealing with whole bones is to observe their external features, such as tuberosities and pits, foraminae and fossae, grooves and ridges. These are often defined, in evolutionary studies, as 'characters'. They are then often treated as 'present' or 'absent'. Some workers recognise that they are really quantitative and use terms like 'small', 'intermediate' or 'large'. The states of these characters are then assessed as primitive, shared derived, uniquely derived, etc., and they are employed in cladistic analyses as though they were genetically determined.

In actual fact, most of these architectural features are truly quantitative, and they are evidence of both the genetic and epigenetic plasticity of bone. Thus some such features may be initially generated by genetic factors with epigenetic contributions, for example genetically produced during early development and mechanically maintained and/or mechanically changed during later development and growth. Sometimes, even, they are produced *de novo* by functional change even as late as in the elderly adult. Some features (e.g. the head of a bone) usually appear in development long before mechanical function is established. However, many features of the head of a bone do not appear unless function commences (e.g. many processes around a

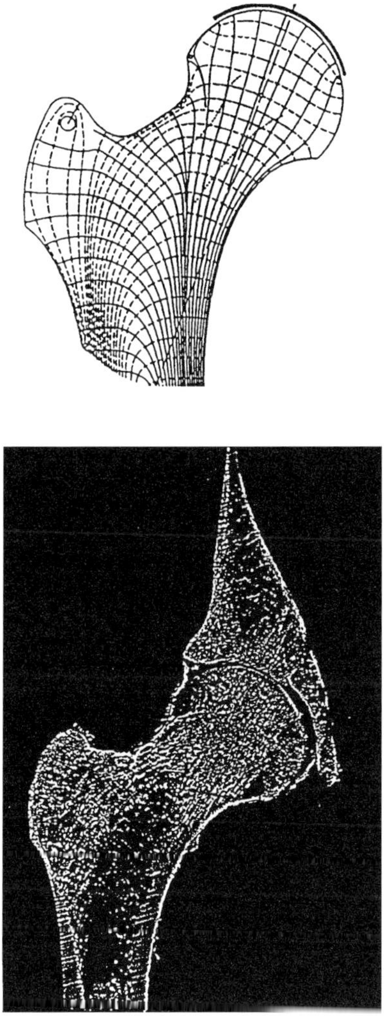

Fig. 3. The similarity between the stresses that can be computed as existing within a coronal section of a femoral head, and the actual spongy bone architecture that is displayed in the section (see Evans, 1957).

bone head for muscle attachment). There are even some features that do appear, only to disappear, if mechanical function is removed (e.g. Washburn, 1947; Avis, 1959; Moore, 1981). Some appear with early function (e.g. raising the head) then disappear or are replaced or changed into other features as function develops (e.g. the curvatures of the vertebral column in the transition from raising the head, to sitting up, to crawling, to standing, walking and running). Some appear only if function is totally changed (e.g. bony ridges in relation to bone healing or surgical bone anastomoses). It is therefore important to know about these biomechanically related morphological features of bone before attempting evolutionary assessments.

The matter is yet more complex. Many such architectural features apparently exhibit opposite arrangements in different bones in the same creature. Why, for example is a whole bone, the incus, present in most mammals, yet a large portion of it may disappear in humans in certain situations (Oxnard *et al.*, 1995). Why are there tuberosities (often very large) on some bones where tendons or ligaments attach (e.g. the patellar ligament to the tibial tuberosity) when there are pits (sometimes very deep) where others attach (e.g. the insertion of the obturator externus tendon at the internal aspect of the greater trochanter of the femur)? Why, sometimes, are there both a tubercle and a pit present at certain tendinous attachments (e.g. the insertion of biceps tendon into the radial tuberosity)? Why are there sesamoid bones embedded in some tendons (e.g. the tendon of peroneus longus as it winds around the cuboid) but not in others (e.g. the tendon of obturator internus as it courses around the ischial spine) (Oxnard, 1993)?

Questions like these were at the heart of a series of initial biomechanical investigations that I have carried out over the years (Oxnard, 1991). I hope that their explication will improve the understanding of bone characters in evolution and illuminate important bone features within medicine.

Some of these mechanical questions seemed to revolve around a (now) quite old question: does overall tension exist in bone (Currey, 2002). Tension clearly does exist in many situations in bones at specific times, especially during reversals between tension and compression in cyclical bending induced by locomotor movements. This has been shown many times (especially in a series of investigations by Lanyon and Smith, 1970). But what happens when tension exists all the time or at least most of the time, or perhaps as some 'average' over time (though surely not a simple arithmetic average), is not clearly understood. Some of this is a matter for this chapter on external bone form, some a matter for the next chapter on internal bone structure.

Is Bone Ever Subject to Tension Overall?

When Currey (1962) first posed the question: "does overall tension exist in bone?" he was initially looking for a whole bone in overall tension. He was also not thinking of any instantaneous state of tension (which happens frequently), but a state of tension over the period of time relevant to architectural change. Indeed, Currey asked me on several occasions if I knew of bony situations where the main stresses might be tensile. He was thinking at that time about the possibility that the arm bones of gibbons might be principally in tension associated with their extreme form of locomotion that includes much hanging and swinging by the arms. He knew that I was interested in brachiation. My answer always was: given that the elbow joint functions during brachiation, net tension ought not to exist in gibbon elbows. I thought there would be compression at the elbow because it is a synovial joint and that type of joint operates under compression. However, Currey's question was intriguing and at that time I started several investigations designed to discover if it were at all likely that tension might commonly exist in particular regions of bones, if not in whole bones. I initially employed simple thought experiments and then simple photoelastic analyses (Oxnard, 1972) in attempting to answer Currey's question.

Zero Dimensions: Tension and Compression at a Point on a Bone

My first thought was to look at the point of attachment on bone of powerful tendons. Surely Currey's idea, overall tension, might predominate at such points (really, of course, rather small areas) when tendons are powerful, when the loads applied by them are large, and when the tendon/bone interface involves collagenous bundles (Sharpey's fibres) that perforate the surface of the bone. These features are found in many places throughout the body. The histological nature of such tendon/bone junctions may be quite complex, involving not only penetrating collagenous fibres, but also some degree of interpenetration of the fibres with hydroxy-apatite crystals, and often, too, an interface containing zones of fibro-cartilage and even, on occasion, mineralised fibro-cartilage.

The processes involved in fossilisation and in the preparation of dried bones for museums generally remove these non-osseous materials. For example, the attachment of the powerful temporalis tendon to the coronoid process of the mandible is first through a cap of fibro-cartilage (that is usually later converted into bone). In some dried mandibles (even in younger adults) this cap is lost. The loss may not be recognised; hence measurement involving the tip of the coronoid process as a landmark may be incorrect.

A similar phenomenon exists in many bony processes associated with ligamentous and tendinous attachment throughout the body. Other examples include the anterior and posterior clinoid processes around the pituitary fossa inside the skull, and the styloid and vaginal processes on the outside of the skull base (see later chapter). The styloid process of the skull is particularly obvious. The length of that process may totally depend upon how many of its four centres of ossification actually become incorporated into that part of the process that is attached to the skull base. Indeed one variation in humans exists where the styloid process completely links the skull base and the hyoid bone, and another can be found where none of the centres of ossification fuse with the result that, in the dried skull, the styloid process appears to be absent.

I originally studied this question of stress at a point at three levels. First, I used arm-chair analyses employing simple theoretical biomechanical concepts (Oxnard 1972). Second, I started to apply experimental investigations using actual biomechanical situations but through simplified photoelastic simulations (Oxnard, 1973). Finally, using FEA encompassing greater degrees of complexity, and with colleagues, I was able to take matters much further (Oxnard, 1983/1984).

The first level, biomechanical thinking, involved the following. Let us assume that the bone where the tendinous attachment occurs is flat. Let us assume, too, that there is a compression along the length of the bone due to the body weight. A tendon attached at close to right angles to the bony surface will be under tension due to contraction of the attached muscle. Somewhere within this structure there must be a junction where the compression in the bone grades into tension in the tendon. The question is: where is that junction located?

If, during the movement that is produced by the action of the muscle, the change in angulation of the tendon to the bone surface is small, then the state of tension may protrude some distance into the bone. This is because the load producing tension at the small region of attachment of a large tendon (of a powerful muscle) is very much greater than the load producing compression resulting from body weight.

If, however, during the movement that is produced by the action of the muscle, there is a large change in angulation of the tendon to the bone surface, then the situation may change. That is, though there will certainly be tension in the tendon some distance from the bone, this will be reduced close to the bone because the tendon fibres are under compression laterally, being constrained at their attachment. The larger the change in angulation during function, the more the fibres are squashed into a smaller cross-sectional area, and the greater the amount of compression of the fibres one against another. Depending upon the degree of this new compression it would be entirely possible for a region of compression to extend upwards into the tendon (Fig. 4).

If, therefore, the early ideas of Frost (1964) and Currey (1962) that overall compression induces bone apposition and overall tension bone resorption are correct, then a pit might be expected to develop in the first case and a tuberosity in the second.

One difficulty for this line of argument is raised by those instances mentioned above in which muscular (tendinous) attachments are to regions that exhibit both pits and tuberosities. For example, the area of origin on the skull of what must have been a very large and powerful temporalis muscle in sabre-toothed cats is undulating. It comprises both approximately circular concavities

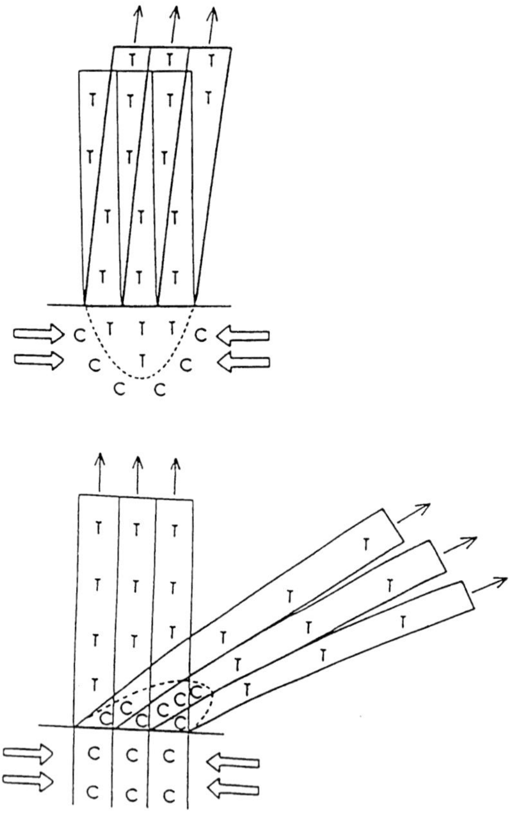

Fig. 4. Arm-chair analysis of the stresses existing in a tendon attached to the surface of a bone, upper frame, when the tendon changes angle to only a small degree during function, and lower frame, when the tendon changes angle to large degree. Arrows indicate loading in the tendon and the bone. T = tension, C = compression.

(pits) on generally raised convexities around the pits. However, though clues might be gained from dissections of present day cats, we do not actually know the precise tendinous relationships of this muscle in sabre-tooths.

Another somewhat simpler example is easier to understand, i.e. the attachment of the tendon of the biceps to the tuberosity of the radius in humans. This tuberosity is actually a complex structure with both a convexity and a concavity. The most ventral part of the tubercle is convex but is quite smooth. It does not have any part of the biceps tendon attached to it. It is overlain by a bursa allowing friction free movement between the free tendon and this part of the bone when the radius is in the wound-up pronated position. It is not in contact with the biceps tendon in the unwound supinated position. A somewhat more dorsal part of the tuberous convexity (a tuberosity) is roughened, however, and bears part of the attachment of the biceps tendon. The dorsal-most part of the attachment of the tendon is actually a concavity (a pit). The biceps tendon (as the temporalis muscle in the prior example) thus attaches both to a tuberosity and a pit.

Let us now apply the same biomechanical thinking to this situation (Fig. 5). When the forearm is in the fully prone position, the biceps tendon is wound completely around the radial tuberosity. Although there is certainly tension in the free portion of the tendon, there ought to be mainly compression in the portion of

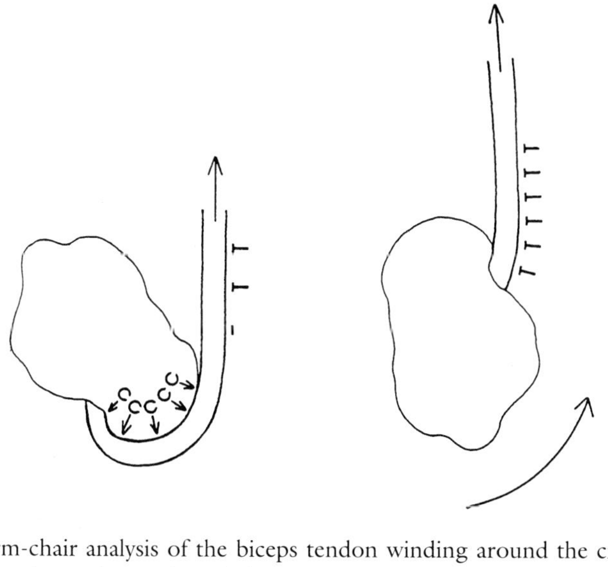

Fig. 5. Arm-chair analysis of the biceps tendon winding around the cross-section of the radius as it attaches to the radial tuberosity. Right frame: in the fully prone position of the forearm; left frame: in the fully supine position of the forearm. Arrows = directions of pull and movement; C = compression; T = tension.

tendon wound tightly around the bone. It is entirely possible that this additional compression will be great enough to outweigh the tension and to make the overall stress at this position compression. As supination occurs, the tendon unwinds from the tuberosity and the compression due to tight apposition against the tuberosity will gradually disappear as the previously wound portion of the tendon comes to move away from the tuberosity. However, the unwinding results in the more ventral fibres of the muscle markedly changing their angle of orientation in relation to the surface of the bone; from being initially parallel to the surface, they are no longer parallel. If the bony surface had been flat it would presumably have been associated with lateral compression of fibres and the development of compression extending into the base of the tendon. This would produce the observed convexity. In contrast, the more dorsal tendinous fibres remain in an unchanged orientation to the bone surface. The fibres do not press laterally against one another. This implies that compression is not generated and therefore the tension already in the tendon as a result of muscle contraction predominates. This could produce the observed concavity. The combination of these two situations could then conceivably lead to a combined sinuous convexity and concavity of the bone surface (which in fact exists).

The arm-chair thinking above has been tested by the second level of analysis: biomechanical analyses using simple photoelastic experimental stress analysis simulations. These employed the photoelastic method (as outlined by Frocht, 1941; Coker and Filon, 1957) and the results (Oxnard, 1972, 1991) tentatively confirmed these ideas.

However, that technique, involving a homogeneous model, is crude compared to the complex heterogeneity of the actual bio-architecture that exists. It is superceded by better simulations that involve the third level of study: finite element analyses (Hirschberg, 1997; Hirschberg *et al.*, 2000; Hirschberg, 2005) using computational methods. These involved testing the ideas computationally using FEA performed with FLAC technology (e.g., see Cundall and Board, 1988; Hoek *et al.*, 1990).

Thus, the compression due to body weight loading in the bone and the tension due to muscular pull on the tendon can be combined. The actual effect depends upon the proportion of these two sets of local stresses. Though generally such loads produce marked changes in stress (many different contours in the diagrams in Fig. 6) along the surface between tendon and bone (whether tubercle or pit), for any given size of tubercle or pit, there are particular combinations of applied forces that produce uniform values along the surface. These are the optimum efficiency solutions (Fig. 7). One might suppose that the final architectural result (height of tubercle, depth of pit) relates to that height or that depth necessary to render the optimum solution for the average set of loads acting upon the system over the response time of bone adaptation.

Even this situation can have further complexity. Consider the case where very large tendons have penetrating fibres. This can also be modelled and shows that, when there are penetrating fibres at the surfaces (of both tubercles and pits) peak tensions actually occur somewhat below the bone surface in both situations (Fig. 8). This fascinating result may explain why many tendon avulsions involve failure of bone beneath the tendon, not at the surface between tendon and bone, nor in the tendon itself. This is a phenomenon that makes it easier to perform orthopaedic repairs in such failure situations. It may be related too, to the phenomenon of cortical excavations that are of considerable interest to paleopathologists.

It is of special interest to note that the original explanation has been greatly extended. FLAC does indeed show that large changes in angulation of tendon fibres produces compression creeping up into the tendon (hence a tubercle). But the additional data about the balance of loads and about the effects of penetrating fibres are new.

One Dimension: Tension in a Linear Crest on a Bone

I also thought of a simple one-dimensional situation: a linear bony crest where muscles arise from opposite sides of the crest. This seems to imply that tension might exist in such a ridge. It would

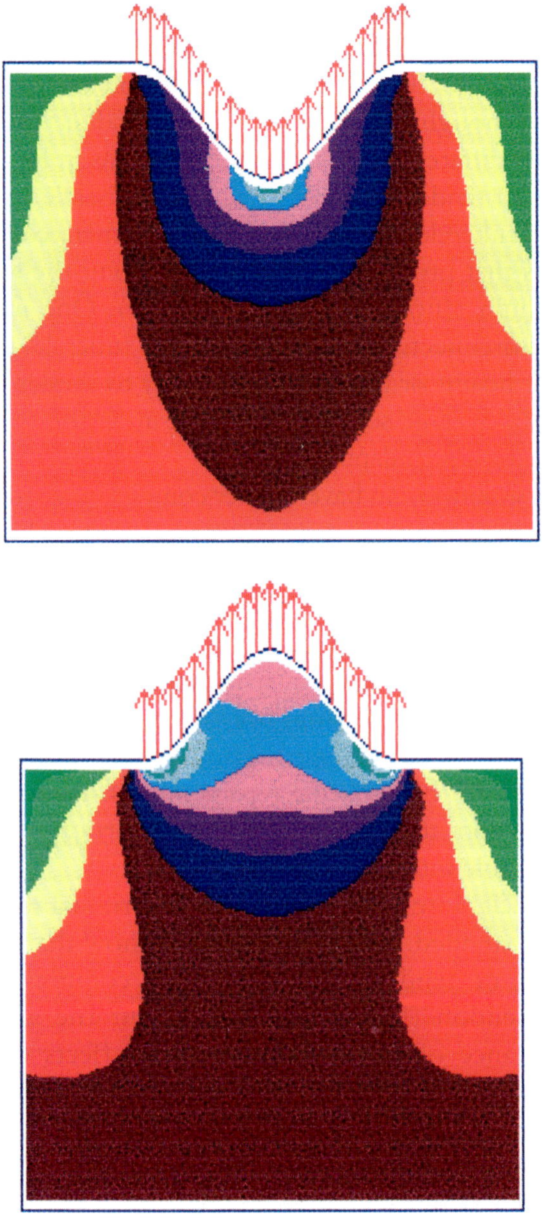

Fig. 6. Stresses as calculated in a tendon attaching to a pit (top frame), and a tuberosity (bottom frame). There is a marked change of gradient (change in colours) across the surface of attachment in each case.

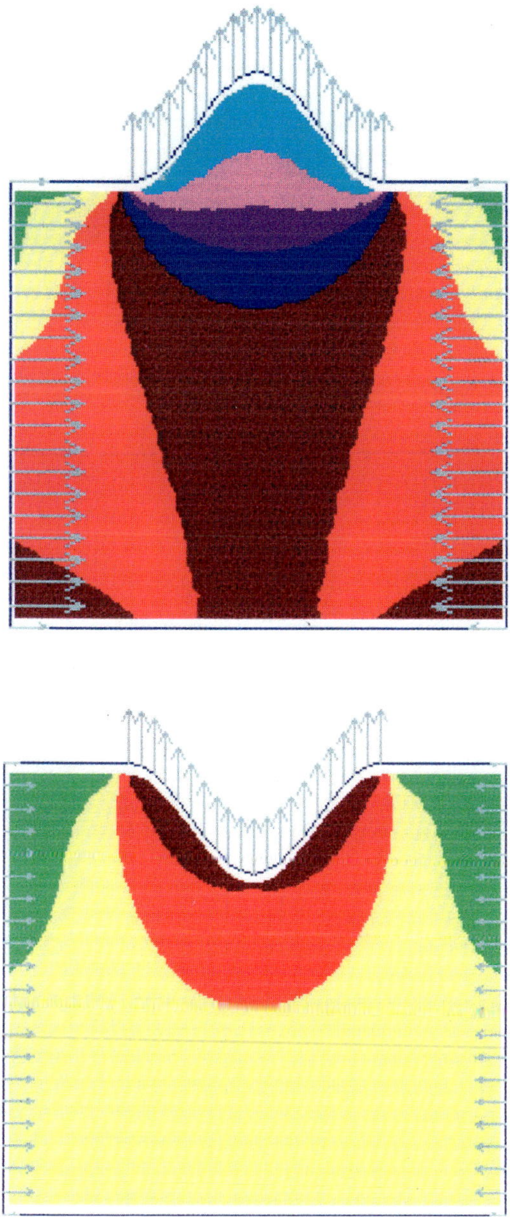

Fig. 7. Stresses as calculated in a particular pit and tuberosity showing that there are combinations of compressive and tensile loads that produce a constant gradient (same colour) over the bony surface (i.e. mechanically efficient).

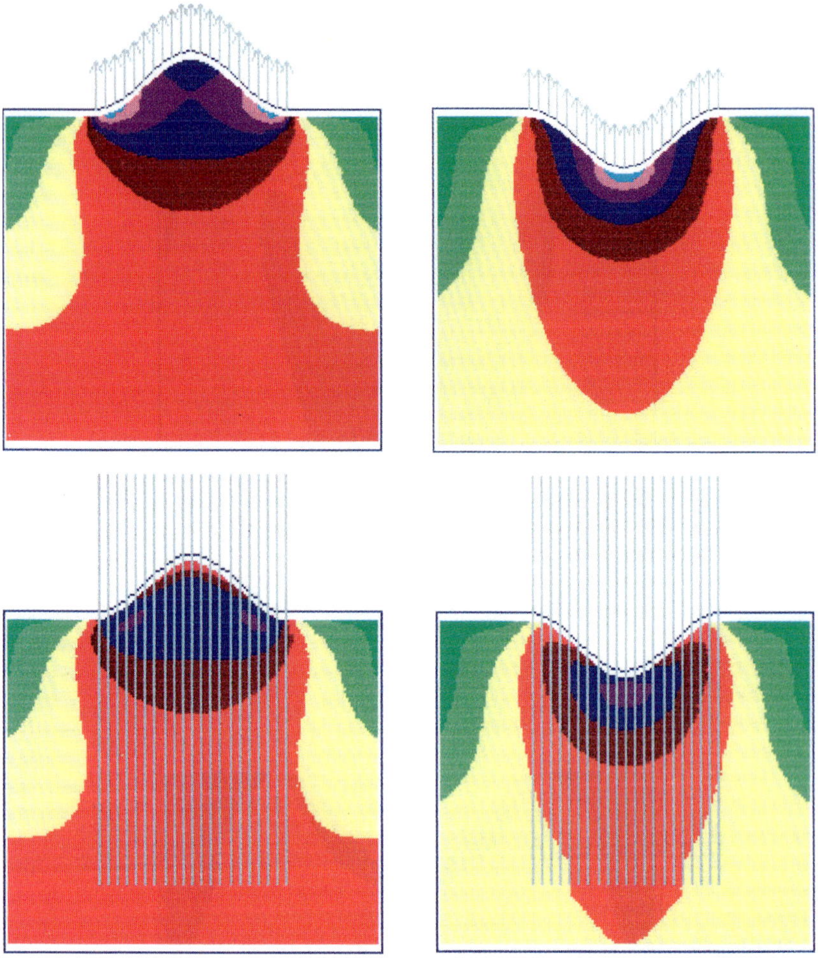

Fig. 8. Colour comparisons of tendon inserting into a pit and tuberosity where the attachment is on the surface (upper frames) and where tendon fibres go into the bone (lower frames). In the upper frames, the maximum stress is at the surface, the tendon bone junction. In the lower frames, the maximum stress is located at some distance into the bone below the surface.

require that the opposite muscles would have to contract at the same time. The midline cranial crests of gorillas, some chimpanzees and many australopithecine skulls have jaw closing muscles (temporalis) arising from each side of the cranium. Of course, though the muscles on the two sides can co-contract during some jaw closing movements, they mostly contract alternately in the usual chewing cycles of these and most other mammals. As a result the predominant effect that they would have on the crest would be tension on one side and compression on the other in one phase of the chewing cycle with reversal in the other phase. A somewhat similar situation exists in the occipital crests of large primates, both living and fossil. These have jaw muscles originating from one side and extensor neck muscles from the other. Again, the effect on these crests would be mostly bending one way and then the other as the muscles function at different times and in different behaviours. The overall result would be that tension would not predominate for any significant period of time.

I was, however, thinking at the time of the crest on the scapula (the scapular spine) that gives attachment to portions of trapezius and deltoid muscles. It is easy to work out that, in the locomotor activities of most terrestrial creatures, these two muscles, like the two temporalis muscles in the masticatory example, do not usually co-contract in locomotion. Trapezius is a protractor of the limb, posterior deltoid a retractor. The scapular spine would be bent first one way and then the other. As in the cranial crests, compression and tension would alternate (Fig. 9). There would not be an extended period when tension alone would exist. If, however, one could find some animal in which these muscles did co-contract in locomotion then tension should exist in the spinal crest (Fig. 10).

It transpires that in some bats, these two muscles do frequently act together because, in acting together, they move the shoulder dorsally as in the upward movement of a wing. This upward movement is not simply produced by upwards air pressure (though that no doubt helps) but requires muscular activity to increase wing velocity to get it back for the start of the next power stroke. This effect of the action of two muscles co-contracting, should, one would think, produce tension in the scapular spine. This is complicated but further supported by the fact that in many bats the scapula is also strongly bent ventrally because it extends far out from the trunk like the first segment of a wing. This, too, might make the dorsal scapular spine act

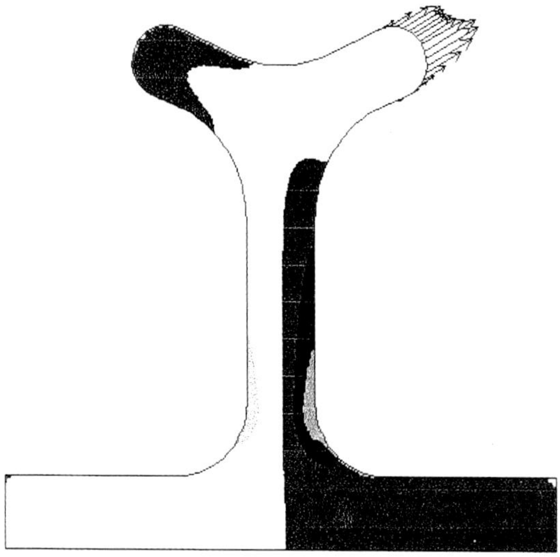

Fig. 9. Stresses in a simulacrum of a cross-section of the spine of the scapula on the scapular blade. A muscle is pulling on only one lip of the crest of the spine. The shaded contours show that the spine is undergoing bending.

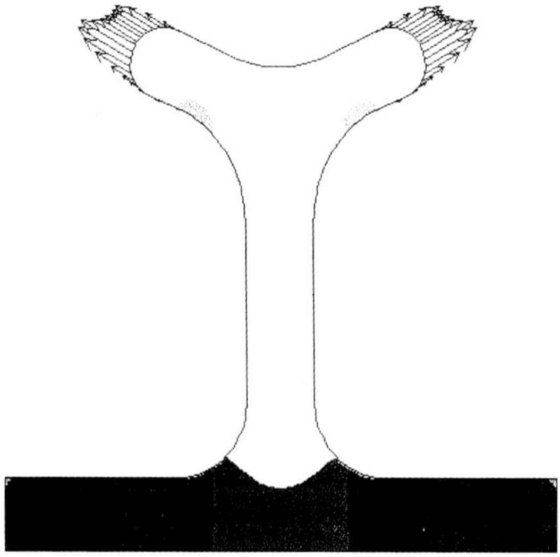

Fig. 10. Stresses in a simulacrum of a cross-section of the spine of the scapula on the scapular blade. Muscles are pulling on both lips of the crest of the spine. The lack of contours show that the spine is in total tension.

like a dorsal tie taking up the tension on the outside of a ventrally bent scapula, rather than acting just as dorsal bony ridge strengthening a flat scapular plate.

I was therefore delighted when inspection of bat scapulae, dissections of bat shoulders, and discussion with my graduate student now Dr Timothy Strickler (who did his doctoral work on bats and flight, 1975) indicated that the bony crest is often replaced by a ligament and a fascial sheet (Fig. 11). In other words, in one situation where it is likely that the crest bears tension for most of the time, the crest is actually a powerful connective tissue ligament not bone.

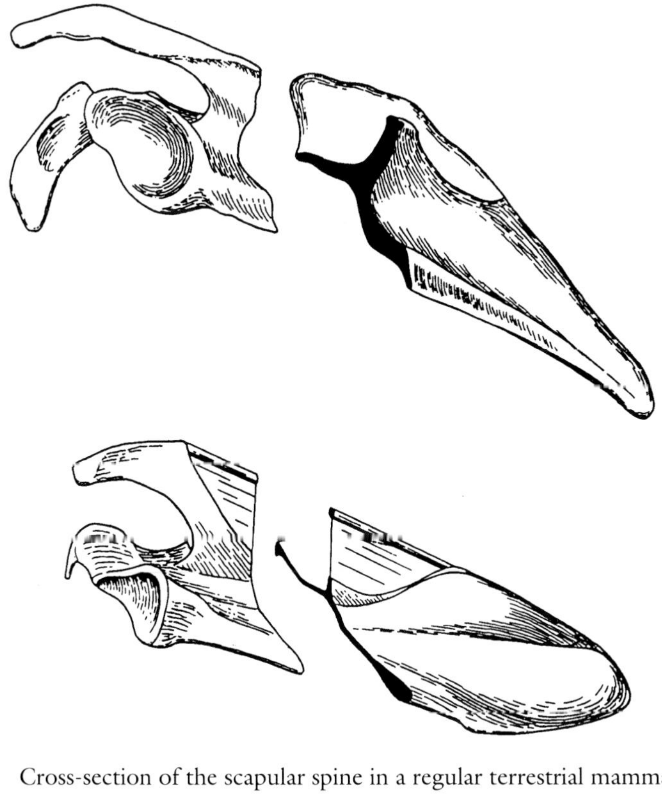

Fig. 11. Cross-section of the scapular spine in a regular terrestrial mammal, above, and in a bat, below. In the bat most of the spine consists of a connective tissue sheet; the crest of the spine consists of a thick round ligament.

Two Dimensions: Tension in a Flat Plate of Bone

I next took Currey's question into two dimensions: a flat two-dimensional bony plate. Again I had the scapula in mind. The arrangement of muscles originating from each side of the scapular blade implies that, at least in special cases where compression from the glenohumeral joint might be overweighed by tension from muscles (supraspinatus and infraspinatus on the one side, and subscapularis on the other), tension might exist in the scapular blade. This could not be fully tested because data about the actual sizes of the various loads were not available. However, using simulated loads, it did seem possible that this might sometimes occur, perhaps especially, in those primates with the largest scapulae like humans and gorillas. In humans the upper-limb does not participate in bearing the body weight so muscular forces producing tension might well outweigh forces due to compressive loads at the shoulder joint. In gorillas where the upper limb muscles are enormously powerful, again the muscles forces might well be the greater.

These species are, as it happens, the very species in which the scapular fossae are at their relative thinnest; holding the bone up to the light readily confirms this! On rare occasion in gorillas and humans a foramen is found in dried bone preparations (Fig. 12). Of course, this is not an open foramen in life. In life it is closed by the fusion of the two periosteal membranes from the two sides of the scapular blade lying across the foramen. It is this periosteal membrane, greatly strengthened by connective tissue fibres, that gives origin to the muscles on the two sides. It seems possible that in these cases there is no compression bearing bone but a tension-bearing collagenous membrane because overall tension exists.

A situation like this also obtains in the pelvic obturator foramen filled in by the obturator membrane and with the obturator internus and externus arising from each side of the membrane. It exists between the double forearm (radius and ulna) and leg (tibia and fibula) bones where the intervening space is infilled by the respective interosseous membranes with muscles arising from each side. In these cases of course, the membrane is permanent (as it were), in the scapula it only

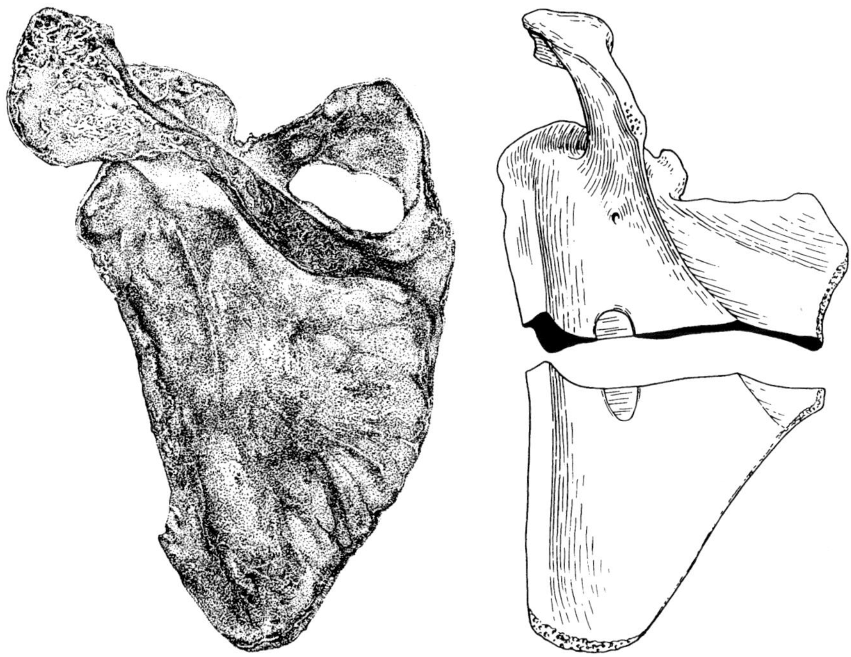

Fig. 12. Holes in bones: adult human and immature gorilla scapulae.

exists when there happens to be a scapular foramen. There are many other examples, even outside mammals (e.g. in reptile skulls), where bony foramina (or even just bony notches) are partially or completely infilled by fascial sheets that give rise to powerful muscles.

Three Dimensions: Tension in a Block of Bone

The problem was finally taken into three dimensions by examining the situation where two large and powerful tendons/ligaments arise from each side of a short more or less cubical bone, suggested again that they might generate, overall, tension in the bone. Take the case of a tendon winding around a bony pulley. It seems clear, on the basis of biomechanical thinking, that bone (a sesamoid) might exist in tendons where the degree of movement was small and a given portion of the tendon was in compression most of the time (upper frame of Fig. 13 where the portion of the tendon labelled 4 and 5 is in compression at both extremes of the tendon movement). It seems equally clear that such bone should not exist in tendons where the degree of movement is great so that no either one portion of the tendon is ever always in compression (e.g. lower frame of Fig. 13 where none of the numbered parts of the tendon are in compression at both extremes of the tendon movement).

These two opposite examples model real situations in the human body. One of these is where the two tendons of the short big toe flexor muscle wind around the under-surface of the metatarsal head but with not much movement along the length of the tendons. Those tendons have sesamoids embedded in the internal surfaces of the tendons. The other is the tendon of obturator internus that winds around the ischial notch of the pelvis. The various movements of the femur on the pelvis dictate that the excursion of this tendon around the ischial notch will be very considerable so that no one part is in compression all the time. This tendon has no sesamoid.

An interesting intermediate case is the tendon of the peroneus longus muscle. This winds around the lateral malleolus at the ankle and again around the groove in the cuboid as it passes to its insertion under the foot. In both of these sites where the tendon changes direction by winding around a bony pulley the tendon is thickened and there is sometimes found a fibrocartilaginous pad (looking rather like a sesamoid bone but made of cartilage instead) embedded in the internal surface of the tendon and rubbing against the pulley. Sometimes, indeed, this fibrocartilaginous pad may actually be ossified so that it is truly a sesamoid bone. This ossification was at its largest in the foot of a Celebes monkey that had walked on two legs all of its life! It may well be that this is at the interface of the mechanical situation producing a sesamoid.

However, again, more extended analysis by Jens Hirschberg using FEA and FLAC was applied to the more specific case of the patella, a large sesamoid, but one which is pre-programmed in development. Yet there is also a biomechanical ontogenetic element. Even when the

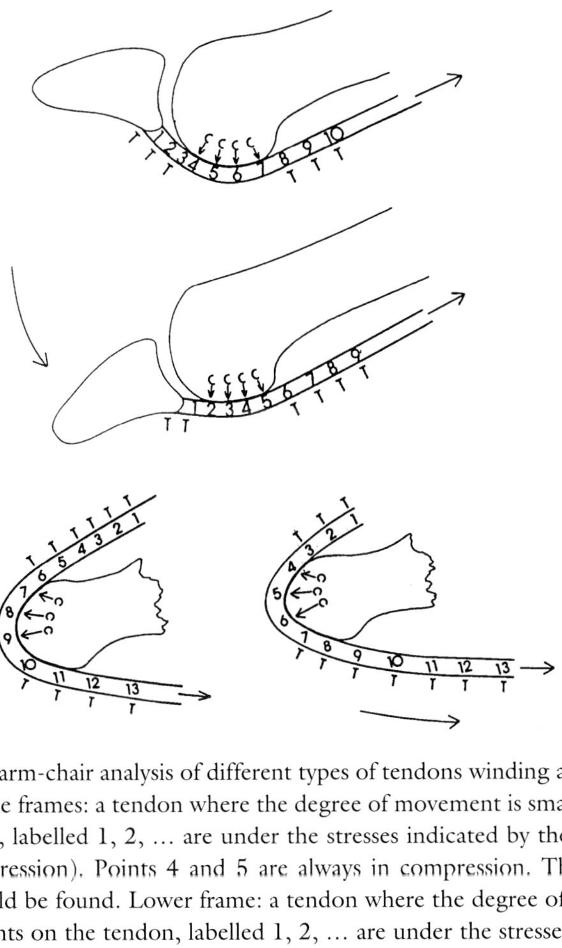

Fig. 13. An arm-chair analysis of different types of tendons winding around bones. Top and middle frames: a tendon where the degree of movement is small. The points on the tendon, labelled 1, 2, ... are under the stresses indicated by the Ts (tension) and Cs (compression). Points 4 and 5 are always in compression. This is where a sesamoid should be found. Lower frame: a tendon where the degree of movement is large. The points on the tendon, labelled 1, 2, ... are under the stresses indicated by the Ts (tension) and Cs (compression). There are no points that are always in compression. A tendon like this should not have a sesamoid.

patella is removed surgically a new bony structure is formed in the patellar position.

The patella has a powerful quadriceps tendon attaching cranially and a powerful patellar ligament attaching distally. In this case, analysis to see whether perhaps tension might be present in the patella indicates fairly conclusively that the situation is primarily compressive

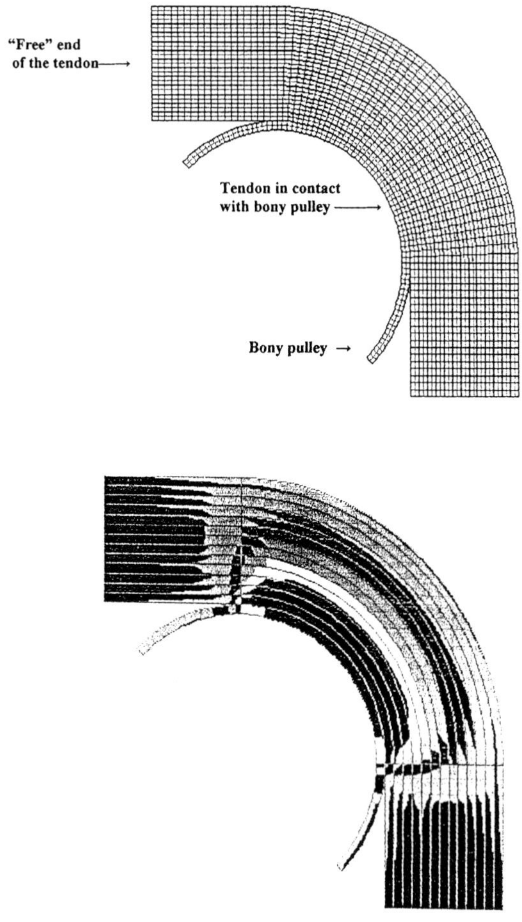

Fig. 14. FLAC of a tendon winding round a bone. The inner eleven contours are in compression, the outer four contours in tension. The compressive portion is where a sesamoid is found.

(Fig. 14). Such tension as exists is mainly located along the superficial surface of the bone; this surface, in life, is covered by a strong collagenous sheet that is the extra-patellar extension of the most superficial layers of the quadriceps tendon and patellar ligament. In the case of other smaller sesamoids, the bony pip is often totally embedded on the under-surface of the respective tendon. In other words, collagen, not bone, exists where it can be shown that tension predominates; bone only exists where it can be shown that compression predominates.

The Case of the Disappearing Incus

Further answers to this question arose through a chance meeting with an ear, nose and throat surgeon, Francis Lannigan who had, at that time, only just moved to Western Australia. He approached Paul O'Higgins (now at the University of York, UK but then at the University of Western Australia) and me with a request regarding the mechanical function of the human incus. How did we think that the human incus worked during hearing? Why did the human incus undergo resorption in certain situations? What might be the mechanical implication for bone grafts or artificial replacement of the incus in these clinical situations during life?

The best analysis of this situation, obtaining the real strains in the incus during function, obviously cannot be carried out. And in any case I was not about to make any great investment in the problem unless I had a little more evidence. I therefore made initial crude estimates by an arm-chair thinking examining the mechanical implications of the movements of the middle ear bones during function. Lannigan supplied me with accurate shape information in two dimensions for the incus ear ossicle. My assessments implied a paradoxical result: that a large portion of the bone must be under tension during both cycles of movement (Fig. 15). This is paradoxical, of course, because, in all other bones so far examined, if a major portion of the bone is strained in tension during one phase of a cyclical movement (e.g. during locomotion) that same portion is strained in compression during the opposite phase (e.g. Lanyon and Smith, 1970). To the degree that this last statement is true then overall tension (*sensu* Currey above) does not exist.

This result was so surprising that Paul O'Higgins and I tried a second (still very simple) estimation using a two-dimensional rubber model of the incus on which had been drawn an orthogonal grid. The distortions of the grid during simulated movement also implied that a portion of the incus could be in tension during both phases of its cycle during hearing (Fig. 16).

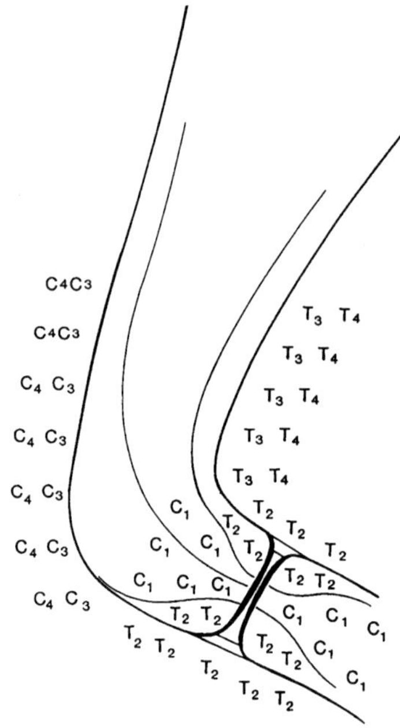

Fig. 15. An arm-chair analysis of the totality of stresses and strains in an object shaped like the lower end of an incus bone. C = compression, T = tension. The outer row of letters indicates stresses and strains when the incus taps against the stapes. The inner row indicates stress and strain when the incus distracts from the stapes. Movements like this do not usually occur in synovial joints.

This meant we felt it was worth O'Higgins carrying out a more extensive study of the matter. This was done using the FEA carried out with FLAC obtained with help of the Geomechanics group as explained previously. That analysis showed what one normally expects, that one portion of the incus (the left-hand edge in Fig. 17) was in tension during one phase of the movement and compression during the other. But the analysis also confirmed that, indeed, there is a particular portion of the incus (the right-hand edge and concave face leading towards the joint between the head of the incus and the stapes in Fig. 17) that is under tension during both stages of the ear drum movements during hearing. In the normal individual, this is obviously compatible

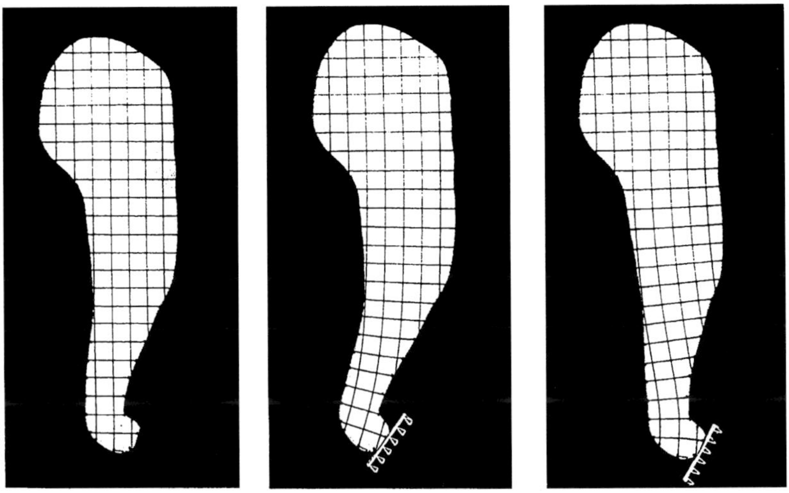

Fig. 16. A rubber model of an incus with a grid inscribed. Left-hand side figure: no loading, the grid is rectangular, i.e. no distortion. Middle figure: loading of the head of the incus when the incus taps against the stapes. The distortion of the grid indicates tension on the right-hand side edge of the incus. Right-hand side figure: loading of the head of the incus when the incus pulls away from the stapes (i.e. in distraction, the incus taps against the stapes). The distortion of the grid again indicates tension on the right-hand side edge of the incus.

with normal structure of the incus. This implies that we have found a situation where tension seems to be compatible with the presence of bone.

At that point Lannigan revealed to us the reason for his questions. It was indeed that it was exactly that portion of the incus that we identified as being under net tension that was sometimes resorbed (Fig. 18). This occurred particularly in older males working with noisy machinery, especially whistling and shushing sounds. It did not occur in individuals in rural environments. Lannigan had further noted that even when a small piece of bone was grafted in to replace the lost incus in these cases, it too, was shortly afterwards often resorbed. All of this had been established visually using scanning electron microscopy of includes in appropriate samples of individuals of different ages and sex. The visual results were

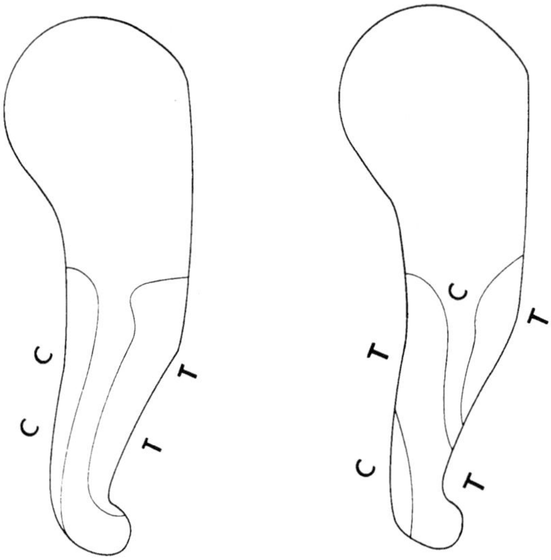

Fig. 17. A FLAC of the stresses in the incus during the tap against the stapes (left) and the distraction from the stapes (right). The lines indicate the contours of the stress differences. The left-hand side edge shows an alternation between tension T and compression C, the expected finding. The right-hand side edge shows tension T in both movements.

confirmed through morphometry of incudes (incus bones — Oxnard *et al.*, 1995).

I must add a caveat here. It could be that the determinant of bone existence or disappearance is not tension *per se* but some threshold level of tension above which bone is resorbed and, possibly some other threshold below which it is not. Under increased stresses induced by abnormal hearing situations (e.g. the continuously varying whistling and shushing sounds possibly existing in some work environments involving machinery, but not existing in normal situations) the degree of tension induced might be outside such thresholds. The bone on the inner curve of the incus might be being resorbed because tension of that higher degree existed there; the enclosing periosteum is still present as a kind of ligament. It is also likely that the height of the threshold differs whether one is going from compression to tension, or *vice versa*.

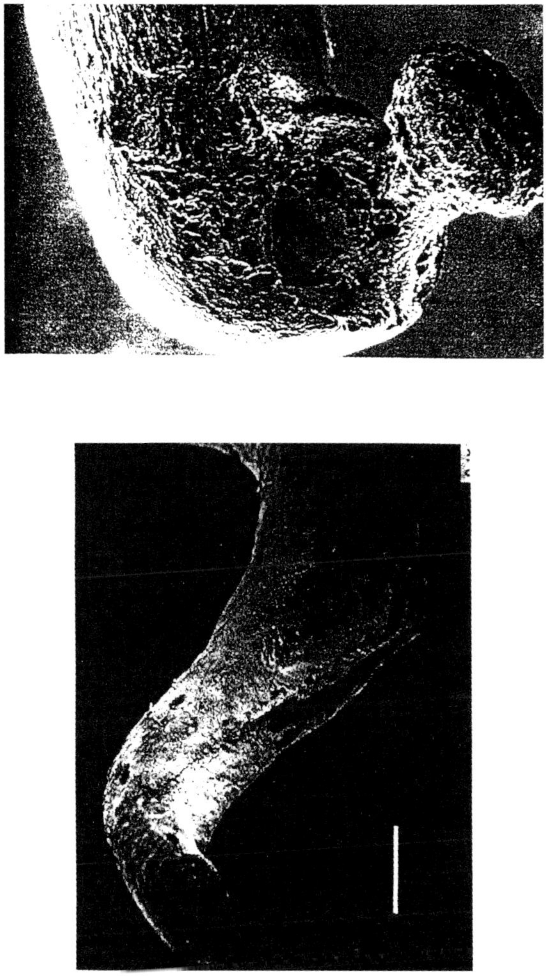

Fig. 18. SEM pictures of the head and neck of the incus showing marked resorption on bone. Upper picture, neck resorbed but head still present. Lower picture, head totally resorbed, large area of resorption of the neck and descending process on the concave side.

This situation in the incus is so far the only one I have found where tension seems to occur in a major portion of a whole bone. However, studies of small portions of bones: e.g. sites of tendon attachments, sesamoid bones within tendons and bony trabeculae in cancellous bones as described above are also possible instances. Much of this work was carried out by Jens Hirschberg employed on my ARC grant at the University of Western Australia.

Are Long Bones Always Hollow?

A final problem that may be worth examining relates to the long bones of the limbs. These are described in most anatomical texts as has having a shaft and two ends. Though the ends contain cancellous bone (like a whole vertebra, the subject of the next chapter), the major portions of the shaft form a tube of cortical bone with a central hollow cavity (the marrow cavity). The biomechanical rationale behind the tubular shaft of a long bone is usually given as strength with lightness for efficiency in bending.

In the standard description of pure bending the highest stresses are at the surfaces of the tube with tension on the one side and compression on the other. There is a neutral axis through the centre where the stress is zero. If the bone were solid there would still be zero stress at the midline; in addition the bone close to the midline would bear very little stress. As a result, the absence of the central material, i.e. the presence of a marrow cavity, would scarcely weaken the bone at all. Certainly the slight increase in strength is not worth the high metabolic cost of maintaining bone in that position. This is the simplistic assumption around the tubular structure of bone.

In fact, this is a simplification of the real situation. In reality a bone is not under pure bending. There is pure bending due to a bone acting as a lever, to be sure, but this is combined with the axial compression of weight bearing. In addition, bones are rarely straight; axial loading of a curved bone produces bending (very few bones are completely straight, though some are almost straight; in humans, for example, the fibula).

Under this more complex circumstance the mechanical efficiency of a tube for bending depends upon an engineering quantity: the second moment of area. The architectural factor in that quantity is the diameter of the bone divided by the wall thickness. This gives information about efficiency in bending and compression, about 'bendability' to coin a term. In theory the strongest column under pure bending with no axial compression is a tube that has an infinite diameter and an infinitesimal wall thickness; think of trying to bend a cooling tower in a power station! Of course, this is an engineer's abstraction. Likewise the strongest column in pure compression is one that is solid that is, has a diameter of twice the wall thickness when the wall thickness is the radius of the column. Equally, this is the practicality of an architect's column in a cathedral.

It is therefore possible to examine this measure in the limbs of a variety of animals (and this has been done by both Alexander, 1983 and Currey, 2002). Figure 19 shows the result. Where bending predominates then this ratio is large. Some pterodactyls have very large values (Fig. 19). In gliding, the wing bones bend, but because the animal glides, the wings are not flapped by muscular activity to anywhere near the degree in (say) birds, so that the compression along the length of the wing due to off-axial muscular activity is much

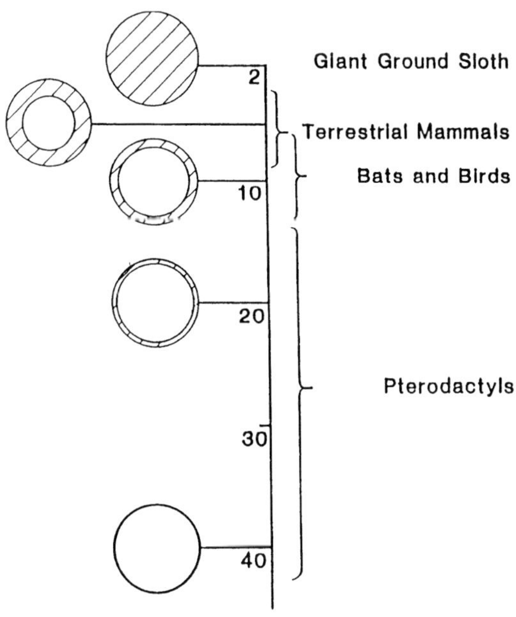

Fig. 19. Plot of values of limb bone diameter divided by limb bone wall thickness for a variety of animals.

smaller than in a true flying creature. Birds have a somewhat smaller ratio because of the powerful muscular forces compressing the wing bones as in Fig. 19 (but their mechanics is complicated by the presence of air sacs within their wings). Bats come next. Those terrestrial creatures that are smallest, lightest and most bouncy come next again; the heaviest creatures (with, therefore, greater degree of axial compression due to body weight though still with much muscular activity) are at the lower end of the axis.

The lowest figure that one can have is 2.0 (diameter divided by wall thickness when wall thickness is half diameter because there is no marrow cavity). The aforementioned columns in cathedrals have a value of 2.0: that is, they are solid. They do not move at all. They support the sometimes extremely heavy roofs of the earliest cathedrals. Are there any animals that are comparable? It turns out that there are: the giant ground sloths of the Californian tar pits have no marrow cavities (Fig. 20). Unfortunately, however, we do not actually know for certain how they moved though it is obvious that they were very large and heavy. If they were also very slow (as seems likely) then the major component acting on the limb bones would be, somewhat like the cathedral column, under great compression (even though muscular activity would produce some bending).

Giant ground sloths, however, are edentates. Could they have this feature for that reason? This is complicated by the fact that two living edentates (also extremely slow creatures, the two tree sloths) also have almost no marrow cavity. In these creatures, the marrow cavity has the dimensions of a needle; their diameter/wall thickness ratio is very close to 2. The fact that their limbs but not their bones bear weight by tension rather than compression does not deny the idea. However, this means that this could be an edentate characteristic and nothing to do with biomechanics at all.

Happily almost the first thing that occurred when I reached Australia was a trip to the West Australian museum.

"Have you any large slow marsupials?" I asked.

"Yes" said John Long, the curator.

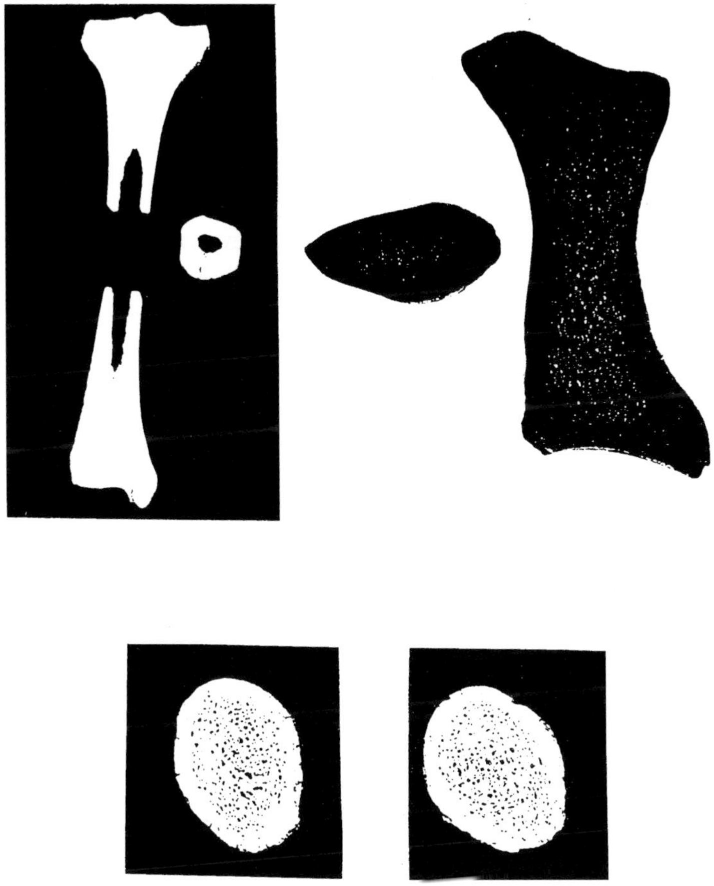

Fig. 20. Upper frames, longitudinal and cross-sections of tibiae of a human and a giant ground sloth. Lower frames, cross-sections of long bones of giant fossil marsupials (*Zygomaturus* and *Palorchestes*).

"Can I cut sections of their long bones?"

"No need for that" said John Long.

He opened a drawer. It contained many long bones of giant marsupials like *Zygomaturus* and *Palorchestes.* Many were broken across. They had no cylindrical marrow cavities (Fig. 20). It must be a biomechanical adaptation. It is not an edentate characteristic (Oxnard, 1993).

Again, therefore, we had evidence relating to the effects of tension on bone.

Question Again: Does Tension Ever Exist in Bone Overall?

After all this do we now have an answer, even if only tentative, to Currey's question: does overall tension (in the sense of tension existing for the period of time influencing adaptation) ever exist in bone? Does this have implications for the mechanical adaptation of bone? The above discussion implies that: where overall tension is unlikely to exist, bone does; where overall tension probably does exist, bone does not. Yet it is clear that this is not the whole story. For further understanding of this problem we have to move on from macroscopic components of bone shape to microscopic elements of bone structure: the subject of the next chapter.

References

Alexander, RMcN, *Animal Mechanics*, 2nd ed., Blackwell, Oxford, 1983.

Avis V, The relation of the temporal muscle to the form of the coronoid process, *Am J Phys Anthropol* **17**: 99–104, 1959.

Coker EG, Filon LNG, *A Treatise on Photoelasticity*, 2nd ed., Revised by HT Jessop, Cambridge University Press, 1957.

Cundall PA, Board M, A microcomputer program for modelling large-strain plasticity problems, in Swoboda S (ed.), *Numerical Methods in Geomechanics*, New York, 1988.

Currey JD, The adaptation of bones to stress, *J Theor Biol* **20**: 91–106, 1962.

Currey JD, *Bones: Structure and Mechanics*, Princeton University Press, Princeton, 2002.

Evans FG, *Stress and Strain in Bones*, Thomas, Springfield, 1957.

Frocht MM, *Photoelasticity*, Vols. 1 & 2, Wiley, New York, 1941.

Frost HM, *The Laws of Bone Structure*, Thomas, Springfield, IL, 1964.

Galileo, G, *Discourses and Mathematical Demonstrations Concerning Two New Sciences Pertaining to Mechanics and Motion*, Macmillan, New York, 1638 (Translated by H Crew and A de Salvio, 1933).

Hirschberg J, A study of the effects of tubercles and pits on the stresses in bone at tendon attachments, MSc Thesis, University of Western Australia, 1997.

Hirschberg J, Simulations of mechanical adaptations and their relationship to stress bearing in skeletal tissue, Doctoral Thesis, University of Western Australia, 2005.

Hirschberg J, Milne N, Oxnard CE, Biomechanics of the tendon/bone interface, *Persp Human Biol* **5**: 55–68, 2000.

Hoek E, Grabinsky M, Diederichs M, Numerical modelling for underground excavation design, *Short Course Lecture Notes on Ground Control Principles*, Ontario, Laurentian University, Geomechanics Research Centre, 17, 1990.

Hsu A-T, Trabecular architecture and finite element analysis of the human calcaneus, Doctoral Thesis, University of southern California, 1989.

Kupczik K, Dopson C, Fagan M, Crompton R, Oxnard CE, O'Higgins P, Assessing mechanical function of the zygomatic region in macaques: Validation and sensitivity testing of finite element models, *J Anat* **210**: 41–53, 2007.

Lanyon LE, Smith RN, Bone strain in the tibia during normal quadrupedal locomotion, *Acta Orthopaed Scand* **41**: 238–248, 1970.

Moore WJ, The Mammalian Skull, Cambridge University Press, Cambridge, 1981.

Murray PDF, *Bones: A Study of the Development and Structure of the Vertebrate Skeleton*, Cambridge University Press, London, 1934.

Oxnard CE, Tensile forces in skeletal structures, *J Morphol* **134**: 425–436, 1972.

Oxnard CE, *Form and Pattern in Human Evolution*, University of Chicago Press, Chicago, 1973.

Oxnard CE, *The Order of Man*, Hong Kong University Press, Hong Kong, 1983, Yale University Press, Princeton, 1984.

Oxnard CE, Mechanical stress and strain at a point: Implications for biomorphometric and biomechanical studies of bone form and architecture, *Proc Austral Soc Human Biol* **3**: 57–109, 1991.

Oxnard CE, Bone and bones, architecture and stress, fossils and osteoporosis, *J Biomech* **26**: 63–79, 1993.

Oxnard CE, Lannigan F, O'Higgins P, The mechanism of bone adaptation: Tension and resorption in the human incus, *Recent Adv Human Biol* **2**: 105–125, 1995.

Runnion CK, Oxnard CE, Robertson WV, Windsor CR, Biomechanical modelling of vertebrae using experimental stress analysis, *Proc Austral Soc Human Biol* **4**: 125–133, 1991.

Strickler T, The functional morphology of the pectoral girdle muscles in the Chiroptera, Doctoral Thesis, University of Chicago, 1975.

Thompson D'A, *On Growth and Form*, Cambridge University Press, Cambridge, 1917.

Timoshenko S, *Strength of Materials. Part I: Elementary Theory and Problems*, 3rd ed., Van Nostrand, Princeton, 1955.

Washburn SL, The relation of the temporal muscle to the form of the skull, *The Anatomical Record* **99**: 239–248, 1947.

Chapter 2

A Fifty Year Love Affair with Spongy Bone

How It Started

The previous chapter has documented how I became involved in biomechanics and bones. It is aimed primarily at my attempts to understand something of the biomechanical adaptations of the form of whole bones and macroscopic parts of bones. It particularly documents how I became involved with numerous engineering colleagues in theoretical learning about the strength of materials and in practical usages of the technologies. Of course, such studies also required an understanding of the architectures involved in bearing mechanical stresses and exhibiting deforming strains. That chapter suggests that the adaptation of macroscopic bone form is related to the precise details of compression and tension in some overall sense (Oxnard, 1971; Oxnard *et al.*, 1995). Where tension exists, a bony plate does not; a connective tissue sheet is present. Where compression exists, a bony plate does; a connective tissue sheet is absent. This seems reasonable; after all bone is strongest in compression, connective tissue in tension.

However, for these ideas to be properly tested, it must also be shown that they work at the microscopic level. The first part of the problem then was to acquire knowledge of the complex internal texture of bone (especially spongy bone which is most evident in sections and radiographs). Studying macroscopic bone form by observation and measurement is relatively easy. Finding out about microscopic bone structure is much harder. This can obviously be done by visualising bone structures through microscopic examination of sections of bones and by radiography looking at the shadows of the interior of whole bones or sections of bones. Today, the information can be much improved by microCT scans and other techniques (though these were not available in those earlier days). No matter how the structures are visualised, however, they need to be rendered quantitatively in order to carry out analyses. In those earlier days quantification was generally achieved using radiographic densitometry or *ad hoc* microscopic morphometry. Such methods provided single measures such as the total amount of bone (densitometry), or the average sizes of arbitrarily defined bony trabeculae (morphometry).

However, for biomechanical purposes, it was apparent to me that techniques that could define the changing nature of trabecular complexity from point to point were needed. Standard microscopic morphometry and densitometry were not sufficient. Sections and radiographs could indeed provide the initial picture though, today, microCT provides better pictures.

My first real attempts at such quantification came, paradoxically, from the work of two geologists (John Davis and John Green) employed by the Kansas State Geological Survey. They were using a 22-foot long optical bench (with a laser and a camera) to photograph the Fourier transform of the complex patterns seen in sections of rocks. The American Petroleum Industry was willing to fund such work to find out how oil flows through porous substrates. This technique could give statistical information about patterns far more definitively than was provide by *ad hoc* measures of individual elements of the patterns. The moment I saw their results, I knew that the technique would be excellent for looking at the 'porous' structure of bone. It should be remembered, of course, that computer generated Fourier transforms were, at that time (the 1960s) far in the future. Indeed one confirmatory computational test that was carried out on the mainframe computers in Kansas took 9 months to generate a single Fourier transform.

The idea of using such a technology was further consolidated when I took a course in optical data analysis provided at the University of Milwaukee by Harold Pincus. It showed me how that 22-foot optical bench could be curled up with mirrors into a small box.

The technique worked; it was possible to characterise a complex pattern or texture whether this arose directly from a picture of trabeculae in a section of a bone, or indirectly from the 'shadows' of many superimposed trabeculae in a radiograph of a bone section or a whole bone.

Given that we could elucidate cancellous bone structure, the second half of the problem was to understand how such a structure, when loaded, bears stress and is strained. Again, in those days, the modern methods of finite element analysis (FEA) were far in the future. The earlier photoelastic methods were extremely useful (Frocht, 1941; Coker and Filon, 1957) in engineering. But in biology, especially in trying to understand how cancellous bone bears stress they were not satisfactory. Making photoelastic analyses of the shape of sections of whole bones was difficult enough. Making photoelastic models of the shape of sections of spongy bone was extremely difficult. They were even more difficult to load. Putting such studies into three dimensions using 'frozen' (though actually heated in an oven) photoelastic analysis was even more difficult; I did carry out a partial study but it was never completed. In any case, it involved examining the three-dimensional model with stresses 'frozen in' by cutting sections, and that, somewhat, defeats the purpose.

In addition, in the days before computer graphics, such studies meant a great deal of work with a drafting table. Yet I did manage one investigation that started to show how cancellous structures bear stress, and what adaptations in cancellous structures might be associated with increased mechanical efficiency in stress bearing. I used a series of holes in a plate as a model of trabecular structure in two dimensions. Of course, the stress caused by a single circular hole in a plate, was a standard student exercise in stress analysis in those days. Multiple circular holes in a plate (rivet holes) had also been studied by German engineers early in the 20th century as they strove to understand stresses in metal plates in ships (especially submarines).

What I did, therefore, was to take a number of circular rivet holes and change their shapes towards those existing in a two-dimensional slice of cancellous bone. First, the circular holes were changed into ellipses; ellipses are a little more like the shapes of the spaces in

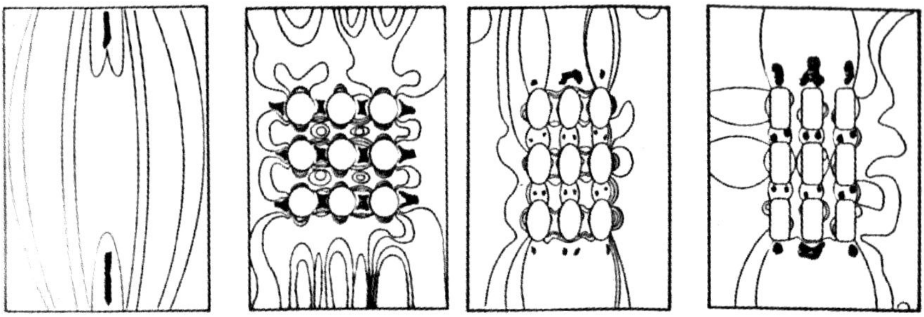

Fig. 1. The stresses in a plate as determined by photoelastic analysis. First, the gradient of stresses (as shown by the contours) in a plate with no holes is low — the contours are few and far apart. Second, the gradient of stresses in a plate with circular holes is high — the contours are many and very close together. Third, the gradient of stresses is lower when the holes are elliptical and arranged in the direction of loading. Fourth, there is a final reduction in the gradient of stresses when the holes are rectangular with rounded corners. Of the various patterns of holes, this last is mechanically the most efficient (Oxnard, 1991).

cancellous bone. This immediately reduced the overall stresses in the structure. Then, the ellipses were changed into rectangles with rounded corners; such spaces are even more like those in cancellous bone. This further reduced the stresses in the structure (Fig. 1). It was immediately clear that modified architectures like these reduced stresses enormously.

In earlier studies of vertebral cancellous structure, working with then graduate student Harry Yang (Oxnard and Yang, 1981), we had found a particular vertebra in which, in addition to largely rectangular spaces with rounded corners, there were angulated facets ground onto (as it were) the rectangles. The facets were all angulated one way. We thought this might be because the loads on that vertebra were angulated. We, therefore, analysed two patterns, one with the correct angulation in relation to the loads, one with the angulation in the opposite direction. It was immediately obvious that the pattern with the correct angulation was much more efficient, mechanically, than the other (Fig. 2). This seemed to be a further confirmation of the mechanical efficiency of spongy bone at this, admittedly rather simplistic, level.

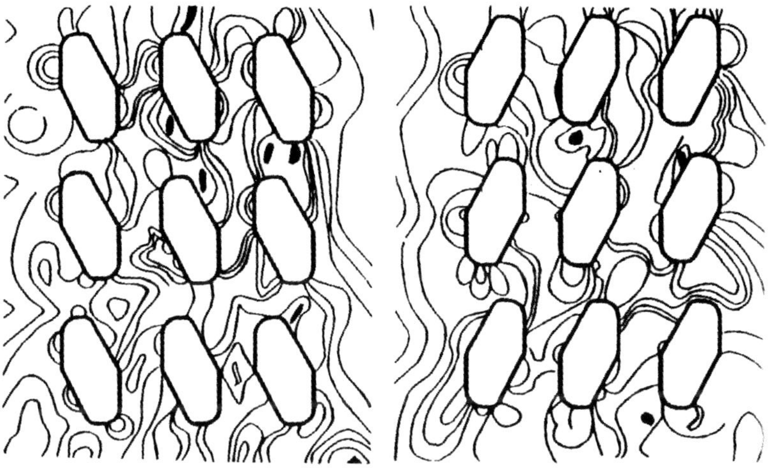

Fig. 2. The stresses in a plate when the rectangular holes with rounded corners have an angulated flat. In the first frame, the angulated flat is directed against the loading direction; in the second, it is directed with the loading direction. The numbers of contours in the second are significantly reduced compared with the first.

More recently, however, such studies have been enormously improved using computational stress and strain analysis. At first this was with standard FEA on mainline computers (in collaboration with engineer and physical therapy doctoral student, Hsu, 1989). Though initially he did not speak English very well, he spoke Mathematics extremely well, and carried out the first FEA of a complex bone in which I was involved. We worked on the calcaneus, the heel bone. Many prior workers had assumed that the major stresses acting on the human heel in walking and running were during heel strike. The very words, 'heel strike' and the picture in the mind's eye that it evokes, seems to indicate a powerful and violent load. In contrast, however, these studies quickly made it clear that by far the largest stresses in the calcaneus in walking and running in humans were at the other end of the cycle, 'foot and then toe off'. It was to this specific set of stresses that the internal spongy structure of the human calcaneus was most correlated (Figs. 3 and 4). This is now well known.

Later, these stress and strain studies came to involve applications of FEA that could be carried out on desktop computers. For me this

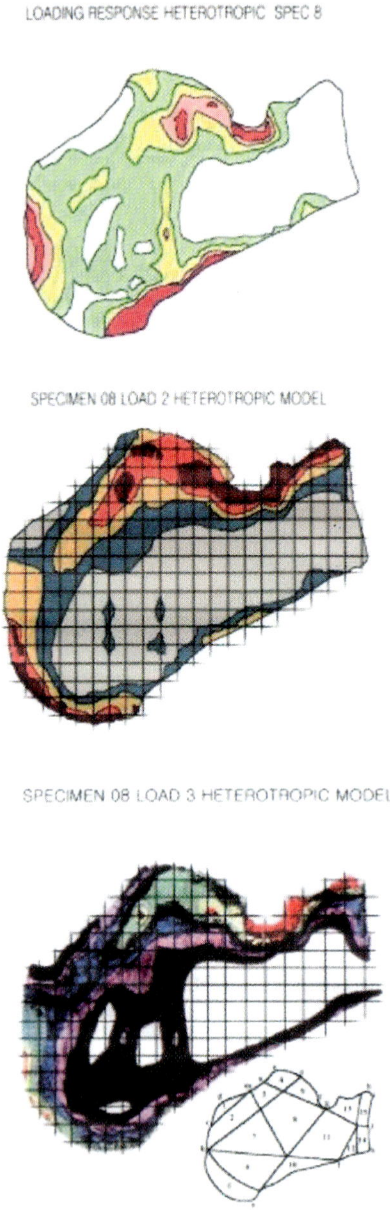

Fig. 3. The stress contours in a calcaneus in three different analyses, the first, when the calcaneus is loaded as at 'heel-strike'; the second, when it is loaded as at 'foot-flat'; the third, when it is loaded as at 'toe-off'.

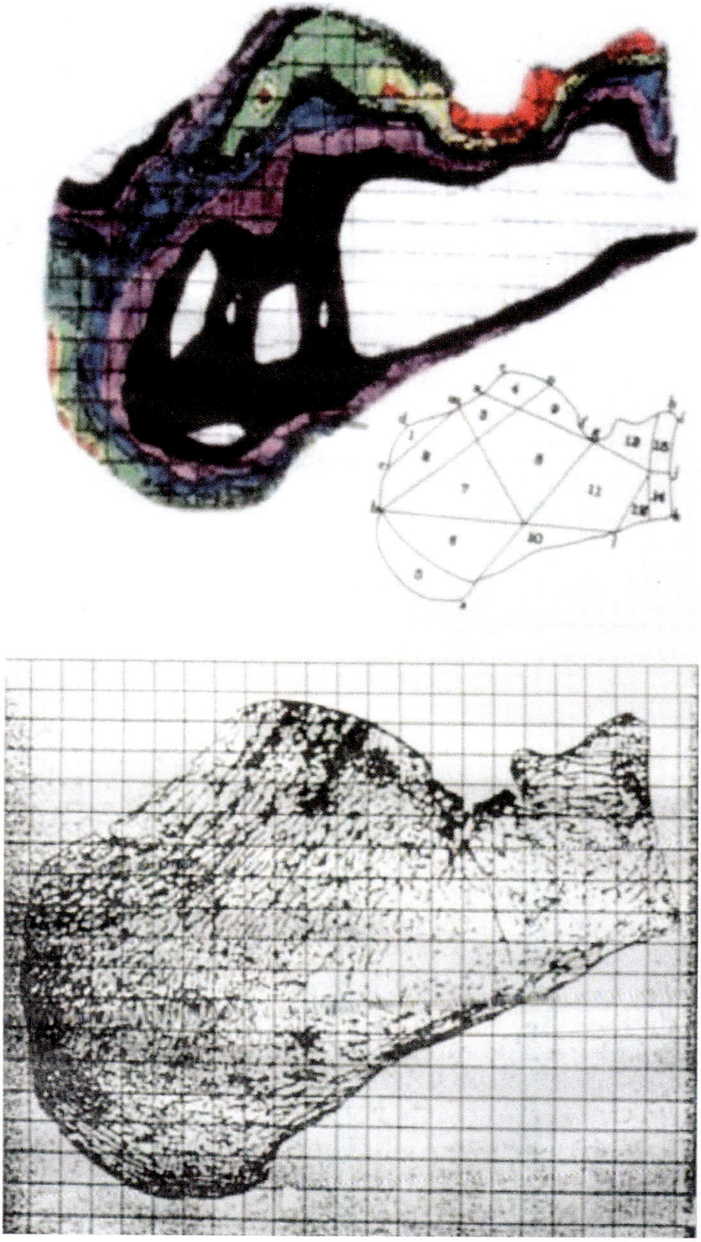

Fig. 4. This figure shows that it is with the 'toe-off' arrangement that the calcaneal trabeculae are most closely related ($P = 0.001$).

involved Fast Lagrangian Analysis of Continua (FLAC) (Cundall and Board, 1988), thanks to colleagues David Windsor and Wayne Robertson in Geomechanics in Western Australia (Runnion *et al.*, 1991).

Today I am involved in bioengineering collaborations at the Universities of York and Hull, UK with Michael Fagan, Paul O'Higgins, Kornelius Kupczic and Catherine Dobson (e.g. Kupczik *et al.*, 2007) and with Jens Hirschberg (e.g. Hirschberg, 1997; Hirschberg *et al.*, 2000; Hirschberg, 2005). Hirschberg was initially a graduate student at the University of Western Australia but is now a senior lecturer at Notre Dame University, Fremantle, Western Australia. Their abilities in two-dimensional (Hirschberg) and later three-dimensional (Fagan, Dobson and Kupczik) stress and strain analysis using finite elements promise to take us far beyond my initial investigations.

The Nub of the Problem

Let us now return to the beginning. My initial interest in the biomechanics of spongy bone was first stimulated (as indeed was also the case for many other investigators) by the apparently very clear relationship between the pattern of spicules (trabeculae) of spongy bone in the head and neck of the femur and the stresses that could be calculated as existing in such a shape when loaded (see previous chapter). That is, there seems to be a clear association between the right-angle network of stresses or strains that can be calculated as operating within a bone during loading, and the right-angle network of architectural elements (trabeculae) that actually exist within that bone (Fig. 5). This is a very old idea that was called 'the trajectorial theory of bone architecture' (reviewed in Murray, 1934). It seems especially confirmed by the study in which the stresses and architectures were calculated in a knee joint from a patient who had had the joint successfully excised, the femur and tibia fused, and who lived for many years thereafter. A two-dimensional analysis of the new stresses in this modified area of bone correlated so well with the two-dimensional pattern of spongy bone revealed at the eventual post-mortem, that the relationship seems cast-iron (Fig. 6).

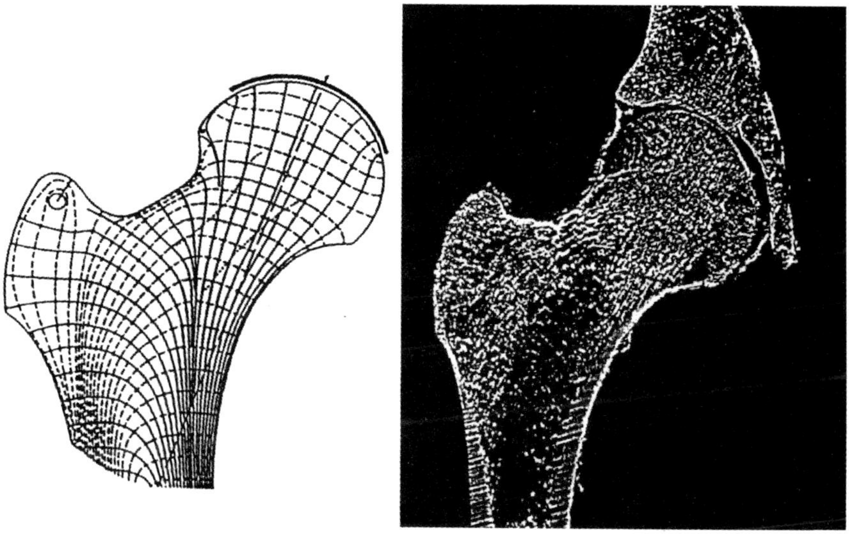

Fig. 5. The orthogonal network of principal stresses calculated as being in a shape of the head and neck of femur compared with a section (in the same plane) of an actual femoral head and neck. The similarity of these two orthogonal networks seems clear.

The association can be expressed as though it represents the reduction in bone mass that is most efficient for bearing the stresses and strains of function. It is as though the bone material were 'frozen' into trabeculae along the lines of principal stresses and/or strains. These principal stress/strain trajectories were then, and are often now even today, assumed to be one compressive and one tensile (not only Meyer, 1867; Murray, 1934; Evans, 1957; Frost, 1964, but even as recently as Geraets, 1994; Currey, 2002). Of course, the actual stresses/strains in a two-dimensional simulacrum of a cancellous network are completely different from those existing in a plane isotropic model of external bone shape (Oxnard, 1993, 1997). Though this idea is most attractive, and has been taught to generations of medical students, more thinking over the years has gradually required me to believe that this is but a pale skeleton of a real, far more complex, association.

Such thinking directed me to examine cancellous architectures and cancellous biomechanics, and the relationship between them, in

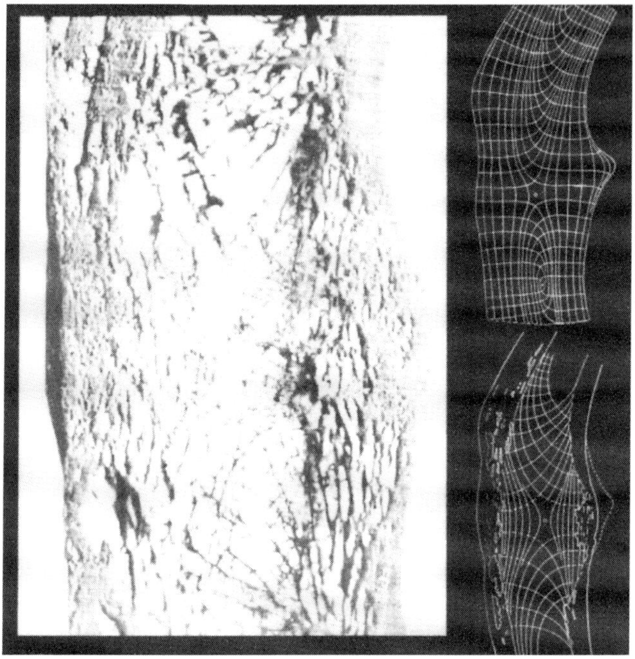

Fig. 6. The bony architecture in a long since healed femorotibial fusion. The actual section on the left shows the cancellous pattern. The lower diagram on the right shows a cartoon of that architectural pattern and the upper cartoon shows the stress pattern. The specimen shows a completely new architectural pattern that has adapted following healing and function. That new pattern is remarkably similar to the new pattern of stresses calculated as now existing in that bone.

more detail. Because, in those days, understanding cancellous architectures required extremely tedious but careful histology of bone, a hard tissue that was difficult to cut, I used radiographs of whole bones. Of course, radiographs, especially in those days, only gave shadows of many trabeculae superimposed. Nevertheless they showed fairly clear patterns and textures. I chose to examine vertebral bodies, partly because they were approximately symmetrical about the median plane, but mostly because it was easy to get consistent radiographs.

There was still the problem of analysing the patterns (better, textures) shown in the radiographs. As explained earlier, I had rejected microscopic morphometry and radiographic densitometry because they did not give detailed local changing textural information about a pattern, just overall statistical summaries. What alternative was there? This is where the relationship with members of the Kansas State Geological Survey came in. It was the accident of finding how they were examining the structure of oil-bearing rocks that provided me with a technology. Once I saw how the geologists did it in oil-bearing rocks I knew the technique was far better for my problem of spongy bone. The method involved using the Fourier transforming properties of a lens to obtain a Fourier transform of a laser beam passed through a transparency of the complex (but still two-dimensional) pattern of trabeculae exhibited in bone radiographs. This provides information about every element of a complex two-dimensional pattern or texture (Oxnard, 1973).

With an MD/PhD graduate student, later surgeon, Dr Harry Yang, at the University of Chicago, we applied optical Fourier transforms in an evolutionary context to whole radiographs of vertebrae of great apes and humans (Oxnard and Yang, 1981). Subsequently, and with the help of graduate student, later Forensic Anthropologist, Dr Alanah Buck at the University of Western Australia (Buck, 1998) and using first this optical method but later doing it computationally (Fast Fourier Transforms (FFTs) by then had been invented), we applied it in a clinical context to radiographs of thick sections of human vertebrae, including sections of vertebrae from individuals with osteoporosis. These sections were graciously provided by Dr David Johnson of Leeds University, UK.

It was the evolutionary study (Oxnard and Yang, 1981) that first alerted me both to the difficulties and the excitement of Fourier transforms. Thus, Fig. 7 shows a lateral radiograph of a chimpanzee vertebrae and its optical transform. It seems evident that the superimposed shadows of all the trabeculae within the vertebra summate to an overall radiographic pattern that is a marked right-angled network. This is clearly confirmed by the Fourier transform which is shaped like a Maltese cross implying that the pattern is indeed primarily a right-angle network. It was confirmed by examination of the section of the chimpanzee vertebra.

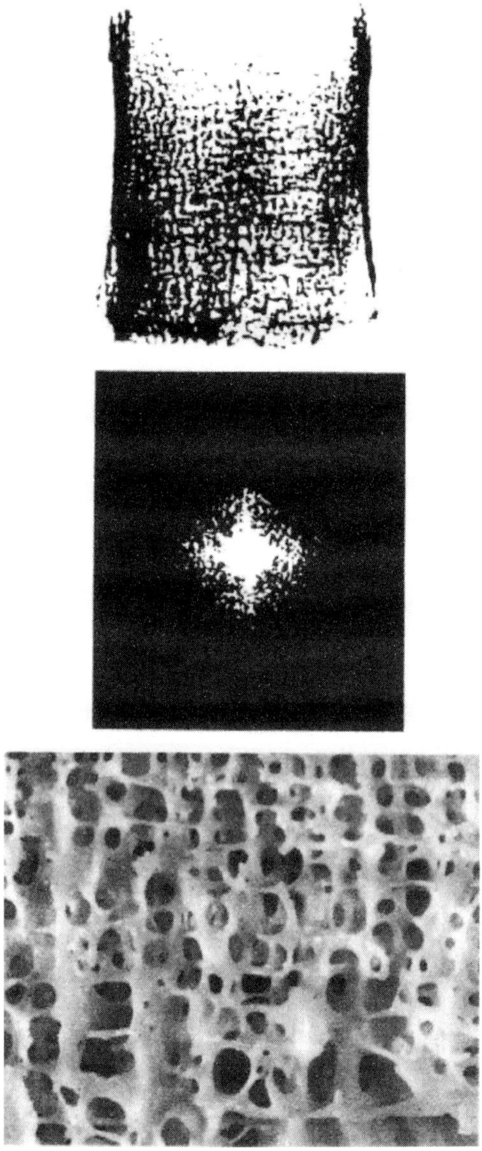

Fig. 7. Above, a lateral radiograph of a chimpanzee lumbar vertebra; centre, the Fourier transform of that radiograph; below, a section (in the same plane). All three show that, whatever minor variations there may be, the major texture of the radiographic shadows is orthogonal, the form of the transform is cruciate (like a Maltese cross indicating general orthogonality), the actual section confirms both of the above.

When, however, we examined a radiograph of an orang-utan vertebra, a radiograph that seems not so very different from that of the chimpanzee, the Fourier transform indicates completely otherwise. The transform has an approximately circular shape with many rays at various angles. This implies that the trabecular arrangement must be at a variety of angles other than orthogonal. The section of the orang-utan confirms that this is so. Its structure is rather more like a honeycomb than a right-angle network (Fig. 8).

Of course, I went on to speculate why this should be. My idea was that chimpanzees (and gorillas and humans, which all have Maltese cross-shaped transforms and therefore orthogonally arranged trabeculae) operated in a biomechanical mode where there truly were, most of the time, fairly consistent and similar stress/strain networks on vertebrae. These, I speculated, might summate to a generally similar overall orthogonal arrangement. In the orang-utan in contrast, I speculated, perhaps because of the extremely varied postures and movements in such a slow climbing acrobatic animal, there would many very different stress/strain networks at many different times. Under such a circumstance, a pattern of trabeculae that resembled polystyrene packing material (to protect a parcel from blows in all different directions) might be what was most adaptive in this animal. I suspect I am right; but the matter has to be taken further. Such studies would allow us, not only to understand bony adaptation to biomechanical forces in living species, but also to assess possible biomechanical adaptation in fossils. In other words, we might be able to proceed from fossil anatomy (which is known) to fossil behaviour (which is not). This would be a most useful evolutionary usage of the methods.

The study of the human materials with first, optical, but later computational Fourier transforms took us in a very different direction. As with the human sample in the prior study, most of the trabecular patterns were judged to be orthogonal, whether by inspection of radiographs, sections or Fourier transforms. However, at this point, the overall problem became much easier. The optical data analysis system using a laser required the maintenance of an optical bench, careful photography, and analysis of the transforms using computational methods that were not easily supported by the computers of those days. When better

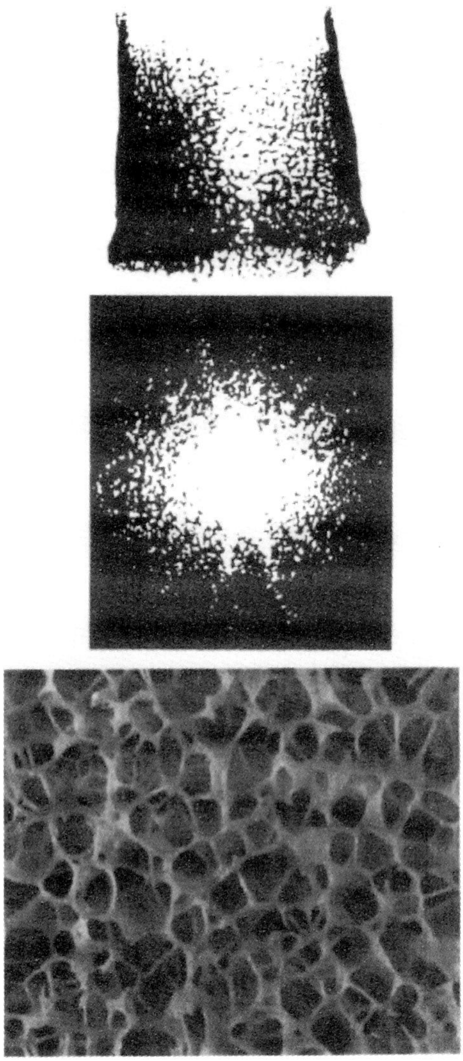

Fig. 8. Above, a lateral radiograph of an orang-utan lumbar vertebra; centre, the transform of that radiograph; below, a section (again in the same plane). The radiograph is very little different from that of the chimpanzee in the last figure. The transform shows, however, that the real pattern is completely different — there is no semblance of a Maltese cross (orthogonality); there is, by contrast, a starburst with many rays at different angles. The section confirms that the trabecular pattern is completely different from that of the chimpanzee. It is entirely consistent with the starburst transform with elements arranged in many different nonorthogonal directions.

computers and FFTs came along Alanah Buck and I were able to dispense with the optical approach in favour of the computational. This allowed us to analyse many more specimens, to get better quantitative information about them, and to apply multivariate statistical methods for their analysis (this was a technique with which I was already familiar).

In summary, we again confirmed things we already knew. Most human vertebrae in sagittal section or lateral radiograph exhibit patterns that confirm the orthogonal arrangements of trabeculae (Fig. 9). But,

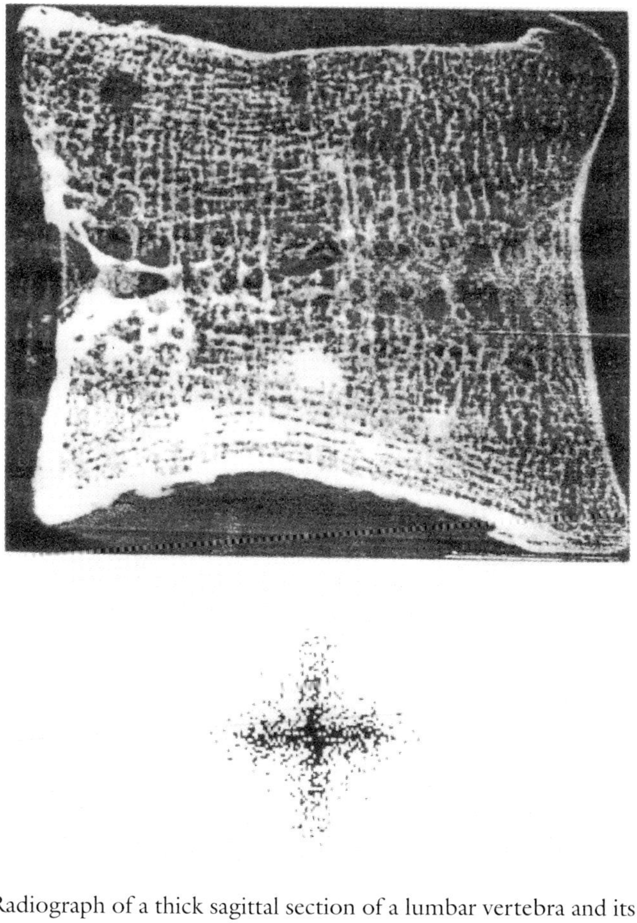

Fig. 9. Radiograph of a thick sagittal section of a lumbar vertebra and its transform in a young male. The cruciate transform confirms the orthogonal architectural arrangement of trabeculae in the section (which is any way quite obvious).

Fig. 10. Radiograph of a thick sagittal section of a lumbar vertebra and its transform in an old female. The rugby ball-shaped transform indicates that the overall orthogonality has been considerable degraded. This is the Fourier 'signature' of osteoporosis.

we also got some surprises. We already knew (from inspection of the sections) that many of the middle-aged and almost all of the older women in our sample had osteoporosis. This was confirmed by the Fourier transforms that indicated that the normal orthogonal network was degraded (Fig. 10). We also already knew that the middle-aged men did not have osteoporosis, but that the very old men had as great a degree of osteoporosis as the very old women. Nothing new here! These transforms did confirm, however, that our technique was sensitive enough to detect these findings. We even found that, whereas our middle-aged (35–55 years old) male sample was fairly uniform, our middle-aged female sample was clearly divided. The division was, perhaps naturally, largely in relation to which individuals were or were not well into the menopause.

We were perhaps much more surprised to find that there was a sex difference between the men and women in our young sample (of individuals aged 16–35, Fig. 11). How could this be? What was the nature of the difference? Further studies showed that in the specimens from young females, though most of the vertebral shadow was essentially like that in the males, one vertebral quadrant only was different (Fig. 12). That quadrant, the third (anterior-superior)

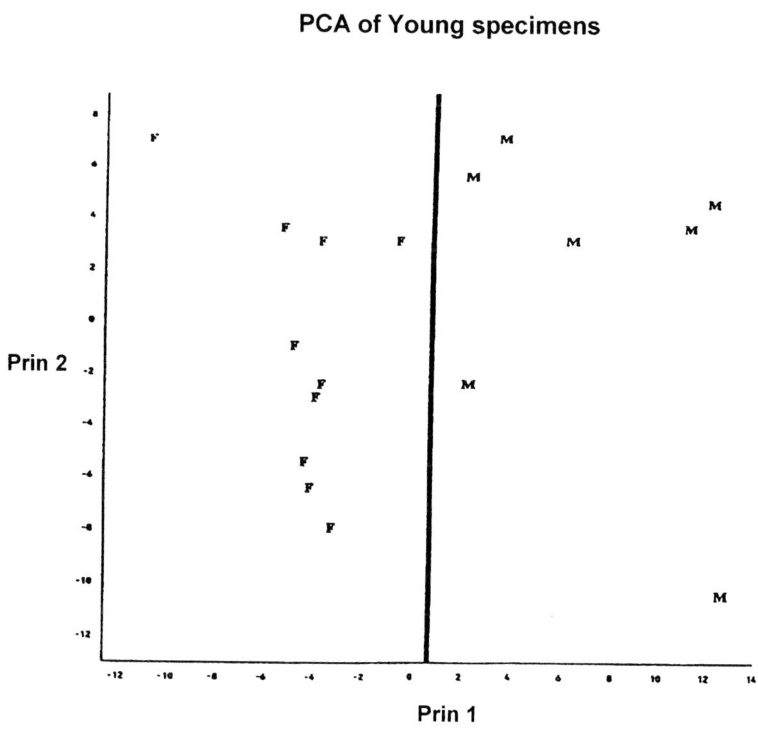

Fig. 11. Principal components analysis of the quantitative data inherent in transforms of young males and females. A line can be drawn completely separating males and females. This implies some difference that cannot be seen in the inspection of the radiographs alone.

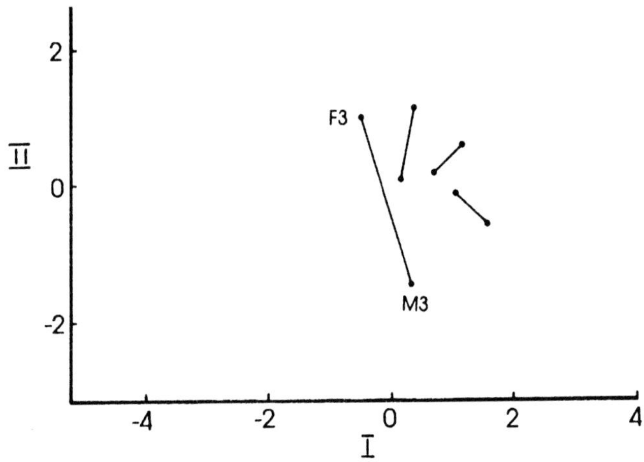

Fig. 12. Principal components analysis of the differences between the Fourier transforms of the individual quadrants of the vertebral radiographs. Whereas males and females are very similar in the first, second and fourth quadrants, females differ significantly from males in the third (antero-superior) quadrant.

quadrant (in lateral view) showed a local Fourier signature like the whole vertebra in women with osteoporosis (Fig. 13). Of course, these young women do not have osteoporosis. But is it possible that they have the hidden beginnings of the condition? There are plenty of factors in earlier life that could give rise to this situation (smoking, lack of exercise, poor diet). It may well be that we have here, a method of predicting later onset of osteoporosis when the condition itself is quite occult. Of course, this could not be done with sections of vertebrae; but it might be possible with images of, say, some other bone that could be imaged in life.

Though this method of analysis has never been accepted and published by any medical journal, it has readily been published in science journals and books. This is mainly because in medicine densitometry is thought to be the gold standard for examining osteoporosis. Indeed, of course, densitometry is very good when the condition is fairly obvious. However, it cannot as easily detect early changes as can Fourier analysis because it is based upon overall averages of density primarily of the bone cortices, while Fourier analysis provides evidence of detailed localised textures of trabeculae.

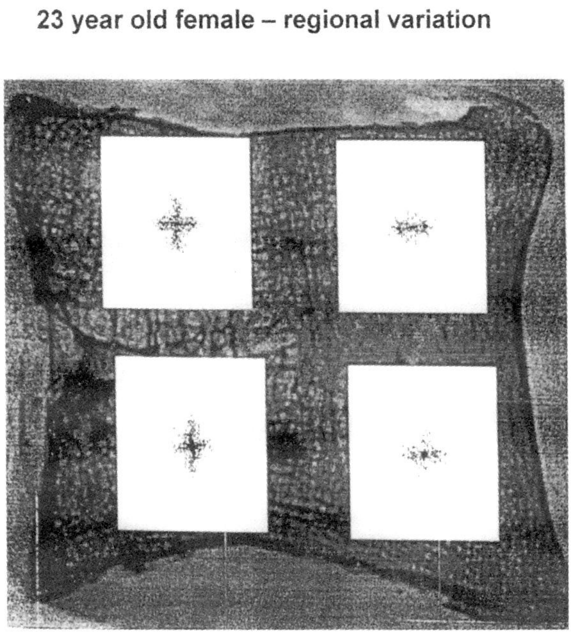

Fig. 13. The actual transforms in each of the quadrants in one of the females of Fig. 12. Three of the quadrants have Maltese crosses (like all four quadrants in males). One of the quadrants (the antero-superior quadrant) has a rugby ball-shaped transform as in old osteoporotic females.

Trabecular surfaces are many times greater than cortical surfaces. The process of calcium loss in osteoporosis is a surface phenomenon. Bone can only be removed at surfaces. The surface of bone is maximised in trabeculae. This, therefore, is where osteoporosis should first be evident. This is, therefore, why it is best revealed by Fourier analysis. Presumably this is why we see early changes in young women even though they do not yet have osteoporosis. Further, the particular site of the earliest lesion, the anterior-superior quadrant of the vertebral body, is significant. When vertebral fractures occur as a complication of late osteoporosis, they often first occur in the anterior-superior quadrant of the vertebra often presenting as a wedge fracture!

I therefore had a lot of very good reasons for wanting to know more about cancellous patterns in bone and their relationship to biomechanical patterns.

The Association Between Biomechanics and Architecture

Let us first think in more detail about mechanics. A load upon any object produces stress and strain at all points within that object. The stress at a particular point can be thought of as the intramolecular resistance of the material of the object to the loads. The strain at a particular point can be described through the molecular deformation produced at that point by the loads. Although this obviously occurs in three dimensions, let us first look at the two-dimensional case: a particular point in a plane.

It is easiest to visualise this by concentrating on the strain, i.e. the deformation because we can draw deformation. Before the object is loaded, we can think about the particular point as being an infinitesimal circle. After the object is loaded that circle becomes deformed into an ellipse. This is shown in Fig. 14. Of course, there are strains in all directions in the plane but two particular strains, one describing the greatest tensile strain and one the greatest compressive strain summarise the overall strain situation. These special strains are called principal strains. One is the maximum elongation of the circle into the

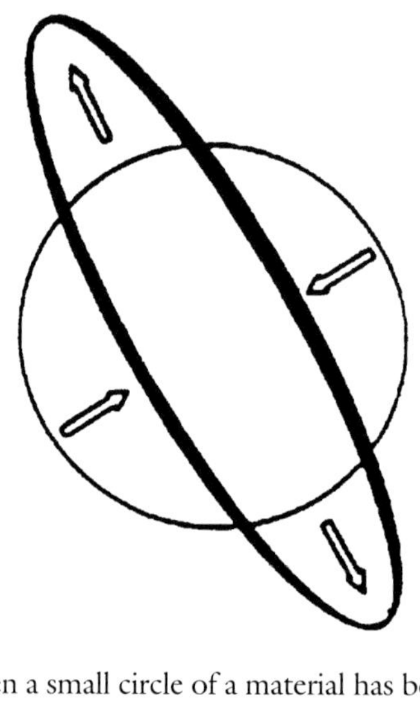

Fig. 14. The strains when a small circle of a material has been strained into an ellipse.

ellipse and is called the maximum principal strain. The other, at right angles to the first, is the maximum contraction of the circle into the ellipse and is called the minimum principal strain.

At any moment the strains at a point in the material can, therefore, be summarised as a cross, the arms of which represent the two principal strains. However, over a period of time, the loads on the object will vary and as a result the crosses will lie in different positions at the point at different times. Let us assume that, though the loads change from moment to moment and this changes the strains from moment to moment, there is some average of the strains that exists over some particular time. This 'average' can be thought of as resulting from various individual strains (Fig. 15).

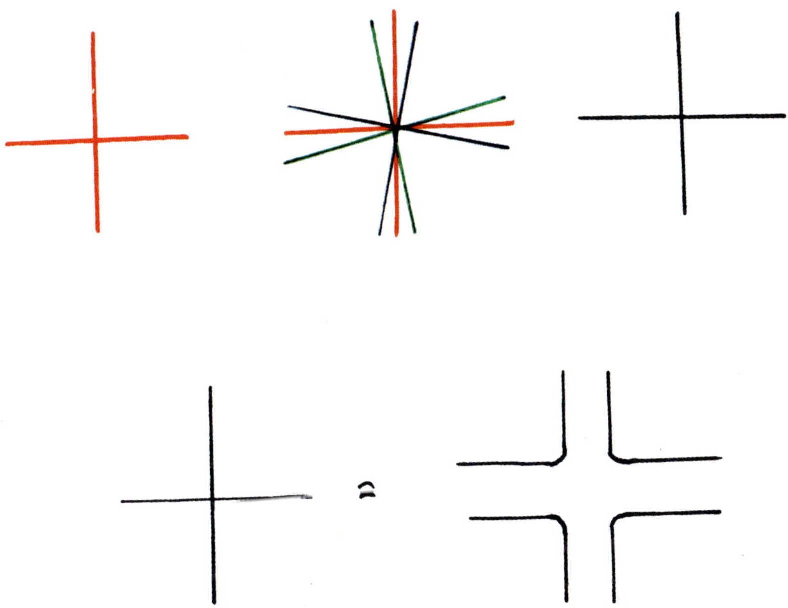

Fig. 15. The overall average orthogonal linear principal strains resulting from changing linear principal strains from moment to moment, and their relationship with the local orthogonal architecture. Red cross (left) = single strain; red, green and black crosses (centre) = several different strains at different times; black cross (right) = average of all strains from centre. Black cross (left below) repeats that average strain and implies that it is borne by crossed bars (right below), the architecture that might best bear the strain.

The Simplicity of Two Dimensions

Figure 16 (upper frame) gives a geometric mechanical engineering description of the strains at many points in a two-dimensional structure. The shape of the element is that of a calcaneus, though that fact is irrelevant to the present discussion. The upper frame shows that, at any specific point in the shape, one can think of the points in the material as being very small circles. When the shape is loaded the circles deformed into ellipses. One diameter of each ellipse is shortened (compression) while the other is elongated (tension) and these changes are linear.

Figure 16 (middle and lower frames) shows the picture when we get rid of the circle and ellipses, leaving only the diameters. The middle frame shows the locus of all maximum diameters of the

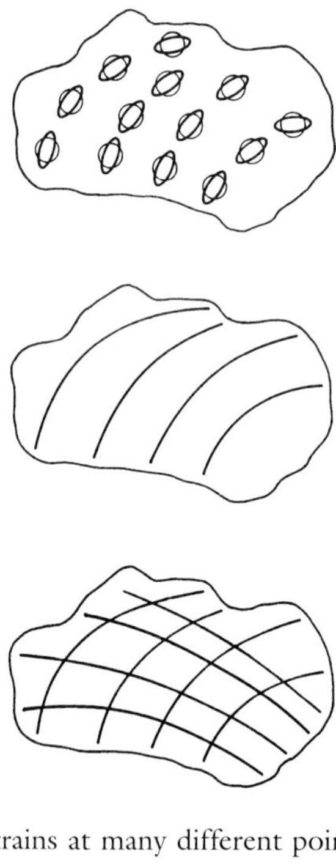

Fig. 16. The pattern of strains at many different points in an object generates an orthogonal network.

strained ellipses. The lower frame shows that the maximum and minimum diameters of these ellipses together carve out a series of lines that cross at right angles. These are the linear principal strain trajectories of the material within the shape. (This is also true of the stresses: intramolecular resistance to the loads, but because I have described the phenomenon in terms of deformation I am specifically talking about strains, the changes in shape of the circles.)

Figure 17 shows a small portion of such a network of orthogonal principal linear strains (continuous lines). It also shows that at 45° to this network is a second network of principal shear strains also at 90° to each other (dashed lines). Of course, there are linear and shear strains at all angles around the ellipses; it is just that the major and minor diameters of the ellipses are those radii of the original circle that have changed in a linear manner, hence principal linear strains. There are also shear strains in all directions; it is just that those that are at a maximum, the principal shear strains, are at 45° to the linear strains and lie in the positions where the linear principal strains are zero

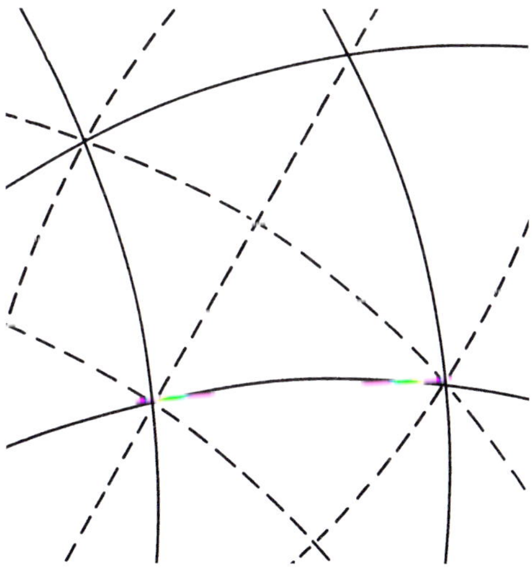

Fig. 17. A small portion of the orthogonal network of linear principal strains showing that there is another orthogonal network rotated through 45° of shear principal strains.

(i.e. those diameters of the circle and ellipse are the same). In other words, the principal shear strains are where there is no linear strain (no change in length of the radius) but greatest shear strain (change in angle of the tangents). This is especially clear when we look at a single such point, its original circular shape and its strained elliptical form.

In order to be able to use this simple idea in a more complex milieu (see later), we must first look at another way of graphing these changes in linear and shear strain in two dimensions. Figure 18 shows the circle deformed into an ellipse. Arrows within it show a few of the deformations of the original radius of the undeformed circle at different angles. What it is now useful to do (but the usefulness will not become apparent until a little later) is to plot these strains in a diagram where the value of the strain is plotted at twice the angle it occupies in the ellipse. This is called a Mohr's circle (Timoshenko, 1955). It is shown in Fig. 19 but needs a little explanation.

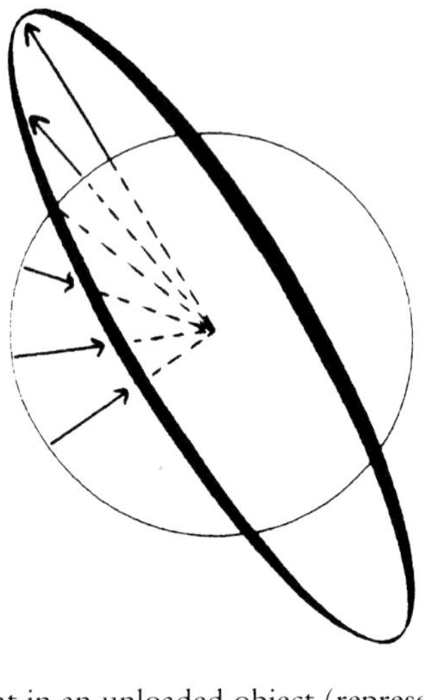

Fig. 18. At a single point in an unloaded object (represented by the circle) the act of loading the object deforms the circle into an ellipse and one can now see that the linear principal strains (represented by the maximum and minimum radii of the ellipse) are merely representative of an infinite number of strains at all different angles around the circle.

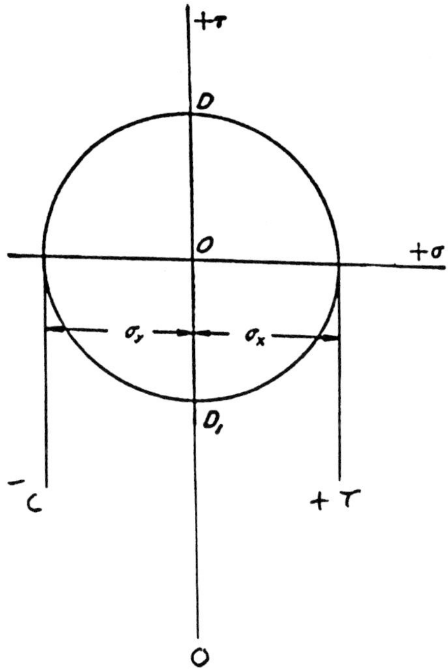

Fig. 19. The 'Mohr's circle' obtained when the values of various strains are plotted at twice the angle they occupied in Fig. 18. For full explanation see text.

Thus, the extreme right-hand end of Mohr's circle in Fig. 19 represents, by its measure on the *x*-axis (the units on this axis measure linear strain), the greatest diameter of the ellipse. This is the maximum linear deformation or strain (called a tensile strain; by convention tension is positive). The extreme left-hand end of the circular plot represents the least diameter of the ellipse called the minimum linear strain (compressive, negative by convention).

Shear strains are plotted along the *y*-axis. At the points of the maximum and minimum linear strains the shear strains are zero. If you look at the ellipse–circle comparison, the tangents in the ellipse and the circle at these two points are parallel, i.e. there has been no change in angle, i.e. shear is zero. At zero linear strain (where there is no difference between the radii of the circle and ellipse, i.e. at 45° to the principal linear strains) the tangents between the circle and ellipse are maximally angled (largest positive and negative values on the *y*-axis which measures shear strain). As a result, the plot of linear strain against shear strain gives a semi-circle (in an isotropic homogeneous material). The *x* and *y* coordinates of the semi-circle are the locus of the linear and shear strains in all directions within the strained two-dimensional ellipse. Of course, because of the $2x$ transformation in angle as one goes from the strained ellipse to the Mohrs strain plot, one has to go round the strain circle twice to get all angles around the ellipse just once.

Why do this? It is certainly something that biologists often have difficulty understanding. The answer is that an extension of it allows us to understand better what happens in three dimensions. Drawing a three-dimensional circle (which is a ball) and straining it into a three-dimensional ellipsoid (which is a kind of sausage) and seeing the three-dimensional nature of all the strains from the ball to the sausage allowing for the perspective of three-dimensional objects is exceedingly difficult. It is difficult to draw, never mind to understand. But plotting these three-dimensional strains around three Mohr's circles can be achieved on a flat page. It reveals interesting relationships about strain in three dimensions.

The Complexity of Three Dimensions

There is no doubt that the right-angle network of principal strains in two dimensions (Fig. 15) is equivalently replicated in three dimensions as a three-dimensional right-angle network. This is obvious. However, if the simple association between the right-angle networks of trabecular architecture (also Fig. 15) were to be similarly replicated in the third dimension, then bony trabeculae ought to form a similar three-dimensional orthogonal arrangement. Though, a three-dimensional architecture is easily constructed in buildings, it is not the case in any bone of which I am aware (Fig. 20). It is, therefore, necessary to look more carefully at the three-dimensional situation in bone.

This is where Mohrs (strain) circles are useful. They can be drawn for three dimensions on a two-dimensional page. There are several possible arrangements for Mohrs circles. Figure 21 shows one example.

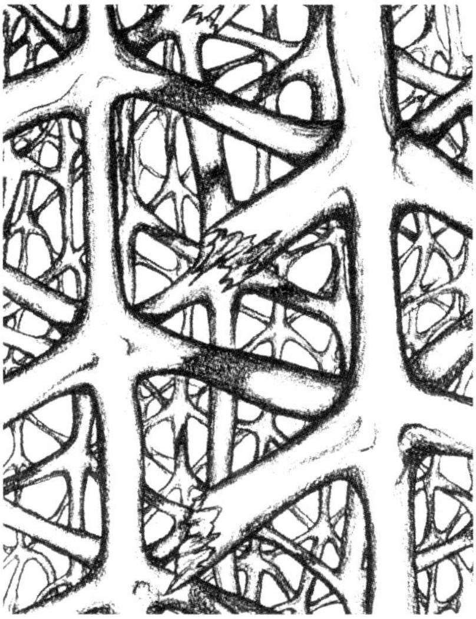

Fig. 20. Direct extrapolation from the orthogonal network of strains in three dimensions would seem to imply that the corresponding architecture of bone might be like this picture (modified from MC Esher).

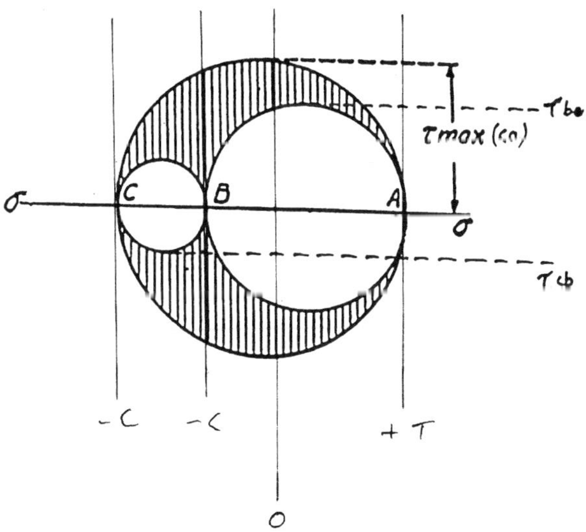

Fig. 21. The arrangement of Mohr's circles at each of the orthogonal directions in three dimensions. For full explanation see text.

Each circle shows the strain solution for a two-dimensional slice of the three-dimensional structure. Two of the circles resemble the circle in Fig. 19 in that they have one compressive (labelled c) and one tensile (labelled t) principal linear strain. It is these two circles that are associated with a right-angled architectural network in each respective plane.

The third circle, however, must (logically) be different. It can only have principal strains that are of the same sign; in this particular example, one circle has two principal linear stresses that are both compressive (though one is, algebraically, larger than the other). What, then, is the architecture that is associated with this circle?

Figure 22 hypothesises what the architectural arrangement might be: two right-angles networks and one with the elements in all

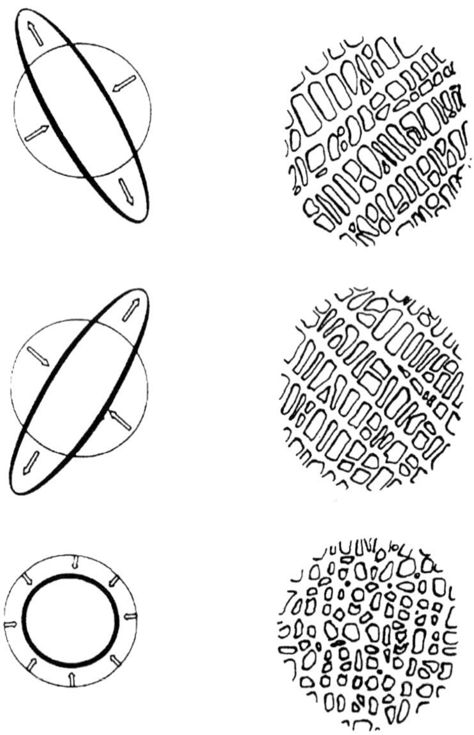

Fig. 22. Three strained ellipses in the three-dimensional case of Fig. 21, and the architectures that might be hypothesised as being associated with them.

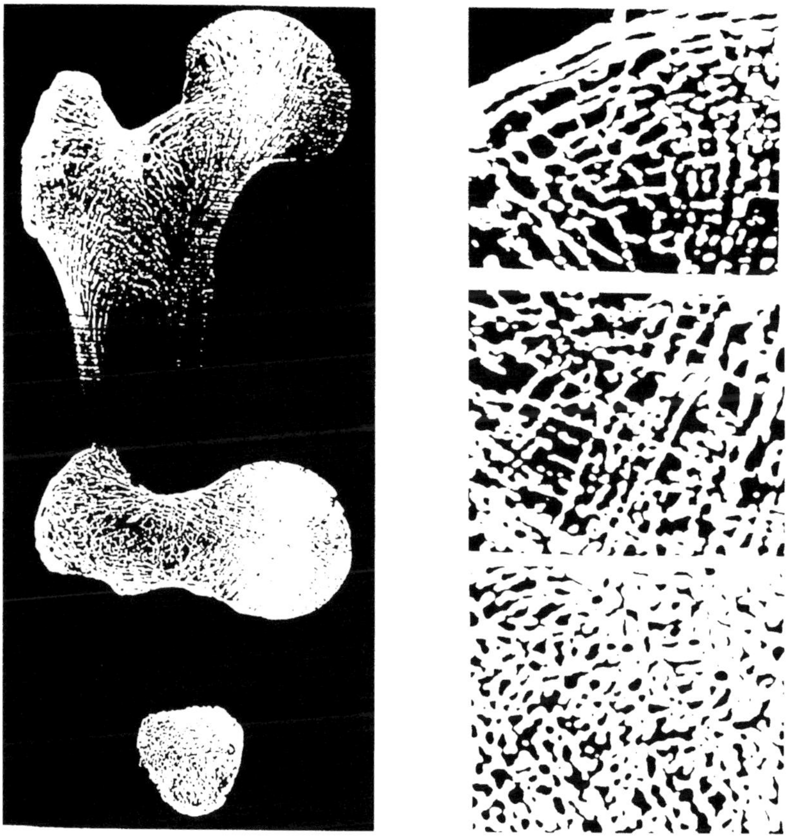

Fig. 23. Three actual architectures that exist in the neck of the femur related to the three different Mohr's circles of Figs. 21 and 22.

directions. Figure 23 shows the actual cancellous patterns in three orthogonal sections of a real femoral neck. The trabecular arrangements in two of the sections are both, indeed, orthogonal. But the arrangement in the third is not. If anything the third is honeycomb-like, i.e. approximately similarly arranged in all directions. There is, thus, still a clear association between strain and architecture. It is not, however, simply a direct relationship between similar orthogonal patterns.

For this to be clearly true, however, it is necessary to examine other possible arrangements of Mohr's circles. For example, it is

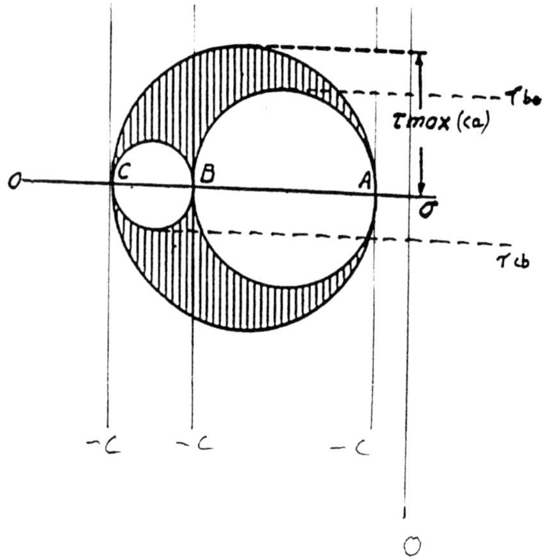

Fig. 24. The Mohr's circles when they are all compressive.

entirely possible for all three linear strains to be compressive. This is shown in Fig. 24. Could this ever occur in bone in practice? A moment's thought about stresses in abyssal fish, with tons of water pressing upon them in all directions, indicates that, yes, this does occur. Does it ever occur in ordinary terrestrial animals? Again a few more moments thought implies that, yes, it might occur near the centre of the ball in a ball-and-socket joint where, at different positions of the joint, loads are impressed upon it in all different directions. The head of the femur is such a structure. Figure 25 shows the prediction. Figure 26 shows the actual situation in all three sections of the ball-shaped femoral head. In each section, the trabeculae are arranged in all directions. No section shows a clear orthogonal arrangement.

There cannot, then, be a simple relationship between orthogonal strains in three dimensions and architectural elements. It is clear that there is a strain/architecture relationship; it is just more complex than implied by a simple reading of the trajectorial theory of bone architecture.

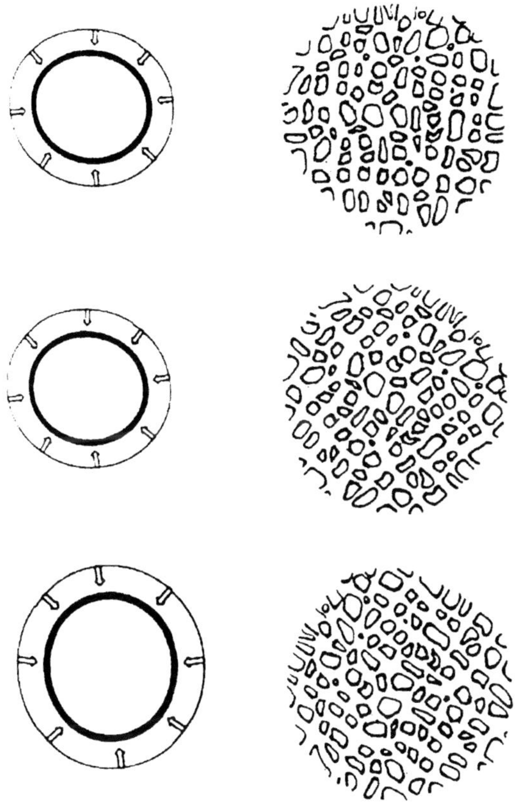

Fig. 25. The predicted trabecular networks in the circumstances of Fig. 24.

This idea can be further tested by examining yet other possible combinations of Mohr's circles. One of these is the situation where all the Mohr's circles contract down to a single point. Can this happen? What does it mean when it does? The Mohr's circles involved are shown in Fig. 27. In this case, it is easier to go first to the biological example and work back to the Mohr's circles. Thus, Fig. 28 shows the biomechanical analysis for a two-dimensional section of the upper end of the tibia in a cow. The technology, here, was the older two-dimensional photoelastic method (Smith, 1962). The result however, is not in doubt. The various principal linear strains are figured as the usual orthogonal network. However, there is one point in that network,

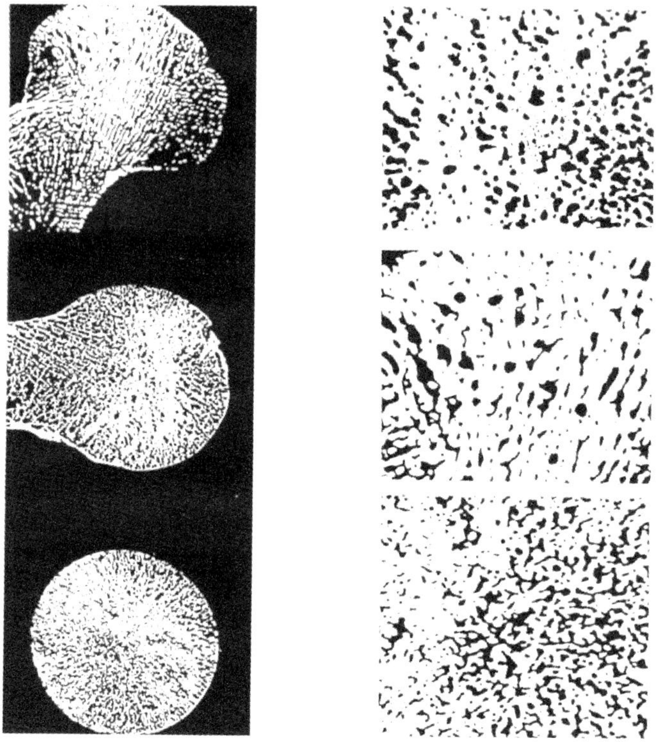

Fig. 26. Three actual architectures that exist in the head of the femur related to the three different Mohr's circles of Figs. 24 and 25.

indicated by the black dot, that all the trajectories avoid. In other words, this is a region where all stresses and strains, linear and shear, are zero in all directions. There are always such points in the analyses of complex shapes. They are known as isotropic points.

During different phases of function, however, with different loads at different alignments during movement cycles, this point would normally occupy different regions within the structure. It just happens that the combination of the specific shape of the cow tibia, with the reduced movements that occur in that fairly large and fairly inactive (for most of the day) creature, that that particular isotropic point does not move around much. In this animal, therefore, the isotropic point is located around the same small locus most of the time. How is

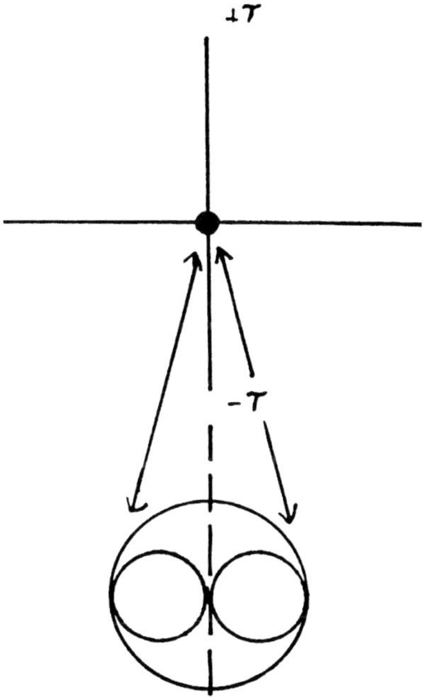

Fig. 27. The Mohr's circles in the situation where all three principal strains are zero.

this small locus of zero stress and strain associated with the architecture? Figure 29 shows the trabeculae in various parts of the picture. Around the position corresponding to the isotropic point, there are no trabeculae at all. In fact, there is a small fluid filled sac that must bear no stress and is not strained. This example is extreme. There are, however, many instances, usually well known to radiologists, where trabeculae are absent in local regions of bones (e.g. in both the neck of the femur and the neck of the calcaneus).

There are yet other combinations of Mohr's circles that are possible. For example, they could be all tensile. I imagine that this might occur, for example, perhaps during the movements of some aquatic creature such as a jelly fish, or in the developing blastocyst, or in the heart wall during the heart beat, or in some exotic alien in space. The heart wall example is especially interesting because, though there is no bone within the hearts of most creatures, in really large animals there

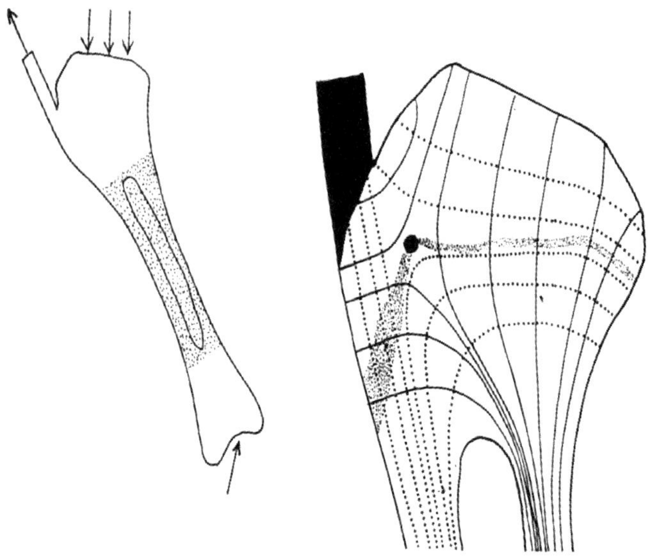

Fig. 28. A biomechanical analysis of the upper end of tibia in a cow (Smith, 1962).

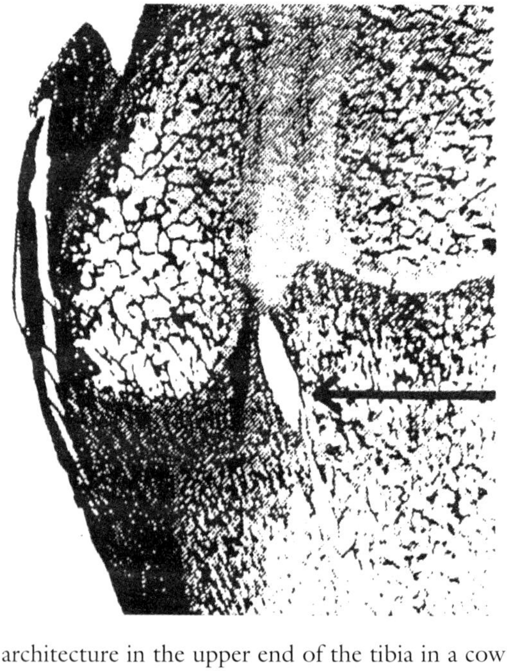

Fig. 29. The architecture in the upper end of the tibia in a cow (Smith, 1962).

may well be a piece of bone in the membranous part of the septum between the main chambers where bone exist (the so-called os cordis or 'heart bone'). What is happening, mechanically, here? Again, however, I leave it to my younger colleagues to work these further examples out more completely.

What is the importance of all this? It relates to at least two overall sets of problems.

The first is that of working out what might be the functional significance of particular trabecular networks in different situations. These include: the functional adaptations of structures of organisms in evolution (e.g. vertebral cancellous patterns in different apes), estimations of function in structures whose functions are unknown (e.g. fossils) or assessment of architectures in clinical conditions (e.g. osteoporosis).

The second relates to understanding how architectures change in relation to changing biomechanics: the mechanism of adaptation of bone. These include: changes during normal development and growth, and changes in clinical conditions.

Curvature in Three Dimensions

Even this is not the whole story. The situation is further complicated on the mechanical strain side by the fact that the orthogonal network from point to point is not necessarily flat. In fact it is highly likely not to be flat. Of course, in a sagittal section of a vertebra the orthogonal network should be flat, i.e. in that plane (and it is). In other parts of irregular objects, such as vertebrae or pelves, the 'planes' containing the orthogonal principal strains may themselves be curved.

Thus, it follows from the last section that when the orthogonal linear strains in the curved planes are opposite (i.e. one tensile and one compressive) then that curved plane contains the orthogonally arranged trabeculae. Of course, such a curved plane cannot be 'found' by cutting flat sections. Sections, by definition are flat, so that the curved plane containing the orthogonal network flows in and out of the flat surface. As a result such flat planes often do not show orthogonality. Neither can such curved planes be seen through regular

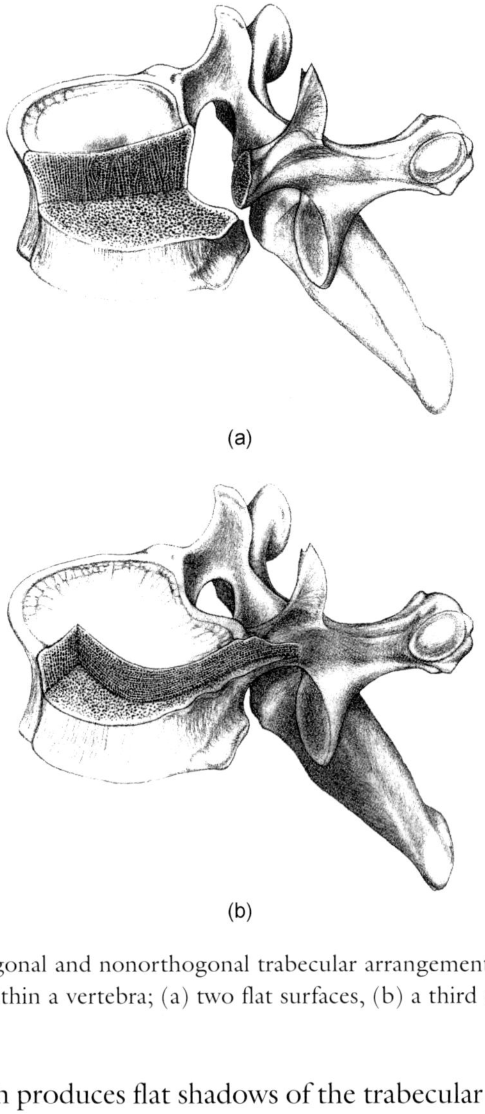

(a)

(b)

Fig. 30. Orthogonal and nonorthogonal trabecular arrangements within different curved surfaces within a vertebra; (a) two flat surfaces, (b) a third curved surface.

radiology which produces flat shadows of the trabecular network along the direction of the ray. They can be revealed in the old-fashioned way by 'dissecting' the cancellous bone with a dental drill (Fig. 30).

In contrast, however, the three-dimensional complexity of the cancellous network can be much better revealed by CAT scans (nowadays these, MicroCT scans, are sensitive enough to show them). Even so, the curved surface containing the orthogonal network is not easy to find. It is continuously confused by the other surfaces containing other arrangements. If the CAT scan image is appropriately rotated, one may be able to follow the three-dimensional orthogonal structure. Such work is being carried out by Dr Michael Fagan and his group at the University of Hull.

A Return to Two Dimensions: The Question of Bone Adaptation

These new ideas of the more complex relationship between strain and architecture mean changes in understanding how the mechanics affects the architecture during adaptation. The chapter on bone

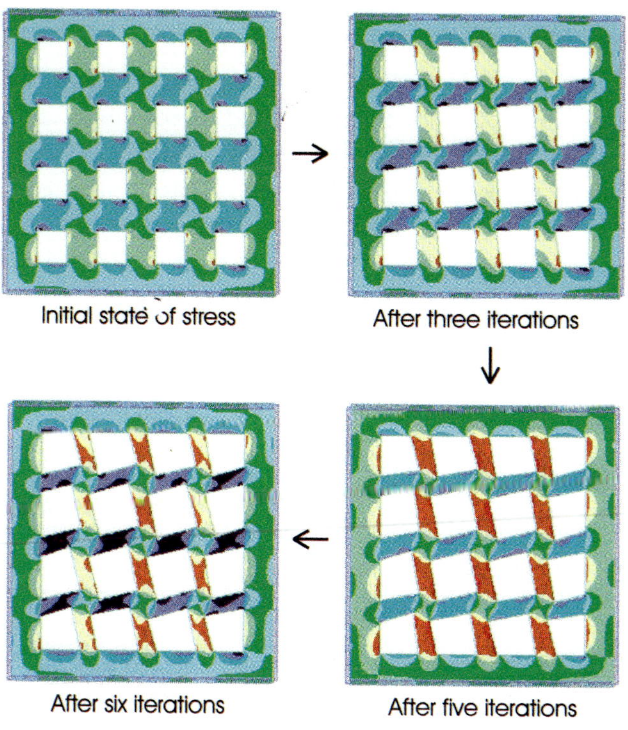

Fig. 31. An orthogonal architectural network 'adapting' to angulated loads by changing its angulation.

shape provided a key piece of information. It was that in studying a tendon attached to a bony pit and a bony protuberance, respectively, there were two particular combinations of loads in the bone and the tendon at which surface stresses are at their minimum. If our hypothesis of bone adaptation related to these particular combinations of tension and compression, then these particular pits and protuberances would be the optimal architectural arrangements.

We therefore decided to attempt to model adaptation in cancellous networks. This is probably best done using FEA. Hirschberg started with a model of a two-dimensional orthogonal architectural network that is efficient under a given orthogonal network of stresses. That is, he started with an orthogonal trabecular network that lay at the same angle as the two orthogonal applied loads. The directionality of the orthogonal loads was then changed so that, though still orthogonal, they now lay at a different angle to the original orthogonal direction. That is, there was a new pattern of loading that was eccentric to the architectural network. This produced much higher and more complex levels of stress in the architecture. The model was then 'required' to realign its architecture to reduce these stresses to the greatest extent possible (Hirschberg, 2005; Hirschberg and Oxnard, work in progress). This resulted in a new diagonally aligned set of trabeculae (Fig. 31). This result looks promising.

Of course, it may not be the orthogonal network of linear principal strains to which the bone is adapting. It could just as well be the orthogonal network of maximum shear strains that lie at 45° to the maximum linear strains. This would make real sense — bone is much weaker in shear than in either tension or compression. Perhaps the adaptation is to not having bone in the 45° shear directions as much as to having bone in the 90° linear directions.

And it is also possible that the adaptation relates to strain energy density, a component of strain that is squared so that there are no negative numbers. This is the parameter most used by engineers and immediately provided in all finite element programmes. However, strain energy density in its raw form does not differentiate between tension and compression, and as we have seen, this differentiation seems to be important. There are even other possibilities to consider,

Fig. 32. The architectural network that is produced by requiring a standard orthogonal network to adapt under a different loading regime. The changed thicknesses of the vertical and horizontal trabeculae are very similar to the trabeculae in a typical osteoporotic vertebra.

for instance, the cytoplasmic network that exists among the osteocytes within the bone. This matter is not yet settled.

We especially examined the cases where the initial model was asked to remodel in relation to different loads in each orthogonal direction loads but with no change in directionality. Under this circumstance, the vertical columns of trabeculae thickened and lengthened and horizontal columns thinned and became shorter. This is a result entirely similar to what is actually found in trabecular bone in osteoporotic vertebrae (Fig. 32, Oxnard, 1997).

Finally, we looked at a case where a single trabecula was broken (fractured — a microfracture). This produced a series of adaptive changes in architecture that were transmitted throughout the entire cancellous structure. Again, this resembles what can be found in osteoporosis (Fig. 33).

The Question of Bone Adaptation in Three Dimensions

In the above studies, the orthogonal architectural networks that have been required to remodel are two-dimensional. They are like the

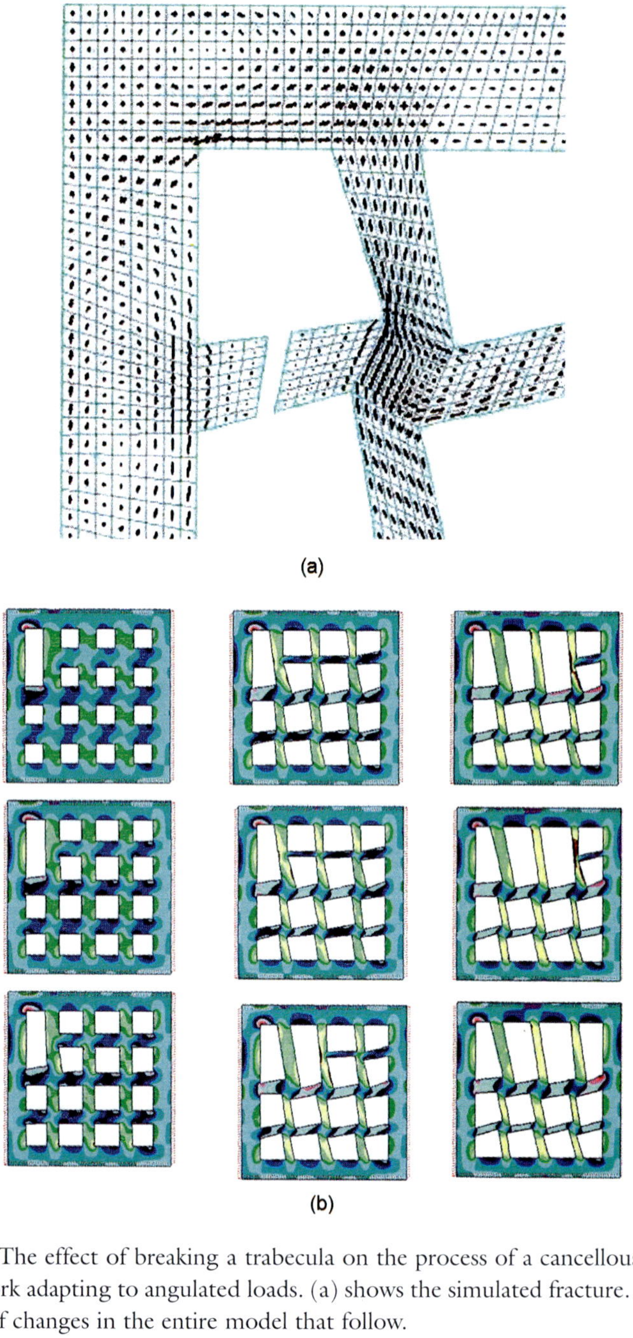

Fig. 33. The effect of breaking a trabecula on the process of a cancellous architectural network adapting to angulated loads. (a) shows the simulated fracture. (b) shows the series of changes in the entire model that follow.

two-dimensional patterns in many plane sections or radiographs of a bone where, according to the prior ideas, the principal strains are one tensile and one compressive. What is now required are similar analyses but of adaptation in a three-dimensional architecture that is orthogonal in only two of its dimensions.

Because physical dissection is time-consuming and physically damaging of specimens the dissections of the regions above that I have performed are limited. It would be better if the three-dimensional architectures were computed and imaged with computer imaging modalities. The computer models could then be 'dissected' computationally. Such work is in progress in Dr Michael Fagan's laboratory in the University of Hull, UK.

Another way of looking at the three-dimensional problem might be to apply the reverse engineering method of the two-dimensional case: computationally 'requiring' a particular three-dimensional architecture to 'remodel' in such a way as to reduce the surface strains to a maximum degree (or perhaps to an optimum). How would three-dimensional trabecular networks remodel when the loading regime is changed? How similar would the result to the actual architectures that exist. Especially, how similar would the changes resemble what has been found resulting from surgical interference?

This is quite a complex problem. It requires a fully three-dimensional method for displaying the results of a finite element stress/strain analysis — this is now available. It requires animated displays to show resulting changed architecture — we are in the process of adapting our two-dimensional situation to the three-dimensional one, though it will not be easily done. It may have to involve an interface between the three-dimensional computational programmes and an anatomist operator to allow coincident three-dimensional display of finite element analyses with three-dimensional (e.g. cyberglove) control of CT segmentation.

I expect that such work will be carried out by Dr Michael Fagan and his colleagues at the University of Hull, UK. As an Honorary Professor there, I hope to see it completed. The implications for evolution, development, growth, and clinical science are immense.

Such studies would be invaluable in attempting to understand the basic mechanics of both the form and remodelling of trabecular networks. They would especially provide information about the nature of the relationships between the stress/strain trajectories and the architectural arrangements (some of which are trajectorial and others not). They would help elucidate problems of mechanical adaptation in the skeleton in extant species in evolution. They would be most useful in the reverse attempt to assess, from trabecular networks in fossils, the nature of the loading patterns that may have existed in them during life. They might even enhance the understanding of changes in architectures that occur as the result of diseases such as osteoporosis and surgical interference in mechanical function in bone (and due to prosthetic devices).

References

Buck AM, An investigation of vertebral cancellous architecture along the thoraco-lumbar column in humans using fast Fourier transforms: An anatomical study, Doctoral Thesis, University of Western Australia, 1998.

Coker EG, Filon LNG, A Treatise on Photoelasticity, 2nd edn. Revised by HT Jessop, Cambridge University Press, 1957.

Cundall PA, Board M, A microcomputer program for modelling large-strain plasticity problems, in Swoboda S (ed.), *Numerical Methods in Geomechanics*, New York, 1988.

Currey JD, *Bones: Structure and Mechanics*, Princeton University Press, Princeton, 2002.

Evans FG, *Stress and Strain in Bones*, Thomas, Springfield, 1957.

Frocht MM, *Photoelasticity*, Vols. 1 & 2, Wiley, New York, 1941.

Frost HM, *The Laws of Bone Structure*, Thomas, Springfield, IL, 1964.

Geraets WGM, *Computer Aided Analysis of the Radiographic Trabecular Pattern*, VU University Press, Amsterdam, 1994.

Hirschberg J, A study of the effects of tubercles and pits on the stresses in bone at tendon attachments, M.Sc. Thesis, University of Western Australia, 1997.

Hirschberg J, Milne N, Oxnard CE, Biomechanics of the tendon/bone interface, *Persp Hum Biol* **5**: 55–68, 2000.

Hirschberg J, Simulations of mechanical adaptations and their relationship to stress bearing in skeletal tissue, Doctoral Thesis, University of Western Australia, 2005.

Hsu A-T, Trabecular architecture and finite element analysis of the human calcaneus, Doctoral Thesis, University of southern California, 1989.

Kupczik K, Dobson C, Fagan M, Crompton R, Oxnard CE, O'Higgins P, Assessing mechanical function of the zygomatic region in macaques: Validation and sensitivity testing of finite element models, *J Anat* **210**: 41–53, 2007.

Meyer Hv, Die Architektur der Spongiosa, *Arch Anat Physiol* **34**: 615–628, 1867.

Murray PDF, Bones *A Study of the Development and Structure of the Vertebrate Skeleton*, Cambridge University Press, London, 1934.

Oxnard CE, Tensile forces in skeletal structures, *J Morph* **134**: 425–436, 1971.

Oxnard CE, *Form and Pattern in Human Evolution*, University of Chicago Press, Chicago, 1973.

Oxnard CE, Mechanical stress and strain at a point: Implications for biomorphometric and biomechanical studies of bone form and architecture, *Proc Australas Soc Hum Biol* **3**: 57–109, 1991.

Oxnard CE, Bone and bones, architecture and stress, fossils and osteoporosis, *J Biomech* **26**: 63–79, 1993.

Oxnard CE, From optical to computational Fourier transforms: The natural history of the cancellous structure of bone, in Lestrel PE (ed.), *Fourier Descriptors and Their Applications in Biology*, Cambridge University Press, London, 1997.

Oxnard CE, Lannigan F, O'Higgins P, The mechanism of bone adaptation: Tension and resorption in the human incus, *Recent Adv Hum Biol* **2**: 105–125, 1995.

Oxnard CE, Yang HCL, Beyond biometrics: Studies of complex biological patterns, *Symp Zool Soc London* **46**: 127–167, 1981.

Runnion CK, Oxnard CE, Robertson WV, Windsor CR, Biomechanical modelling of vertebrae using experimental stress analysis, *Proc Australas Soc Hum Biol* **4**: 125–133, 1991.

Smith JW, The structure and stress relations of fibrous epiphyseal plates, *J Anat Lond* **96**: 209–225, 1962.

Timoshenko S, *Strength of Materials. Part I: Elementary Theory and Problems*, 3rd edn., Van Nostrand, Princeton, 1955.

Chapter 3

Ghosts of the Past: Muscles and Bones

From Bone to Muscle

Many of my first studies involved estimating the evolution of bone form in primates. Because, however, bone form is so dependent upon the biomechanics of function, that work was almost always based upon prior knowledge of muscles. Muscular anatomy elucidates bone form in a number of different ways.

First, muscles contain information associated with the postures and movements of bones. Thus, through knowledge of the sizes, structures and attachments of muscles, through information such as relative muscular leverages, relative action direction, relative muscle mass and muscle fibre cross-sectional area and so on, it is possible, using anatomical inference, to gain some understanding of the function of the bone-joint unit. Such information, for instance, was a necessary prelude to my first investigations of the primate shoulder skeleton. It was only later that anatomical inferences became testable (and were sometimes denied) by experimental methods such as field observations, cinematography, electromyography, dynamics and kinetics.

Second, muscles provide information about the mechanical efficiency of bone shape and architecture. This is a function of muscles that is often forgotten: the effects of muscles and ligaments in relieving bone of stresses and strains. I first studied these problems using theoretical arm-chair analyses (see prior chapters) and later practical experimental stress analysis analogues. It was only later still that better stress and strain studies became possible with computed finite element strain simulations and true strain analysis *in vivo* using engineering strain gauges.

Third, muscles (tendons and ligaments) give information about the way in which tensegrity (compressive integrity through tension bearing) affects the biomechanics of bones (Buckminster-Fuller and Marks, 1973). Thus, though weight bearing by the vertebral column depends upon compressive loads supported by relatively rigid struts such as vertebral bodies and processes, cushioning intervertebral discs and opposing surfaces of facet joints, a considerable portion of this compression is borne through tension in appropriately aligned elastic ties such as: fasciae, ligaments and muscles with their tendons (these last being special adjustable ties).

For all these reasons, almost all our studies of bone form were preceded by soft tissue dissections of numbers of specimens of as many primate species as possible. This helped us understand the nature of the relationship between soft tissue variations and hard tissue form and architecture.

It is not, however, at all easy to procure numbers of cadavers necessary for such studies. However, an accident early in my career (being the first anatomist on the scene at a fire in a local pet store that unfortunately suffocated more than 1000 primates and many birds and other mammals as well) set the stage. It meant that I was able to obtain — I paid £100 for all primate carcasses — and later dissect samples of monkeys and apes large enough to permit quantitative analysis of soft tissue features. Thus, between 1958 and 1962, I dissected the shoulders and arms of more than 60 individual primates, between 1962 and 1966, the hips and thighs of more than a 100, and in 1968 and for several years following with the help of a technician, the arms and forearms of more than 130 specimens. I often had samples of as many as six to eight individuals of each species usually from the same locality. Of course all this needed the help of colleagues, students and technicians (and especially my colleague, the late TF Spence) and these I gratefully acknowledge (Ashton and Oxnard, 1963; Oxnard, 1983/1984; Zuckerman *et al.*, 1973).

It would be extremely difficult to do all this nowadays. The soft tissue materials, the time for dissection, the experienced assistance, the interest in whole organism structure, and the research funds especially, are not available today. In any case, in these days of emphasis on molecular investigations, whole organism study just seems so old fashioned. Most students, assistants and graduates go into other scientific areas. Research funds are nowadays only rarely channelled to whole organism work (though I personally cannot grumble, my work has been very well supported in three continents and over five decades). Yet the anatomical investigations that were carried out with such good samples provided a mint of information about the bone-joint-muscle system in primates that allowed us to interpret our morphometric results more readily. As we shall see later this chapter, however, they have very recently proven especially relevant to the interface between developmental, biomolecular and evolutionary biologies.

Dissecting many specimens lead me to the vast literature on anatomical variations that fill the pages of the zoological and anatomical journals of the 18th, 19th and early 20th centuries (Ruch, 1941). They helped me understand many soft tissue variations that I found. The literature interpretations of many of these variations were often associated with ideas about evolution (those old guys of the 18th and 19th centuries knew what they were doing). They knew that variations must involve heredity, growth and function, though the precise nature of the mechanisms was hidden from them. Today's knowledge of genetic development and biomechanics was not then available. Nevertheless, as far back as Darwin and, indeed, much further back, workers were very clearly aware that anatomical 'anomalies' as they were often called, might give information about development and evolution.

Many of these variations were especially evident in human anatomy (because many more human cadavers have been dissected) and these were often, likewise, interpreted in the light of evolution. Yet many were also recorded simply because they were there; many more because they had clinical implications. In the early and middle of the 20th century this started to give this aspect of human anatomy a bad name. It was denigrated by many biological scientists as 'anatomical stamp collecting'. Of course, knowledge of such features has always been useful, indeed, important in clinical medicine, especially for surgical procedures, and especially before modern anaesthetics when surgery had to be performed extremely quickly. Today, when surgical procedures are so much more complex, so much more frequently guided by imaging methods, and so often done 'scopically', information about such anatomical variations is becoming very important indeed. Today, also, however, knowledge of such variations can now be seen in a new scientific light, the evidence of morphological pattern through functional, genetic and developmental processes.

Patterns in Muscles

My own attempts to understand the biological importance of muscular features looked first at their statistical frequency in different species. I was able to do this by combining data from my own reasonable samples with the information in the literature. Thus, working with the late Eric Ashton, we documented that, though the caudal limit of the origin of pectoralis major was on the 8th rib in 78.5% of our 26 specimens of Old World monkeys and 95 specimens described in the literature, nevertheless in 17.4% of cases it was on the 7th rib, and in 4.1 cases on the 9th rib. These levels contrasted with the range in apes (much smaller samples of 6 and 40, respectively) being from 5th (10.9%) to 8th (13.0%) ribs (Ashton and Oxnard, 1963).

Our assessment was that differences like this were probably of functional import in relationship to overall locomotor patterns, reflecting differences in the orientation of muscle pulls. Thus, in most Old World monkeys, the pectoralis musculature originates very low down on a long narrow rib cage resulting in its fibres being aligned primarily cranio-caudally in such a way as to aid retraction of the limb, the power stroke, in fourfooted locomotion. This contrasts with its more cranial origin on a more barrel-shaped, therefore wider, rib cage in apes and prehensile-tailed New World monkeys (Atelines). This latter arrangement results in the muscle being aligned more medio-laterally when the limb is dependent, but cranio-caudally when the arm is above the head. It thus helps retract the raised arm during tree climbing, branch hanging and arm-swinging postures and movements, and helps support the body (prevent it dropping off the arm, as it were) during such postures and movements. It may possibly be important in respiration in the generally more orthograde postures of atelines and apes. Such statistical information seemed to be a measure of functional adaptation in that muscle. Similar findings were, of course, evident in most other shoulder muscles. Later studies of the muscles of the arm and forearm, and of the pelvis and thigh showed similar quantitative information of apparent functional import.

Second, however, in those later studies other kinds of muscular variations became evident. These often implied that variations could be related to differences between animals that, as far as could be determined at that time, were functionally rather similar. For example, in the forelimb, the deep head of coracobrachialis (coracobrachialis profundus), usually absent in humans, was present in four of eight specimens of owl monkeys (dourouCoulis, *Aotus*) and in all six specimens of titis (*Callicebus*). In contrast, this muscle was usually absent (four out of five, and three out of five) in uakaris and sakis (*Cacajao* and *Pithecia*). It was absent in five of six specimens of capuchins (*Cebus*) but in only one of six woolly monkeys (*Lagothrix*). Likewise, in the hip, gracilis muscle had only a single belly in all the douroucoulis and titis. There were two bellies to this muscle in all the uakaris but in only one of the sakis. There were two bellies in every specimen of capuchins and woolly monkeys. These kinds of muscular variations seemed not to be obviously functional but might have been interpretable as characters adding a little to information about the genetics, the evolutionary relatedness, the phylogeny, of the various species.

Yet a third kind of feature was typified by a specific finding that I made on the deltoid muscle. In most land vertebrates: amphibia, reptiles and non-placental mammals, the deltoid muscle consists of only two parts (a deltoideus acromioclavicularis, and a spinodeltoideus). In contrast, in most primates, the deltoid consists of three parts originating, respectively from the clavicle, acromion and scapular spine. But the primate muscle is supplied by only two branches of the axillary nerve. One branch supplies what is known as posterior deltoid (originating from the scapular spine) and the other branch supplies the other two parts (anterior and middle deltoid originating on the clavicle and acromion). Why should the innervation pattern differ from the muscular arrangement? Or does it? Perhaps the primate innervation pattern is representative of the older reptilian muscular anatomy?

This situation was partially clarified through accidental findings made during a separate study of vitamin B^{12} levels in rhesus monkeys (see another chapter; Oxnard, 1963) where a large series of monkey foetuses of different ages, especially many older foetuses, became available for dissection. This revealed that, though the deltoid appears first as one pre-muscular precursor, it is later evident as three parts. The developing axillary nerve, likewise, has three branches that seem to be aimed towards these three parts. The adult arrangement seems to have been established. The three parts in the embryo surely relate to three parts in the adult. At this point, therefore, embryologists had stopped looking.

Dissection of even older foetuses revealed, however, that one of the parts of deltoid disappeared and so did that branch of the nerve reaching towards it. One of the other parts of deltoid then divided into two (anterior and middle) thus returning the number of heads to three. One of the two remaining nerve branches likewise split to innervate the two new muscular heads. While it has long been known that nerve supplies are useful in helping to determine muscular homologies, this is an especial example of just how useful they are. But it also implies that late changes may occur in development and these are not always recognised.

The importance of nerve supply in determining muscular homologies (Wood, 1867; Testut, 1884) is equally evident in many other regions. Thus, the supraspinatus and infraspinatus muscles in humans take origin from the developmentally dorsal part of the pectoral girdle, the blade of the scapula (Matsuoka *et al.*, 2005). It is evident from comparative anatomy that these are nevertheless really ventral muscles. In opossums (*Didelphys*) they arise from the coracoid (a developmentally ventral element of the girdle) and in reptiles they are actually known as the supra-coracoideus muscle. Are they really ventral muscles that, in placental mammals, have migrated dorsally, or are they a new dorsal development in placentals?

It is evident from a careful dissection of their nerve supply that the former is likely to be so. They are innervated in mammals by the suprascapular nerve. In humans, this takes origin from the brachial plexus at the point where the ventral primary rami of the fifth and sixth cervical nerves join as the superior trunk. This is before the level at which that trunk splits into its dorsal and ventral divisions and it is, therefore, not clear whether this nerve is ventral or dorsal in origin. However, in those few cases where the nerve arises a little more distally in the plexus at the root of the lateral cord, it is evident that it arises with fibres which are destined for the lateral cord, a ventral component of the plexus. Fibres can also be traced proximally by intra-neural dissection into the ventral parts of the fifth and sixth cervical nerves. This implies that the suprascapular nerves, and therefore the muscles, are ventral derivatives.

A fourth type of feature involving muscular variations seems clearly related to functional adaptation. Thus, the dorsoepitrochlearis muscle is present in all non-human primates. Its form and attachments, however, vary in ways that seem related to functional differences. For example, in species that are primarily quadrupedal in habit, this muscle, arising in relation to the tendon of latissimus dorsi muscle, passes down the arm and inserts into various structures (mostly fascial) in the forearm (Fig. 1). With such a course and attachments it is capable of extending the shoulder at the same time as extending the elbow. Together these movements retract the forelimb and are part of the power stroke that helps drive the body forwards in quadrupedal locomotion.

In apes and prehensile tailed New World monkeys which can move using under-branch arm-hanging, swinging and climbing, the lower attachments of this muscle extend only into the fascia of the arm, not passing below the elbow (Fig. 2). In this position, the muscle assists retraction (adduction) of the arm without impeding flexion of the forearm. Together these movements are the power stroke in under-branch hanging, swinging and climbing. These muscular facts and functional evaluations have long been known. This muscle, generally absent in humans, is nevertheless present as a variation in about 5% of human cadavers. When present, it has the form that is found in those apes and monkeys capable of extended arm-hanging and climbing. This is one of the many pieces of evidence for an arboreal climbing ancestry for pre-humans. Is the variation in humans a 'ghost' of *a prior* functional condition?

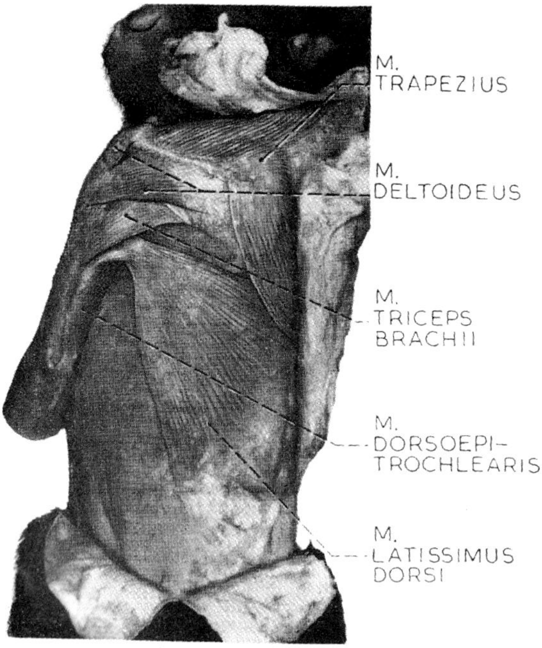

Fig. 1. The dorsoepitrochlearis muscle in a baboon. The muscle crosses both shoulder and elbow joints, thus helping retract the limb in the power stroke of moving on all fours.

Trilobite Eyes and Physicists

These thoughts are now leading me in new directions. They started, however, with much earlier considerations of Levi-Setti's findings (1975) on the compound eyes of trilobites. His book was going through the University of Chicago Press at the same time as my own first book. Thus, Levi-Setti provided information implying that the compound eye of trilobites has evolved independently a number of times. Scanning electron microscopy of trilobites that had been fossilised at different times after death and 'dissected' by taphonomic processes (processes damaging the body after death) revealed the details of the internal structure of the ommatidia of their compound eyes. Some of the ommatidia were preserved at the surface (thus showing the shape of the outer surface of the cornea), others at different

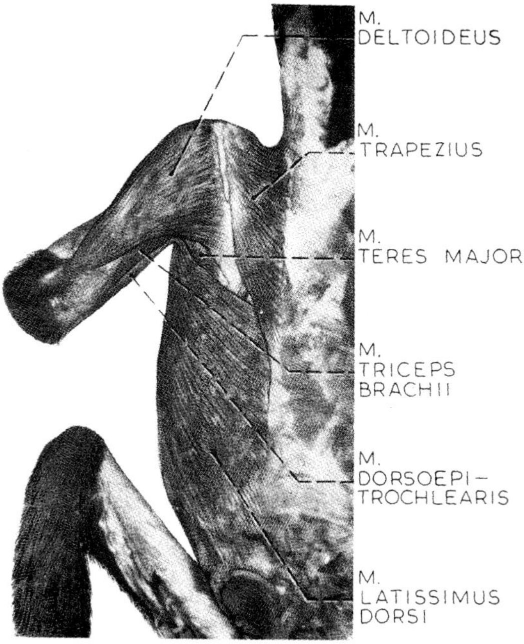

Fig. 2. The dorsoepitrochlearis muscle in a gibbon. The muscle crosses the shoulder joint but not the elbow. It, thus, helps adduct the shoulder while not preventing flexion of the elbow, movements important in hauling the body up by the arms in climbing.

levels in their interior (including the compound lenses with which light was channelled to the nervous elements at the base). The various surfaces of the compound lenses could, therefore, be discerned through surface electron microscopy of the fossils. This showed that different trilobites had evolved two different types of compound lenses as solutions to the light collecting problem posed by a dark muddy environment under water. These two optical solutions, worked out separately as optical problems by Huygens and Descartes, had, therefore, evolved hundreds of millions of years before these two gentlemen, and of course, had evolved in parallel.

In contrast, the closest living relative of trilobites, *Limulus*, the horseshoe crab, has no lens in its ommatidia. Light collection is carried out by a new optical solution (also evolved recently by astronomers for collecting faint light from distant stars). This involves parabolic reflective walls in each ommatidium thus concentrating the light at the light-sensitive point. Yet careful electron microscopic examination of the omatidia in *Limulus* shows that, though there is no compound lens present, there are thin wisps of connective tissue lying in exactly the position that the components of the compound lens would occupy if present. Are these remnants indicative of connective tissues covering a compound lens that has long since disappeared? Are the lenses the product of a different developmental process than the connective tissues with which they are covered? Does this imply that, like the trilobites, horse-shoe crabs once had such lenses, but lost them in evolving a better solution for light-concentrating: the parabolic mirror? Are these connective tissue wisps also 'ghosts' of the past?

These thoughts stimulate new questions. When functionally redundant muscles disappear, is there a double developmental process at work? When, for example, muscles disappear but fascial sheets remain, is that because connective tissue elements are produced by different developmental processes to muscles? In other words, persistent fascial sheets may tell about muscle loss.

Development: The Head Domain and the Temporalis Muscle

Though it might seem that the days are gone when new information may be revealed by dissection, in fact, this is not so. It is true that bones, muscles and nerves have been very well studied. They are easy to observe and dissect. However, the anatomies of ligaments and aponeuroses (flat ligaments in the form of collagenous sheets) are considerably more poorly known. The anatomies of connective tissue sheets lying around and between muscles and bones (fasciae) are scarcely known at all. These connective tissue structures are so much more difficult to dissect than bones, muscles and nerves, particularly in fixed specimens. However, it seemed to me that the comparative anatomy of fascial sheets might provide information about evolutionary changes in adjacent muscles.

The first hint of this arose through an accidental finding as I was dissecting a chimpanzee head at the behest of the late Lord Zuckerman in the Department of Anatomy, University of Birmingham. I was looking to see how the temporal muscles were attached in relation to the cranial crests evident in some ape skulls. As I viewed the temporalis muscle I was suddenly puzzled. The glistening silver tendon of the temporalis muscle so obvious in humans was absent (Fig. 3). Where had it gone? The answer was that it had not gone anywhere; it was still there. I made a transverse incision through the muscle of the chimpanzee. There in the depths of the incision was the glistening silver tendon. It was covered by a thick superficial head of temporalis that was almost entirely fleshy (Fig. 4). Humans do not have the superficial muscle. Remarkably, within a few days, during medical student teaching, I found a human that also displayed this condition (Fig. 5).

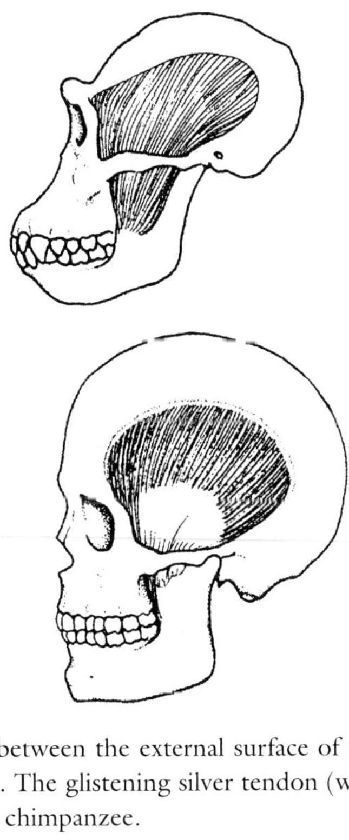

Fig. 3. The difference between the external surface of the temporalis muscle in a chimpanzee and a human. The glistening silver tendon (white region) of the human seems to be absent in the chimpanzee.

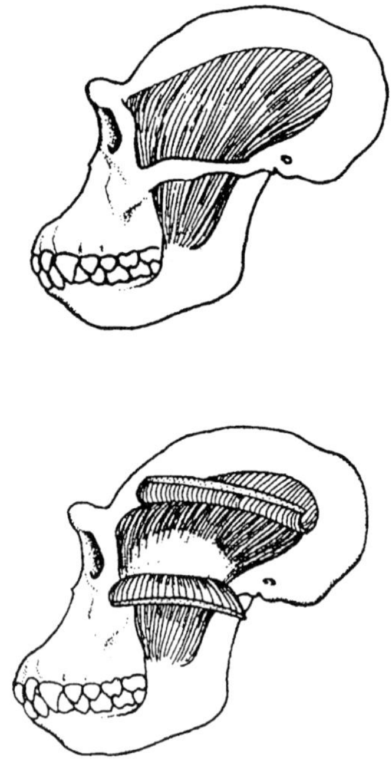

Fig. 4. Cutting into the temporalis muscle in the chimpanzee revealed the tendon lying deep to a thick superficial portion of temporalis not present in humans.

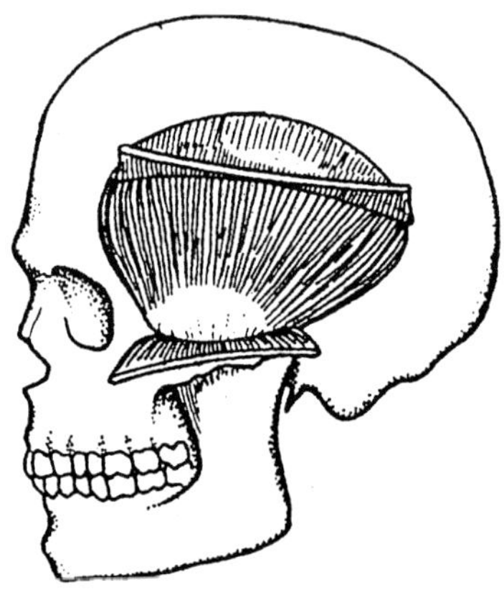

Fig. 5. A variant in humans with a superficial head just like the chimpanzee, albeit one that was very thin. Several others in various stages of development have since been found, a total of about 35 in over 300 cadavers.

As a result of this, over the following years, I made observations on the temporalis muscles in many humans cadavers used in teaching. Though the initial observations were made in Birmingham, and later in the various anatomy departments in the United States and Australia in which I have worked, I was also fortunate to have available observations made by Professor W. James Moore (now retired but originally from Birmingham Anatomy and later from the Department of Anatomy, University of Leeds, UK) and Dr Kenneth Chan (now retired but then of the Department of Anatomy, University of Hong Kong). Professor Paul McMenamin at the University of Western Australia has also been interested in these matters, inviting me to present research lectures to medical students on this topic. To be especially thanked are also the many medical students in these medical schools who have also alerted us to the presence of anomalous superficial heads of the temporal muscle in humans. These variations were found in 35 cadavers from amongst more than 300 that were surveyed. In addition, to check on the situation in non-human primates, actually already fairly well known, I dissected four chimpanzees, four rhesus monkeys, two baboons, two vervets, two colobus monkeys and two langurs (Oxnard and Franklin, 2008a).

The temporal muscles of all the non-human primates dissected here concur in every way with those presented in primate anatomy texts (e.g. Hartman and Straus, 1933; Berringer *et al.*, 1968; Swindler and Wood, 1973). In all species, there is a fully fleshy superficial head to the temporalis muscle. What next turned out to be important, were the fascial sheets associated with the muscle.

This superficial head of the temporalis muscle takes origin from the skull area between the superior and inferior temporal lines (or ridges or crests, depending upon species, sex and size). It is covered by the external investing layer of deep fascia (as is always the case for superficial layers of muscles excepting those muscles that develop in the superficial fascia: e.g. in the head, the muscles of facial expression and in the trunk, the muscles of the panniculus carnosus). The superficial head also has an internal investing layer of deep fascia on its under-surface.

The deep head of the temporal muscle in non-human primates arises from the lateral surface of the cranium below the inferior temporal line, ridge or crest. This head is initially fleshy but gives way to the aforementioned glistening silver tendon which inserts on the coronoid process. The deep head is also covered by deep fascia that arises, as does the deep fascia lining the under surface of the superficial head, from the inferior ridge. However, the fact that these two layers are separate implies that, at least on occasion, these two muscles contract independently, even though, perhaps, they often act together. On the under-surface of the deep head, there is, of course, a further internal layer of the deep fascia arising from the adjacent portion of the periosteum of the temporal fossa.

There are, thus, two muscles and four layers of fascia. Their relationships in chimpanzees are shown diagrammatically in Fig. 6.

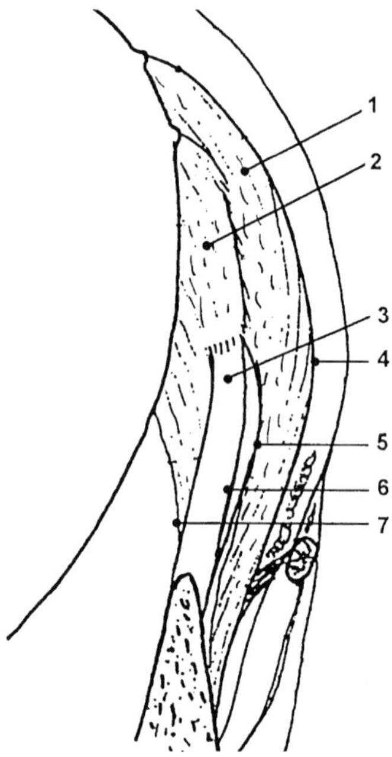

Fig. 6. A diagrammatic exploded section through the temporalis muscle and its associated fascial layers in the coronal plane in a chimpanzee. The left-hand side curved line is the outer surface of the cranial wall; the speckled projection below is a section through the coronoid process; the small round speckled circle near the tip of the coronoid process is a section through the zygomatic bone. 1 is the fleshy superficial head of temporalis muscle, 2 and 3 are the fleshy and tendinous parts of the deep head. The dark lines labelled 4, 5, 6 and 7 are the investing layers of fascia over and around the two muscles.

There are, of course, various other heads of temporalis (e.g. anterior head) but this discussion does not immediately concern them.

In humans, the temporalis muscle is like the deep head of non-human primates. It arises from the cranium at and below the inferior temporal line (a thin line in humans, rather than a strong ridge or even crest as in apes and monkeys). The glistening silver tendon of this muscle is clearly evident through the 'temporal fascia' because there is no superficial head. The implication is that, either the superficial head of the muscle of apes and monkeys has never existed in humans (which we will judge unlikely) or that at some stage in human evolution it has disappeared.

It is evident from the anatomy of various fascial sheets that the superficial head must have disappeared. Thus, in humans, the so-called 'temporal fascia' is a complex structure, not at all equivalent to the external investing temporal fascia in apes and monkeys (Fig. 7). The human layer is made up of three fascial sheets, two on each side of the non-existent superficial head and one overlying the deep head. It can, with care, be dissected fully in the normal situation in humans. That this is true, however, is more clearly revealed from the dissections of 35 human cadavers that had superficial heads of temporalis. These anomalous muscle heads varied from cases in which they were entirely similar (though very much thinner) to those of apes and monkeys, to cases where only portions of the muscle were present. In each of these anomalous cases, the fascial sheets were much more readily identified than in normal humans and were similar to those in apes and monkeys (Fig. 8).

Has there never been a superficial temporal muscle in humans, or was it once present and has it disappeared? Is this complex fascial structure the remnants of an absent muscle and its coverings? These findings say yes. How and when might this loss have happened? Development and fossils may supply an answer.

Back to Bones: The Skull

Answers to these questions may come from the very recent biomolecular work of Stedman *et al.* (2004). Realising that a primary

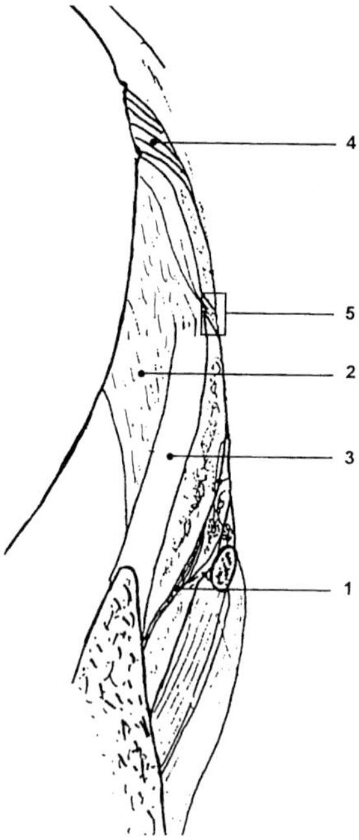

Fig. 7. The same section and conventions as in Fig. 6 but in a normal human. 1, 2 and 3 are the same muscles as in chimpanzees except that the superficial head is extremely small, often missed in dissection, and hidden behind the coronoid process. The dark lines labelled 4 and 5 represent the various layers of fascia that can be dissected with care. They are the equivalents of 4, 5 and 6 in Fig. 6 and include fibro-fatty materials in the position of the absent superficial head.

difference between humans and all other primates is the very much reduced size and power of the masticatory apparatus, Stedman *et al.* suggest that this might be due to a change in the gene encoding for myosin heavy chain (MYH). This change involves a frameshifting mutation that inactivates MYH 16. In the head domain, this particular MYH is responsible for the format of the masticatory muscles. Others are associated with other developmental muscle groups: e.g. MYH 13 for muscles moving the eye.

A specific protein inactivation of MYH 16 is associated with marked size reduction of the entire muscular battery responsible for mastication. Among primates, this inactivation exists only in humans. The result is a

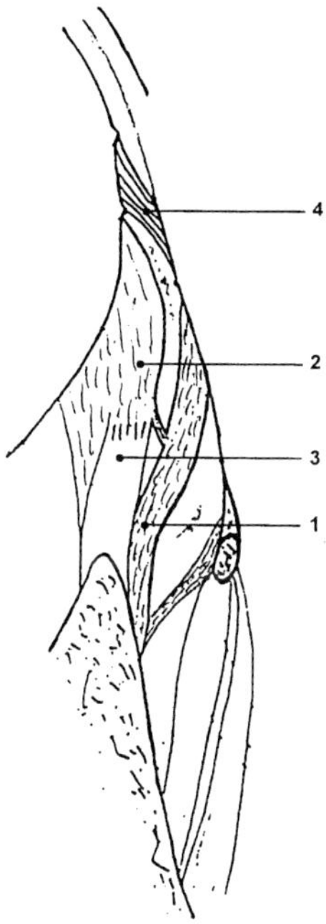

Fig. 8. The same section and conventions as in Fig. 6 but in a human with a superficial head. 1, 2 and 3 are the two muscles and the tendon as in chimpanzees. The dark line labelled 4 represents the upper part of the compound fascia. The remaining dark lines are the investing fascial layers, exactly as in the chimpanzee, investing the superficial and deep muscular heads.

pattern of gracilisation in the muscles of mastication in humans that is not found in any other living primate. Any individual ape or monkey that possessed this mutation would have so much reduced masticatory power, that, with the normal diets of apes and monkeys, it would be immediately inviable. Only the adoption, by humans, of new foods (e.g. meat from extensive scavenging involving groups of individuals) and possibly new food preparation techniques (e.g. by the inventions of butchering, preparation, marination, hanging, burning, and later, cooking) may have rendered small jaw muscles still viable. It is entirely likely that the muscular difference, also unique to humans, that is outlined here, the loss of the large superficial head of the temporal muscle, is associated with this overall masticatory muscle gracilisation (Kidd and Oxnard, 1997).

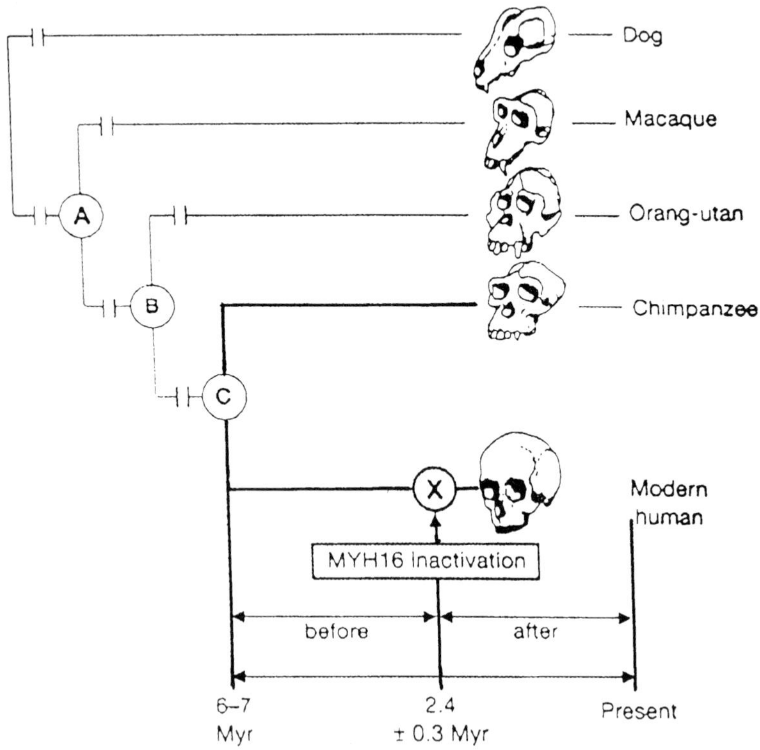

Fig. 9. Phylogenetic analysis of similarities in MYH 16 for a number of species. Only humans possess the MYH 16 inactivation. Using the chimpanzee common ancestor as 6–7 million years ago, then the inactivation occurred about 2.4 million years ago and (according to Stedman *et al.*) ancient humans are included but australopithecines excluded.

Stedman and colleagues went on to show, using the coding sequences of the myosin rod domain, that this protein inactivation may have appeared approximately 2.4 million years ago (Fig. 9) and been present, therefore, in fossil humans (such as *Homo erectus* and *H. ergaster* but not the earlier australopithecines). This 2.4 million year figure is based upon an assumption of a human/chimpanzee divergence time of 6 million years ago. Of course, if the divergence time of humans and chimpanzees was even earlier (e.g. 7, 8, 10 or 13 or even more millions of years, see a later chapter) then the date of this change would also have been earlier, possibly as much as 3–4 million years ago. This would mean that other earlier fossils could be included (but presumably only those that show reduced masticatory muscle attachment areas and a degree of gracilisation of the face and cranium). The various australopithecines, with large masticatory muscle attachment areas, powerful dental batteries, heavy jaws and robust faces and skulls would continue to be excluded.

The question, therefore, arises: can we check in any other way (e.g. from dimensions of the skull) whether or not the temporal muscle is reduced? It turns out that this is possible. Thus, the depth of the temporal fossa at the level of the zygomatic arch (relative to its antero-posterior dimension) is an approximate surrogate for the relative thickness of the temporal muscle, this muscle and its bony insertion being the main large structures that occupy this space. The larger this proportion, the thicker is the muscle (Fig. 10). Thus, in modern

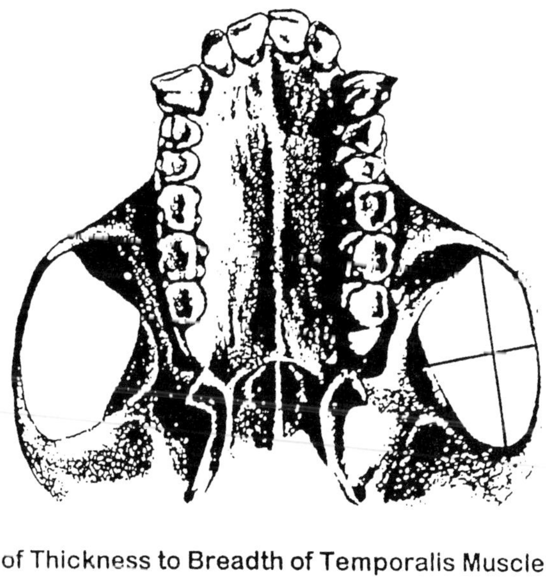

Ratio of Thickness to Breadth of Temporalis Muscle

Reflected in ratio of Depth to Length of Temporalis Fossa

Fig. 10. The temporal fossa ratio that relates, at least in part, to the thickness of the temporalis muscle.

Table 1. The dimensions of the temporal fossa in some primates.

Extant species	Fossil species
Known heavy temporal muscles	*Presumed heavy temporal muscles*
Gorilla 0.88	*Australopithecus afarensis* 0.78
Orang-utan 0.83	*A. africanus* 0.73
Chimpanzee 0.78	*A. robustus* 0.78
Bonobo 0.74	*A. boisei* 0.80
	Homo floresiensis 0.82[a]
Known light temporal muscles	*Presumed light temporal muscles*
Modern humans 0.60	*Homo erectus* 0.64
	H. heidelbergensis 0.66
	H. neanderthalensis 0.63

[a] For *Homo floresiensis*: see later chapter.

apes (with a heavy superficial head to the temporal muscle) that proportion averages 88% in gorillas, 83% in orang-utans, 78% in chimpanzees and 74% in bonobos. This contrasts with the human average of about 60%. Humans have a very thin and light temporalis muscle *in toto* and an absent superficial head in particular (Table 1).

It is also possible to replicate this measurement in various fossils (Table 1). Thus, in three human fossil specimens, *Homo erectus*, *H. heidelbergensis* and *H. neanderthalensis*, the values are 64%, 66% and 63%, respectively; little different, in other words, from modern humans. Can we presume that the reduction of temporalis muscle had already taken place at this point in time (extending back, with the Nariokotome Boy, as much as 2 million years or more)? In contrast, four specimens of australopithecines (*afarensis, africanus, robustus* and *boisei*) have values of 78%, 75%, 78% and 80%, respectively. Can we presume that whatever the age of these specimens, the reduction had not taken place in their lineages? (It is particularly interesting to note that in the recently discovered skull of the Flores fossil, the 'hobbit', this proportion is approximately 80% as estimated from the published pictures of the basal view. Does this mean the Flores fossil had, relative to the size of its tiny cranium, powerful muscles of mastication like australopithecines and apes? However, this is a subject for another chapter.)

Stedman and colleagues also suggested that this reduction in masticatory muscles might have produced increases in cranial volume and therefore increases in brain size. I suspect that any direct link with an increased brain is unlikely. However, despite rejecting the idea of a direct link to brain growth in humans as compared to other primates, there are possible indirect effects upon cranium size that are interesting. Thus careful experiments in the growing rat (Moore, 1958) show reduction in cranial length and height with reduced temporalis muscle activity. This goes with a concomitant reduction in face and mandible size with reduced masseter activity.

Two counterintuitive findings from these experiments, however, were significant increases in the widths of the cranium and jaws combined with significant decreases in the thickness of the cranial and jaw walls. These increases are not small (of the order of 10.5%). Their real effect is significant reduction in thickness of the lateral wall of the cranium and the ramus of the mandible in the experimental animals. Cranial and jaw thicknesses at these points are the biomechanical equivalent of wall thickness in long bones. They relate to the second moment of area (see earlier chapter, Currey, 2002; Oxnard, 1990, 1991). The thickness reductions in these experiments imply much reduced mechanical stress from temporalis and masseter/pterygoid muscles.

The equivalent change in skull and jaw thicknesses in humans are likely partly related to lessened stress from reduced masticatory muscles. In addition, the effect of reduced temporal muscles in Moore's rats also produces smaller temporal lines, a more globular cranium and an increase in internal lateral diameter of the skull (though no increase in intracranial volume was reported presumably because the antero-posterior and dorsoventral diameters were reduced). The equivalent reduction from temporal ridges and crests high on the vault (even reaching to the vertex) in apes and monkeys to lines low on the temporal squama in humans is likely related in part to equivalent reduced biomechanical stress from smaller temporal muscles.

All this implies decreased robusticity of the crania, faces and jaws. The change in the myosin heavy chain (MYH 16) modification producing, in humans, reduced masticatory muscles overall may well have been associated with reduced robusticity of the cranium, face and jaw during evolution.

It is, thus, possible that the more gracile crania, faces and jaws of the newly found but much older fossils of the genus *Homo* may also be morphological markers of this masticatory muscle reduction. Certainly these features are considerably more gracile in these species than they are the much more robust present day great apes and fossil australopithecines (whether robust or gracile).

A second way of answering these questions stems from scanning electron microscopy (SEM) of bone surfaces. This is just another way of looking at the muscle/bone interaction. Thus, in species (non-human primates) possessing a superficial head of the temporalis muscle, the area of bone from which the superficial head arises, the area between the temporal ridges, is relatively smooth as befits a fleshy muscle arising from underlying periosteum. In humans, with an absent superficial head and a strong and complex temporal fascia, the area between the temporal lines is associated with rough vertical striae and deep pits for the attachment of Sharpey's fibres in the fascia (Oxnard and Wealthall, 2003; see also Wealthall, 2000). This is what one would expect if this area were for the attachment of the strong collagenous fibres of a powerful complex temporal fascia.

If any of the fossils were well enough preserved, such SEM studies might well show which of these situations was present (e.g. Kupczik *et al.*, 2007). This has not yet been done. In any creature (e.g. male gorillas and orang-utans, robust and hyper-robust australopithecines) where there are cranial crests because opposite temporal muscles have come together in the midline we would not be able to discern these features. In such specimens, however, crests imply very powerful superficial muscles which arise by strong linear tendons where the area between the two temporal lines has become the lateral lips (one on each side) of a cranial crest. This is true whether crests be temporal or occipital crests such as one finds in the skulls of some apes and australopithecines, or whether the crests are those found on totally different bones (such as the crest of the spine of the scapula). Such a feature automatically implies very powerful oppositely directed muscles.

In summary, then, the information provided by molecular biology about masticatory muscle reductions in humans as compared with all non-human primates, is consistent both with the soft tissue findings of loss of the superficial head of temporalis and the overall reduction of the masticatory muscles in general. However, these features are only obtainable from extant species. The molecular information is also consistent with the changes in bone architecture related to the muscular reductions. The bone measures have been replicated on the small number of fossils available and imply that the muscle reduction was evident in the *Homo* specimens available but not in the australopithecine specimens. SEM has not yet been undertaken on fossils (and indeed may not be possible if the surfaces of the fossils are too damaged). This all seems to suggest that the reduction in masticatory muscles was due to a molecular event, that it was confined to the genus *Homo,* and that it arose about 2.4 or more million years ago. As Nariokotome boy, with a human-like value for the relative dimensions of the temporal fossa, is dated at about 2 million years ago then this is consistent with the molecular estimate of Stedman and colleagues. Both the molecules and the morphology may be ghosts of the prior condition.

All this leads to more general questions. Are there other bodily regions in humans where similar evolutionary loss of muscle bulk may have occurred? One answer may be the upper limb.

Development: The Forelimb Domain and the Dorsoepitrochlearis Muscle

In humans, in contradistinction to all other primates, the upper limb is not used in powerful locomotor movements (because humans do not use their upper limbs in weight-bearing locomotion). Has there been a concomitant reduction in upper limb musculature? Is there any primary evidence of this from persistent muscular anomalies and fascial sheets? Is there any secondary evidence in upper limb bones from changed mechanical effects?

The beginnings of answers may lie in consideration of the already mentioned and well-known difference between humans and other primates, the dorsoepitrochlearis situation. We have already looked at the dorsoepitrochlearis muscle briefly. The variations in this muscle in non-human primates seem clearly mechanically adaptive in terms of shoulder and elbow function in locomotion and posture. In humans, of course, this muscle is usually absent. It is, however, sometimes present in humans (in 5% of cases, Bergman *et al.*, 1984) even though we do not use our upper limbs for weight-bearing locomotion. In humans it has also been named latissimo-condyloideus, accessorius tricipitis, coraco-condyloideus, gleno-tricipitis, and the fourth or fifth (depending upon how they are counted) head of triceps (because some fibres may insert into triceps tendon or the fascia overlying triceps.) Generally, however, its arrangement in humans, when present, is precisely as in lesser apes, great apes and prehensile tailed New World monkeys. That is, its attachments are such as to produce only adduction of the shoulder while not impeding flexion of the elbow. Though this could be interpreted as humans sharing a character primitive for apes and humans, considerations of functional anatomy imply that it is more likely present because humans arose from creatures that had similar arboreal adaptations as apes and prehensile-tailed monkeys. This is an old story well documented in the literature.

What is less well-known is the arrangement of the fascia in this region. Though the muscle is not present in the great majority of human specimens, careful dissection reveals that a 'fascial ghost' of the muscle is present in all humans. Bundles of connective tissue fibres pass from the neck of the scapula, the adjacent deep fascia on the subscapularis muscle, the tendon of latissimus dorsi, and sometimes the capsule of the joint and subjacent bone, and pass distally into the deep fascia of the arm. It is only on the rare occasions when there are muscular fibres present in this fascial structure that a dorsoepitrochlearis muscle is recognised. Medical students never notice these fascial arrangements; in their zeal to clean the muscles that they are expecting to find they remove the fasciae. This fascial structure seems to represent the combination of the investing fascia and internal connective tissue components of a muscle once present; it exists even when the muscle does not. As such it seems to be evidence that once humans (or pre-humans) did have such a muscle.

Given, however, our experience with temporalis muscle above, this finding suggests a more general conclusion. If developmental mechanisms have lead to the loss of the dorsoepitrochlearis muscle (perhaps modification of some muscular factor, equivalent to MYH16 for the muscles of mastication in the head domain) they may not have involved the connective tissue elements (due to different connective tissue factors). Contractile tissue, muscle, may have disappeared, but non-contractile yet tension bearing fascial coverings and internal connective tissue elements may have remained.

Is it possible that this is the beginning of a much broader theme, relating to the comparison of all shoulder muscles in humans as compared with non-human primates. Are other upper limb muscles also reduced or eliminated? Are muscular anomalies present indicating where muscles have been lost? Are the appropriate fascial elements still present? This might be especially evident if the elements in the upper limb domain were compared with their serially homologous counterparts in the lower limb domain. Such comparisons would combine information across various primates (comparative evidence), within humans (variations or anomalies), and between limbs (serial homologies).

Development Again: The Forelimb Domain and Other Upper Limb Muscles

There are many muscles in the limb domains in primates. Let us concentrate upon only those muscles at the upper end of the forelimb domain and compare them with equivalent muscles (where present) in the hindlimb domain. Let us particularly examine them in the developmental groups to which they belong. The comparisons that we shall make are comparisons between developmentally similar muscle groups, in serially homologous elements (the hip and thigh, and the shoulder and arm) and between humans and non-humans (mostly our closest living relatives, the apes, especially chimpanzees, bonobos and gorillas (Oxnard and Franklin, 2008b)).

From Hip to Shoulder: Cranial Ventral Muscles

Two groups of embryologically ventral muscles acting at the hip arise from the embryologically ventral bony elements, the pubis cranially and the ischium caudally, of the pelvic girdle. (Though the ischium is anatomically dorsal in location in the primate adult, it commences as a ventral element in the foetus, and only later becomes dorsal in position through the twist of the limb that occurs during development.)

Of these muscles, one group, the adductor muscles, are cranial in position originating from the more cranial pubis. They are primarily uniarticular muscles acting upon the hip. There are several muscles in this group ranging from the shortest, adductor brevis, through adductor longus and the adductor head of adductor magnus to adductor gracilis. They may, each, have as many as two or even three heads in different species. They splay out like a fan passing downwards and laterally from the pubis onto the medial aspect of the femur. There are also various small superficial bellies, some with long tendons, and these pass more distally into the fascia of the leg (the lower limb segment below the knee). These muscles are similarly complex and big in both humans and non-human primates (Fig. 11).

The equivalent muscular block in the forelimb domain of non-human primates is the coracobrachialis muscle sheet. This is also mainly a uniarticular muscle acting across the shoulder. It arises from the cranial ventral skeletal element equivalent to the pubis in the lower limb, that is, the tip of the coracoid process of the scapula. As with its lower limb serial homologue the different components splay out like a fan in most primates. There is a short uppermost deep head (coracobrachialis profundus) often with two or more components, a middle head (coracobrachialis intermedius) again with as many as three components and a long superficial head (coracobrachialis superficialis) also often with several components. Each of these passes down the limb inserting into the humerus at correspondingly more distal points. As with the adductors in the lower limb, there are also additional fine muscular slips that pass into fascia and inserting beyond the elbow joint into fascia in the forearm. The disposition of this muscle group in the forelimb of non-human primates can be seen to be

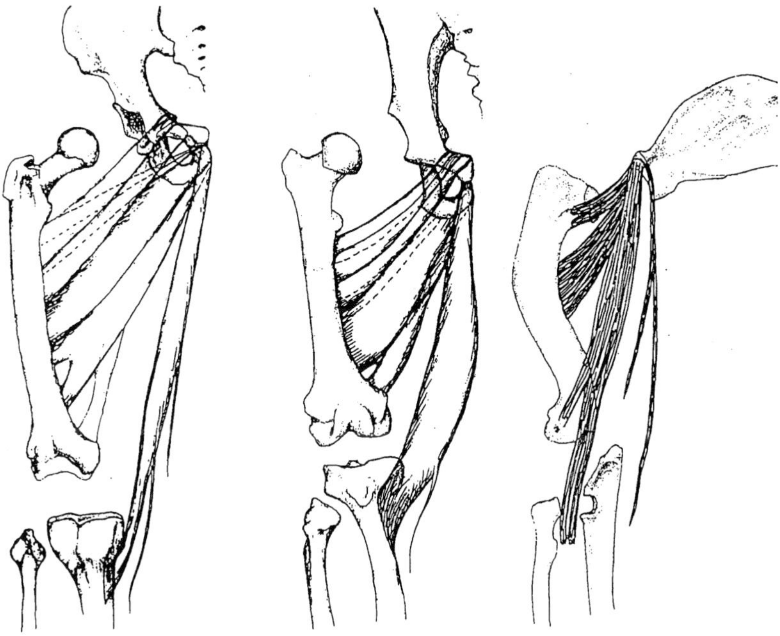

Fig. 11. The cranial ventral muscles around the hip in humans (left frame), a generalised non-human primate (middle frame) and the shoulder in a generalised non-human primate (right frame). The anatomy looks a little curious because the bones and joints have been 'exploded' in order to reveal more clearly the overall muscular arrangements. The overall serial similarity in pattern between lower limbs in both humans and non-humans and upper limbs in non-humans is obvious.

enormously similar to its serial homologue, the adductors, in the hindlimb (also Fig. 11).

There is, however, a major difference between humans and other primates. In humans this muscle sheet is greatly restricted to just one component (or sometimes two) of what in other forms is called the coracobrachialis intermedius (Fig. 12). In humans, the other components are usually absent and the whole muscle is therefore called the coracobrachialis. Variations are found on occasion in humans: small deep and superficial heads, extra intermediate heads, and sometimes, even, one or more very superficial bundles passing far down onto the humerus and with some fibres into fascial sheets as far distally in the

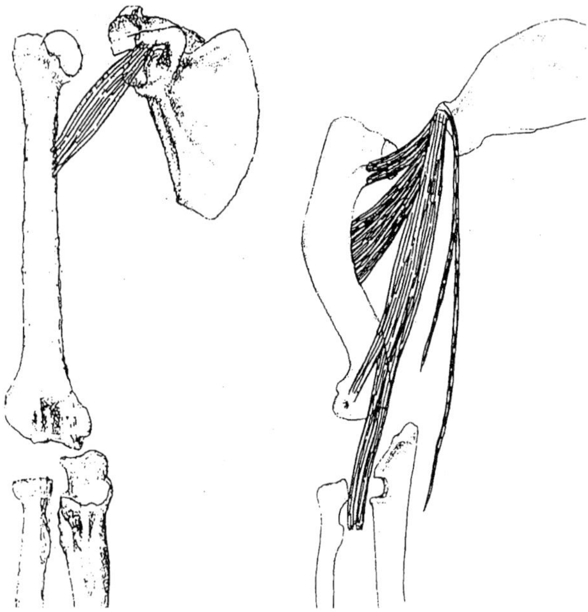

Fig. 12. The cranial ventral muscles around the shoulder in humans (left frame) and the same generalised non-human primate as in Fig. 11 (right frame). The anatomy looks a little curious because the bones and joints have been 'exploded' in order to reveal more clearly the overall muscular arrangements. The marked reduction in humans is obvious.

forearm. When present, these remnants are very small but they clearly mimic the main parts of the non-human muscle. They are, of course, accompanied by their investing deep fascial sheets.

If these variations and their related fascial sheets are ghosts of what the arrangement was in the past, then during human evolution, and human evolution alone, (that is, subsequent to separation from apes) this muscle group has become very much reduced but the fascial sheets are still present.

From Hip to Shoulder Again: Caudal Ventral Muscles

We find a similar story when we come to look at the caudal ventral muscles. In the hindlimb these arise from the ischium; in the forelimb from the base of the coracoid and its related upper part of the glenoid cavity. These bony origins are, again, both embryologically ventral parts of the limb girdles.

In non-human primates, the hindlimb muscular components of this group (the hamstrings) are splayed out as the hamstring head of adductor magnus, biceps femoris, semimembranosis and semitendinosus. The forelimb components are two (sometimes three) heads of biceps, often two separate muscles, and the two (sometimes three) heads of brachialis, also often two separate muscles, together with other occasional slips passing more superficially into the forearm fascia. In each limb all these muscles cross the middle joint (elbow and knee, respectively) and many are also biarticular, crossing both the proximal joint (hip and shoulder, respectively) to insert on each side beyond the distal joint (knee and elbow, respectively).

The situation in the hindlimb is similar in both humans and non-human forms and this similarity is carried over into the forelimb non-human species (Fig. 13). In contrast, again, in humans the arrangements in the forelimb are much simpler: there are only two muscles (a single biceps with two heads, and a single brachialis, Fig. 14). In addition, the two human muscles are much smaller than those in the non-human species. Again, however, variations occur in humans in which both biceps and brachialis may be partly or even completely divided. In such cases, the picture looks more like the forelimb of non-human primates (though, still, the muscles are much smaller). If these muscular variations (and the fascial sheets that accompany them) are ghosts of what the arrangement was in the past, it would again appear that in humans this forelimb muscle group as a whole has become much less complex and very much smaller.

From Trunk to Limb: Ventral Muscles

The group of shoulder muscles that originate from the trunk and cross the shoulder to be inserted very close to the shoulder joint is the pectoral muscle block (Fig. 15). This is organised into superficial and deep layers. These muscles are not represented in the hindlimb.

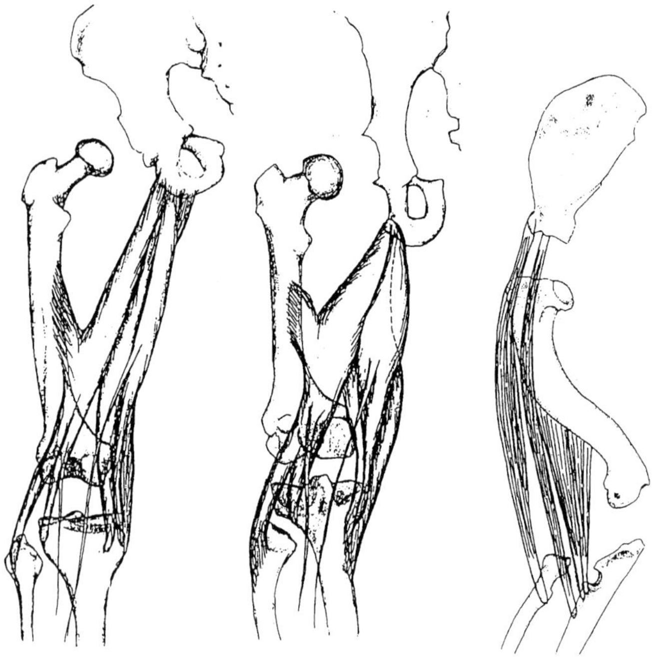

Fig. 13. The caudal ventral muscles around the hip in humans (left frame), a generalised non-human primate (middle frame) and the shoulder in a generalised non-human primate (right frame). The anatomy looks a little curious because the bones and joints have been 'exploded' in order to reveal more clearly the overall muscular arrangements. The overall serial similarity in pattern between lower limbs in both humans and non-humans and in the upper limbs in non-humans is obvious.

In non-human primates pectoralis major comprises several different components including an entirely superficial capsular part (often rather small) arising from the sterno-clavicular joint, a clavicular part proper arising also from the sterno-clavicular joint and a considerable portion of the clavicle and a sternal part arising from the major length of the sternum. This latter may be divided into two or more portions. The deeper layer of the pectoral musculature includes, first, subclavius arising from the first costal cartilage and inserting into both the clavicle and the clavipectoral fascia (and thereby also crossing the shoulder joint). The deep layer also includes: pectoralis minor arising from the sternum and ribs medially and inserting mainly into the

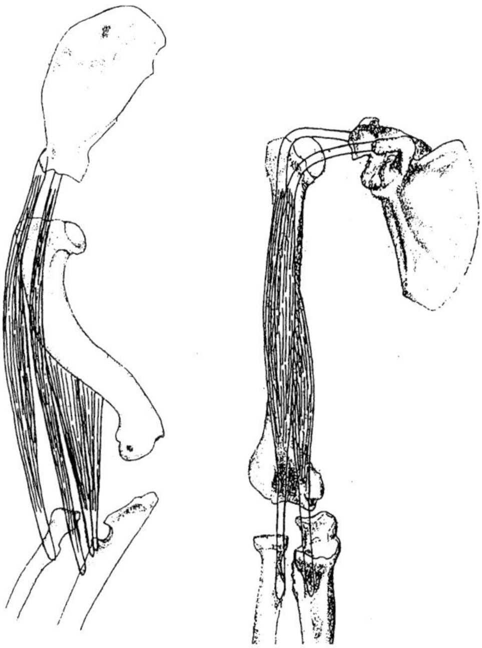

Fig. 14. The caudal ventral muscles around the shoulder in humans (left frame) and the same generalised non-human primate as in Fig. 13 (right frame). The anatomy looks a little curious because the bones and joints have been 'exploded' in order to reveal more clearly the overall muscular arrangements. The marked reduction in humans is obvious (though somewhat less than) was the case for the coracobrachialis sheet (Fig. 12).

capsule of the shoulder joint, and pectoralis abdominis, arising more distally on the trunk from fascial sheets of the abdominal wall and inserting with pectoralis minor. These muscles may be organised into two or more clearly separable fascicles. A final component may include part of the panniculus carnosus muscle. This muscle is primarily a muscle of the superficial fascia and therefore, technically, not a limb muscle; but part of this muscle inserts into a tendinous band lying deep to the pectoralis minor and inserting with the tendon of that muscle. There are frequently a series of connections between the panniculus carnosus muscle and various other shoulder muscles. Some heads of these various muscles may pass far down the arm and

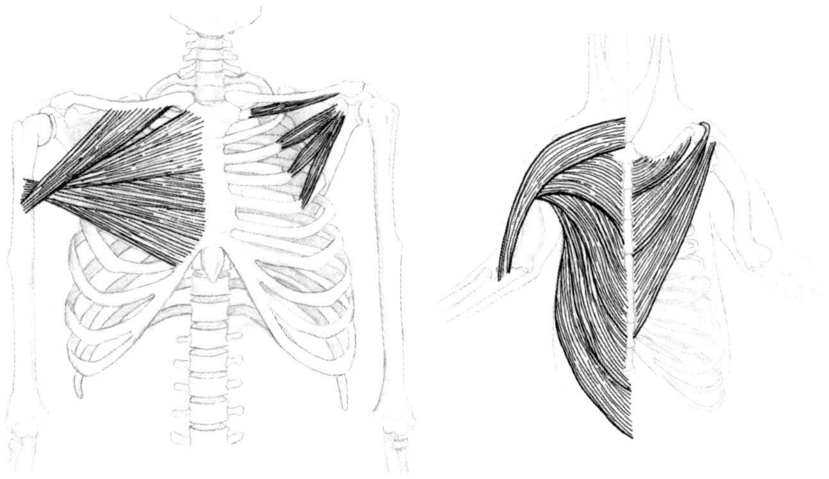

Fig. 15. The ventral trunk to shoulder muscles showing pectoralis major and minor in both a generalised non-human primate and a human. The anatomy looks a little curious because the bones and joints have been 'exploded' in order to reveal more clearly the overall muscular arrangements. The marked reduction in size and complexity in humans is obvious.

even into forearm fascial sheets. This conspicuous development is associated with movements of the limb in creatures (non-human primates) where the forelimb is heavily involved in locomotion (Fig. 15).

In contrast, in humans this double muscle sheet comprises only the pectoralis major, minor and subclavius of traditional anatomical texts (also Fig. 15). The individual muscles pass from the ventral aspects of the thoracic cage, cross the shoulder joint and are variously inserted in the upper end of the arm. They are much reduced in overall mass and in cranio-caudal extent compared to various non-human primates. However, any of the extra muscles described above for non-human primates may be found as variations in humans. The commonest of these is a pectoralis abdominalis although additional digitations of pectoralis major and minor are frequent. Even muscles representing panniculus carnosus connect from the pectoral muscles to the superficial fascia (panniculus adiposus). Are these muscular variations and persistent fasciae ghosts of the complex elements of non-human forms?

Dorsal Muscles in General

Equivalent comparisons of the dorsal musculature in the hindlimb and forelimb can also be made. However, the situation is more complicated because in the forelimb there are extra dorsal muscle groups drawn from both trunk and head that are not represented in the hindlimb.

Thus, a first muscle group at the shoulder is the dorsal trunk musculature that has become attached to the limb girdle: the rhomboid and serratus muscle sheets. The serial representative of these muscles in the lower limb is only the subvertebralis muscle of reptiles and the psoas minor muscle in mammals (and this latter no longer acts across the trunk-girdle joint; there is no such moveable joint in the lower limb). Therefore, a cross-limb comparison cannot be carried out.

A second group at the shoulder comprises those individual head muscles (of branchiomeric origin) that have been drawn into the shoulder over evolutionary time (most of trapezius and sterno-cleido mastoideus). This is a very ancient evolutionary development found in all land vertebrates and indeed even earlier. There is no representation of this group in the hindlimb at all; that limb is too far removed from the head domain. Again, therefore, a cross-limb comparison cannot be carried out.

A third dorsal group at the shoulder consists of limb muscles proper (perhaps a small part of trapezius and certainly the latissimus dorsi, teres major, dorsoepitrochlearis, deltoideus and triceps brachii). These are all represented by serial homologues in the lower limb so that comparisons with the lower limb can be carried out.

Let us examine each of these in turn.

From Trunk to Limb: Dorsal Muscles

The muscle sheet in the upper limb that has migrated from the trunk to the shoulder moves the scapula upon the trunk. These are the rhomboid and the serratus muscle sheets.

In non-human primates (Fig. 16) the rhomboideus sheet is large and has cranial (occipital) as well as several extensive cervical and thoracic heads. Thus, they originate on the skull from the far lateral aspects of the superior nuchal line medially to the medial external occipital protuberance, from the nuchal ligament at the levels of all cervical spines, and from the spine of the last cervical and the spines of the thoracic vertebrae often extending as low as the last. Though originally a single sheet development, there are separations between the parts and they can be identified as rhomboideus capitus, occipitalis, cervicis, minor and major respectively (these last two together also often called rhomboideus thoracis). This sheet often shows further intermediate separations so that there can be as many as six or seven apparently distinct muscles.

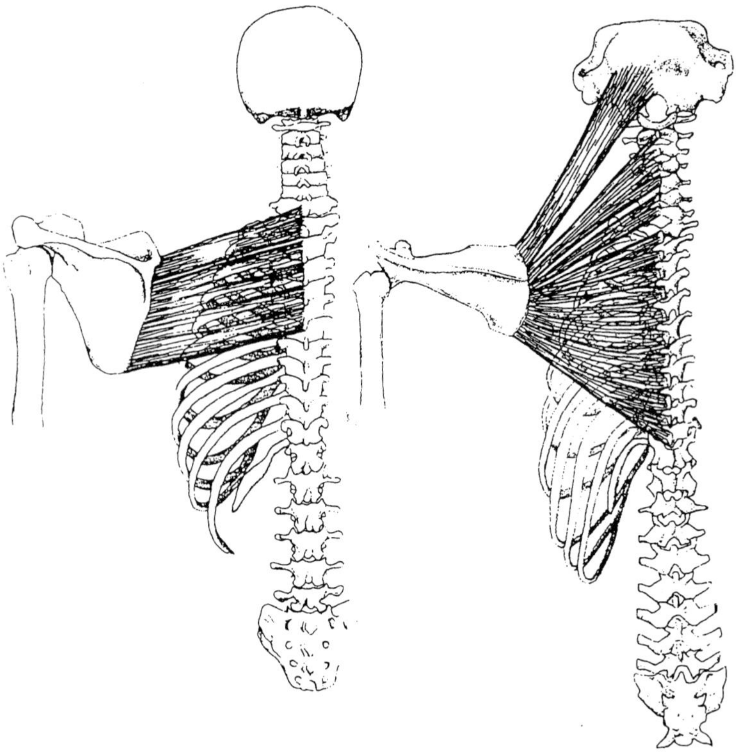

Fig. 16. The dorsal rhomboideus muscles around the shoulder. The anatomy looks a little curious because the bones and joints have been 'exploded' in order to reveal more clearly the overall muscular arrangements. The two muscles in humans are so much smaller and less complex than the extensive muscular components in non-human primate.

In humans, in contrast, the rhomboids are small and usually limited to two muscles (also Fig. 16); rhomboideus minor and major are generally confined to origins from the lowermost end of the ligamentum nuchae, the seventh cervical vertebra, and the vertebral spines of the eighth cervical and first five thoracic vertebrae. In other words, most of the segmental elements are missing. On occasion, however, humans possess small cervical heads (rhomboideus cervicis) and even sometimes occipital heads (rhomboideus capitis).

There is a second muscle sheet in the upper limb that, like the rhomboids, has migrated from the trunk into the limb and is thus able to move the scapula upon the trunk: the serratus sheet (Fig. 17). This sheet contrasts with the rhomboid sheet because it takes origin from lateral derivative of the axial skeleton (the transverse processes and their homologues) rather than medial vertebral spines and their homologues. It is, otherwise, just as widely distributed. At its upper end in non-human primates, the muscle bundles arise as high and laterally on the skull as the mastoid process (often identified as a cranio-scapularis or mastoid-scapularis), from the lateral aspects of the superior nuchal crest (occipitoscapularis), from the transverse processes of the atlas (atlantoscapulares anterior and posterior), from all of the transverse processes of the remaining cervical vertebrae (an extended levator scapulae) and from the lateral aspect of ribs (serratus anterior and serratus magnus) from the first to, often, as low as the 10th rib (and even as low as the lowest rib in some creatures such as bats). These muscles are separately distinguishable hence the profusion of names.

The insertion of this complex serratus group in non-human primates, and even more so in many other mammals, is likewise much extended (also Fig. 17). It includes the lateral end of the clavicle, the acromion, parts of the superior border of the scapula, the superior angle of the scapula, the whole of the medial border of the scapula, and around the inferior angle to encroach upon the lateral border of the scapula. In bats, the form in which this muscle is most extensive, this origin includes additionally almost the whole of the axillary border of the scapula up to the shoulder joint. (The additional arrangements in bats are presumably a further extension of the shoulder musculature attendant upon the fact that

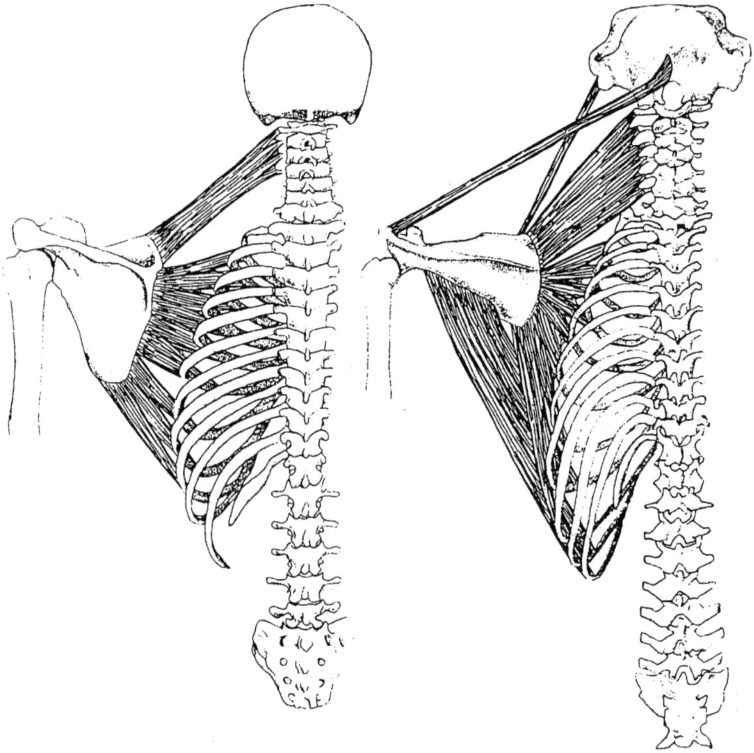

Fig. 17. The serratus muscle sheet around the shoulder. The anatomy looks a little curious because the bones and joints have been 'exploded' in order to reveal more clearly the overall muscular arrangements. The muscles in humans are small and much less complex than the extensive muscular components that exist in non-human primates.

in bats the scapula is not just a bone laid upon the dorsum of the thorax but, in fact, a bone extending so far laterally away from the trunk that it forms a true first segment to the wing.) Again, these are different evolutionary and developmental elements of what commences as a muscular sheet in each body segment.

In humans, in contrast, this muscle sheet comprises only two obvious muscles, a levator scapulae with an origin usually restricted to the first four cervical transverse processes, and a serratus anterior arising from the first to seventh or eighth ribs. Likewise in humans, the insertions of these two muscles are restricted to the vertebral (medial) border and lower angle of the scapula (Fig. 17).

Again, a large number of variations exist in humans. These include: levator claviculae muscles, mastoid-scapular and occipito-scapular muscles, atlantoscapulares anterior and posterior muscles, levator scapulae muscles separated into several sheets and with more extensive origins from as many as the second to the seventh cervical transverse processes, distinct separations of the parallel fibred portion of serratus anterior from the radiating lower fibres to the inferior angle of the scapula, extensions of the origin to as low as the tenth rib, together with various other small muscle bundles linked with other neighbouring muscles.

One particular set of variations include those in which there are distinct gaps in this muscle sheet so that, for example, the serratus anterior portion may present as two or three separate small muscles with gaps between. When this occurs, the gaps are always filled by fascial connective tissue sheets that pass from one component to the next thus infilling the gaps. Whenever these variations are present they are very small and would appear to have little functional importance. When present, however, they mimic, though very much smaller, almost every one of the muscles normally found in non-human primates. When absent the equivalent fascial sheets are present (e.g. a sheet extending between the upper border of serratus anterior and the lower border of levator scapulae). These muscular variations and fascial sheets seem, again, to be ghosts of the lost muscles.

From Head to Limb: Dorsal Muscles

A second sheet of muscle, also not represented in the hindlimb, has migrated from the head into the limb. This comprises head muscles (largely of branchiomeric origin) that have been drawn into the shoulder region: trapezius (and cleido-sterno-mastoid, though we will not consider this latter because it is really a head mover rather than acting on the upper limb). In non-human species trapezius arises extensively along a linear origin that can start as far laterally as the mastoid process, and includes the entire superior nuchal line or crest, the external occipital protuberance, the ligamentum nuchae (and through it, therefore to the upper six cervical spines), the seventh cervical spine and the spines of all the thoracic vertebrae and at least the first two or three lumbar vertebrae.

In humans, the muscle is very thin; its upper origin is restricted to the medial third of the superior nuchal line and its lower limit the 12th thoracic vertebral spine.

The insertion of the muscle, likewise, varies between humans and non-human species. In non-human species, the insertion is extensive and very heavy (especially the upper fibres in the apes). The cranial-most fibres may pass into fascia of the arm and even forearm, other cranial fibres may insert into the region of the deltoid tuberosity on the humerus and may be conjoined or separate from the most ventral part of deltoid. On the clavicle, the insertion can be as extensive as including the lateral two-thirds of the clavicle and again the fibres may be conjoined with the fibres of the anterior deltoid. Indeed, its most superficial fibres may be coextensive with the superficial layers of the anterior deltoid thus forming a cranio-humeralis muscle overlying the clavicle. (This is the usual situation in many animals that have no clavicle, but it is also present in some non-human primates such as pottos.) It also attaches to the acromion, the cranial crest of the spine of the scapula and a large tubercle near the medial end of the scapular spine.

In humans, the insertion is restricted to the lateral third of the clavicle, the acromion and the upper edge of the scapular spine near the vertebral border. Likewise in humans, in contrast to the large and heavy muscle in apes, the trapezius is so very thin that medical students often go right through it as they remove the overlying skin. It shows a much lesser degree of development, both in its overall size, and in its various complexities, in humans as compared with the apes and many other primates.

Human variations in this muscle, though usually very small, nevertheless indicate that humans have the same propensity to develop the many extra ghostly components.

True Dorsal Limb Muscles

Finally, the true dorsal muscle sheet is evident in both upper and lower limbs. In the lower limb, these include various one-joint muscles acting at the hip: the gluteal muscles (gluteus major, medius, minimus and tensor fascia lata) and the iliacus muscle, the two-joint muscles acting on both hip and knee: the rectus femoris and sartorius muscles, and the one-joint muscles acting upon the knee: the other three heads of the quadriceps muscle, the vasti. In the upper limb, the equivalent one-joint shoulder muscles are the deltoid, latissimus dorsi, teres major and subscapularis muscles, the two-joint shoulder and elbow muscles are the long head of triceps and dorsoepitrochlearis muscles; and the one-joint elbow muscles are the remaining two heads of triceps in the arm.

In the hip and thigh, these are all large complex muscles in both non-human and human forms though there are two main differences between non-humans and humans. Thus, those non-human primates with tails have a very large caudofemoralis, and humans have a very large portion of gluteus maximus inserting into the iliotibial fascial tract.

In the shoulder and arm of non-human primates the equivalent muscle units are all large and complex.

In contrast, they are small and much less complex in humans. In addition, in humans there are numerous small muscular and fascial variations (the occasional existence of dorsoepitrochlearis in humans is the commonest example, but so, too, are various fourth, fifth, sixth and seventh heads of triceps; *triceps* was named after the regular human condition). These seem to reflect the normal situation in non-human primates. Again, are they ghosts, perhaps, of prior structures?

In summary, then, and best exemplified in the various figures, humans have greatly reduced size and complexity in shoulder and arm muscles than apes, monkeys and other primates. Likewise, in humans variations similar to those in apes and monkeys occur occasionally. But in humans, all the fascial sheets are arranged as in apes and monkeys

These muscular variations and fascial sheets in humans seem, again, to be ghosts of muscles that have been lost.

Again, Back to Bones: Limb Bones

Do these muscular reductions found only in the forelimb domain in humans have effects on the bones relating to bearing of reduced biomechanical loads? Can we find architectural features of the upper limb bones that are markers for these reductions in muscles in humans as compared with apes and monkeys? Can these markers be seen in fossils to allow us to estimate how far back this change occurred? Are there any molecular markers in living forms? Is the situation in the head domain replicated in the forelimb?

One important measure of loading on long bones is the ratio of the diameter of the bone to the thickness of the cortical wall (Currey, 2002). This ratio is a component of the second moment of area, a mechanical quantity that relates to stress in a column. In creatures in which pure bending predominates over pure compression this ratio tends to be high, 20–40 in the forelimb bones in gliding dinosaurs, for example, where a gliding wing has small loads along its length but larger loads across its diameter (Fig. 18). In quadrupedal animals, the ratio varies from about 7 to 8 in small springy creatures like the smaller antelopes to as low as 3 in elephants and rhinos. The higher figures in animals like antelopes that are small and springy are found

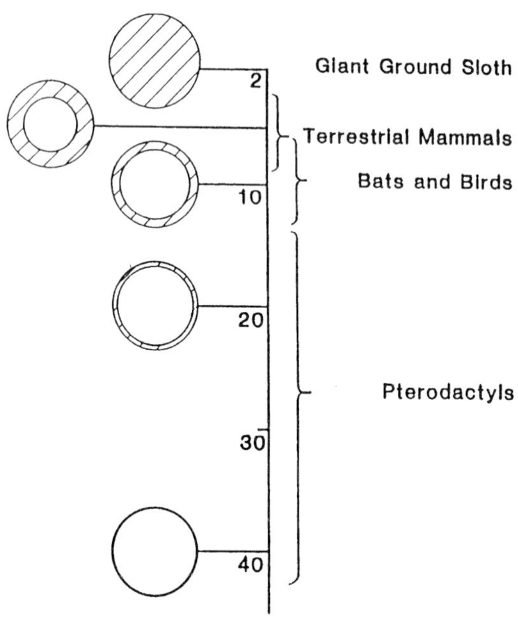

Fig. 18. Plot of bone diameter/cortical thickness in various creatures.

where compression produced by body weight is relatively small, and bending produced by very active muscles is, in comparison, relatively large. This contrasts with larger animals such as elephants where the ratio is smaller because the compression produced by axial loading due to body weight is relatively larger. This ratio is actually at its minimum (it cannot be less than 2 because the wall thickness cannot be greater than the radius of the bone, and the diameter of the bone divided by the radius is 2) in very slow and heavy creatures (Oxnard, 1990). Thus, in the giant ground sloths of North America, and in some giant Australian marsupials such as *Zygomaturus* and *Palorchestes* the figure is 2; there is no marrow cavity; the entire cross-section is bone. The assumption here is that, due to great weight and very slow movement, bending due to muscles is even more reduced in comparison with compression from great body weight. (Even though elephants have great body weight, their movements are relatively fast; elephants have marrow cavities in their long bones.) It is true, however, that we are not actually certain how the various fossils moved.

Among primates, the great apes and larger monkeys have values somewhat intermediate between springy gazelles and galumphing elephants at around 3.0–5.0 for both upper and lower limbs.

Humans differ from all of the above in an interesting way. Humans have different values for the two limbs: 3.3–3.7 for the lower limbs and 4.4–5.2 for the upper limbs (Table 2). This inter-limb difference is likely due to the differential reduction of upper limb muscle mass in humans. It presumably reflects the fact that humans bear weight only on their lower limbs.

This idea is also suggested by experiments in which muscular activity and weight-bearing in the forelimbs were reduced in experimental animals (Lisowski *et al.*, 1961; Doden, 1990). The effect was to thin the walls of the cortices in the bones of the forelimb in comparison to the hindlimb and thus raise the ratio.

A second marker of a biomechanical differential between apes and humans could be expected to be in the relative length of the upper limb compared with the lower limb (Table 3). The large and comparable muscles in the upper and lower limbs of those apes and monkeys that frequently bear the body weight on both forelimb

Table 2. Inter-limb differences in ratio of bone diameter to wall thickness.

Extant mammals	
Small bouncy mammals	Approximately 7.0, both limbs similar
Large yet active mammals	Approximately 2.6, both limbs similar
Large primates: great apes and baboons	Approximately 3.6, both limbs similar
Modern humans	3.4, 3.7, thigh/leg, 4.2, 4.5, arm/forearm
Fossils	
Australopithecus africanus	3.7–3.8, both limbs similar
A. robustus	3.1–3.2, both limbs similar
Homo erectus	3.5, lower limb, 4.4, upper limb
Experimental animals	
Normal dogs (controls)	5.1 — each limb
Dogs (reduced muscle function)	5.9 — each limb

and hindlimb are associated with a fairly close approximation of the sizes of the two limbs (intermembral indices of 110–120 in gorillas, 100–114 in chimpanzees, 97–108 in bonobos, 80–90 in rhesus monkeys and baboons, all close to 100). All these creatures spend considerable time bearing body weight by both limbs. In contrast, the values for modern humans which bear weight only on the lower limbs are much lower, around 67–74, i.e. modern humans have much longer lower limbs.

Of course, longer lower limbs are also found in animals that leap (such as especially many strepsirrhines: tarsiers, bush babies and to lesser degrees many leaping and springing haplorrhines: squirrel monkeys, colobus monkeys and langurs). Thus, some of these creatures superficially resemble humans in this feature. This is, however, clearly a different adaptation. Again, also a different adaptation, animals that are very highly arboreal involved in acrobatics (such as orang-utans, lesser apes and prehensile-tailed New World monkeys) have much longer upper limbs than hindlimbs.

The numbers for fossils are interesting (Table 3). Those australopithecines in which this variable can be estimated have values similar to those of chimpanzees and bonobos, 95–105. Does this

Table 3. Relative sizes of upper and lower limbs as reflected in the ration of upper to lower limb lengths.

Extant species	Fossil species
Locomotor emphasis on limbs approximately equal	
Chimpanzees 100–114	Australopithecines (estimates) 95–105
Bonobos 97–108	Homo floresiensis (estimates) 95.0[a]
Totally non-locomotor upper limb	
Modern humans 67–74	*Homo erectus* 72 and 73
	H. neanderthalensis 67

[a] For further discussion of *Homo floresiensis* see later chapter.

mean that we should presume there is still a heavy locomotor emphasis on forelimb function in these creatures? Lots of other evidences concur, saying that they must have been arboreal in some way, even if they were bipedal as well. In contrast, are the values for *Homo erectus* at 72, 73, and for Neanderthals at 67. All three fossil humans are entirely similar to modern humans. Are we to assume that they had totally non-locomotor usages of their upper limbs? This certainly concurs with all other evidence implying that they were upright. (Also interesting, however, is the fact that the Flores human — the so called hobbit — has an estimated ratio of about 95! We will return to this finding later.)

A third marker of muscular activity in limbs is the relative size of joints. Thus, just as cross-sections of the long bones give a measure of loading in the limb as a whole, so too do cross-sections of bone ends. In this case, however, the measure is rather more complex. The equivalent measure of diameter divided by cortex thickness for long bones ends would be diameter (which is easily determined) divided by the summation of the widths of all the intervening trabeculae modified by some factor for each trabecula depending how far it was from the axis of the bone. This is difficult to determine and I leave it to younger colleagues to work it out.

In the meantime, however, the ratio of the overall transverse dimension of a joint in the upper limb relative to that in the lower limb is an approximate surrogate. Accordingly then, by using the various lower limb joint widths as a common standard for single individuals of human, chimpanzee and orang-utan, it is possible to see how this changes the scaling of upper limb joint widths (of course, in the same specimens). Thus, when lower limbs are rendered similar, the effects on upper limb scaling are as shown in Fig. 19. That is, the human is enormously smaller than the two extant apes.

Can we do anything similar for the fossils? Unfortunately, we rarely have joints from the same fossil individual. Hoping, however,

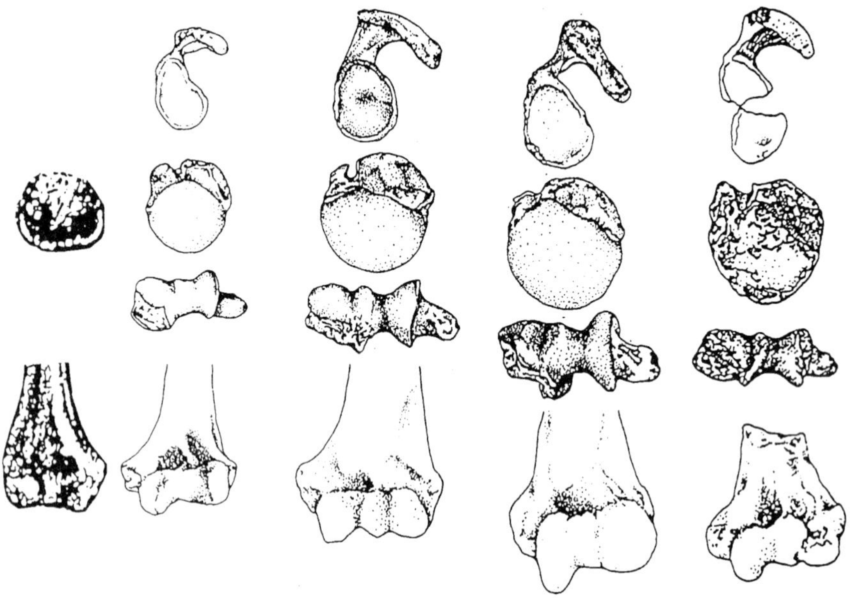

Fig. 19. Sketches of selected upper limb parts scaled on the basis of identical sizes for lower limb parts. The three central sketches are particular human, chimpanzee and orang-utan specimens. The sketches on the left-hand side are of the comparable parts of the Nariokotome boy (scaled in the same way on the basis of his lower limb parts). The sketches on the right-hand side are of different australopithecine parts scaled in the same way on the basis of australopithecines lower limb parts. However, the australopithecine parts are from different specimens and thus, only give an approximate view of the size relationships. With this caveat in mind, it is nevertheless clear that it is with the sizes in the apes that the australopithecines are most congruent.

that the individual fossils are not too far from their averages, a rough comparison can be made. It shows that the upper limb parts of various australopithecines are generally similar to the two apes (also Fig. 19). They are, therefore, relatively much larger than humans. This is again consistent in suggesting that their upper limbs were used in locomotion.

A similar comparison for the Nariokotome Boy (*Homo* species) (which is a single individual) gives the opposite picture, one that is consistent with the upper limb not being used for locomotion.

The differences between apes and australopithecines on the one hand, and modern humans and the known human fossil on the other, are so great that statistics are not necessary.

Implications for Human Evolution

All these findings suggest that there are a number of features of muscles in which humans differ uniquely from other primates. These involve overall reductions in muscle/bone/joint size and complexity in specific developmental domains (Kidd and Oxnard, 1997).

In the head domain, the muscles of mastication in humans are reduced and less complex than in other primates (with consequent smaller bones). A molecular inactivation that produced this reduction would be quickly eliminated in animals still employing masticatory functions like those as in apes. If such inactivations occurred, however, in a species (e.g. humans) in which reduced masticatory muscle mass and complexity were not a disadvantage because of changed diets, etc., then they might well persist because the reduced phenotypes would not be deleterious. That this may have occurred in humans as evidenced by the reduced temporal muscle is enhanced by the fact that other related structures (e.g. the unique temporalis fascia of all humans, the anomalous temporal muscles of a few humans, and apparently related bone differences) are still present. The occasional variations (anomalous muscles) perhaps due to slight persistence of otherwise inactive muscle genetic mechanisms, together with the persistence of fascial sheets that are dependent upon different connective tissue molecular factors, strengthen this idea. As a result, changes in external conditions (e.g. different diets and dietary habits, changed methods of mastication, or additions to mastication such as butchery of foods, or even, at later stages, inventions such as food preparation and cooking, all requiring much less masticatory power) would make such muscle reductions viable. Also a result, the inactivation would not be eliminated, indeed might well be expected to spread, within that species. Whether larger and more globular skulls and the increases in intracranial volume that would also result from the same muscular reductions, would be directly associated with a larger brain, is problematical. These are likely to be largely two separate matters.

Similarly in the forelimb domain it seems that there are smaller and less complex forelimb muscles (with associated less robust forelimb skeleton). The anomalous muscles that occasionally persist in humans suggest that this truly is a reduction. A molecular mechanism for this reduction that is equivalent to the inactivation of MYH 16 in the head domain has not yet been identified. If there were one (or more likely more than one), however, it would also be quickly eliminated if occurring in individuals and species still using the forelimb for locomotion. Individuals with loaded hindlimbs due to true bipedalism and free relatively less loaded upper limbs with reduced upper limb power as in bipedality would remain viable. Such inactivations would therefore not be eliminated; indeed they might quickly spread, within the species.

Equivalent reductions would not have occurred in the lower limb. However lower limb function differs in the different primates, the lower limb always participates in locomotion; it is always loaded.

This requires that muscular variations be indicators of pattern in the developmental basis of evolution. It requires too, that persistent fascial sheets are not eliminated by reducing muscular factors. We know that this is true in the human head domain: myosin heavy chain involves only muscle. Connective tissues are controlled by different molecular factors. Can we assume that some similar muscular factor operated in the upper limb in humans? This is a question for molecular biologists. Certainly there are molecular factors confined, separately, to muscles and tendons in experimental animals.

Because both muscles and fasciae are biomechanical in function, they have their effects upon bone form and architecture. For this reason fossil studies can also give information about relationships and times. Because, also, they are dependent upon developmental factors, molecular phylogenetics can also speak to relationships and times. It is my expectation that, just as the temporal muscle findings implied some underlying molecular factor for reduction of muscles of mastication confined to humans, so too the upper limb findings imply the existence of some molecular factor (more likely several different factors — Sam Cobb, personal communication) for reduction of upper limb musculature (some of it confined to humans). The molecular distinction in the head domain implies that the masticatory change occurred at least at 2.4 million years ago (Stedman *et al.*, 2004).

Will future biomolecular work in the forelimb domain show when the forelimb change occurred? Wouldn't it be interesting if the most potent evidence for bipedality were as much in upper limb losses, as in lower limb gains? Such information as we currently have from the fossils implies that this change had not occurred in australopithecines. How far back it may have been in the various members of the genus *Homo* depends upon information from fossil post-crania yet to be discovered.

It is even possible (see later chapter) that this principle could be extended to apply to the remarkable differences between non-human primate brains and human brains. Brain increase in size and complexity in the developing and growing human individual, produced by new external forms of brain stimulation due to communication and other stimulations applied by the mother, parents, family and community, to the new individual, would produce major changes in the brain. One set of changes would be ontogenetic increase in size and complexity. Such changes might permit gene mechanisms or post-gene factors related to increasing complexity and size to spread.

Are there any such factors available? Indeed there are. They involve the opposite of inactivation, the repetition of one region or domain of the gene in larger brains. Of course, in humans their full expression usually produces major abnormalities of increased brain size and complexity and is totally deleterious. However, that may merely be the deleterious end of a spectrum of possible changes, the smallest of which might just permit increase in size and complexity without problems. This would be rather like the changes that produce a cyclops deformity. Originally thought to be unique, this is now known to be merely one end of a spectrum. At the other end very small changes such as two incisors being fused exist; these produce no noticeable problems in individuals in which they occur. These might be mechanisms which would otherwise not be supported if the new external brain stimulation producing increased size and complexity were not present.

All this implies a concept, unique to evolving humans, that it is external factors, biomechanical functions, methods of ingesting foods, behavioural, social and cultural changes within related individuals that are the real drivers. These all are capable of producing ontogenetic changes of a new kind, that permit the extension within the species of supporting genetic mechanisms that would otherwise be deleterious and rapidly eliminated whenever they occurred. This seems especially clear in the cases of muscle development that I have cited. The matter is obviously much more complex for the brain. In fact, all the situations are almost certainly more complex than I have indicated, probably involving interactions between epigenetic and genetic factors, especially through the cascades of factors, and the backwards and forwards (up and down) interactions between epigenetic and genetic factors that developmental biology now shows is how the individual develops. It presumably could apply to lots of other features of humans. It seems to represent a new mode of evolution only possible in creatures like humans. Again, (see later chapters), it suggests that we may have entered a new classificatory group of life (at present claiming only one set of related species, humans and their immediate forebears) that justifies Huxley's invention of the Psychozoa.

References

Ashton EH, Oxnard CE, The musculature of the primate shoulder, *Trans Zool Soc Lond* **29**: 553–650, 1963.

Bergman RA, Thompson SA, Afifi AK, *Catalog of Human Variation*, Baltimore, Urban and Schwartzenberg, 1984.

Berringer OM, Browning FM, Schroeder CR, *An Atlas and Dissection Manual of Rhesus Monkey Anatomy*, Tallahasse, Anatomy Laboratory Aids, 1968.

Buckminster-Fuller R, Marks RW, Dymaxion World of Buckminster Fuller, Doubleday, 1973.

Currey JD, *Bones: Structure and Mechanics*, Princeton University Press, Princeton, 2002.

Doden E, Effects of hypergravity on rat and dog bones, in Jouffroy FK, Stack MH, Niemitz C (eds.), *Gravity, Posture and Locomotion in Primates*, IL Sedicesimo, Firenze, 1990.

Hartman CG, Straus WL Jr, *The Anatomy of the Rhesus Monkey*, New York, Hafner, 1933 (1961 reprint with additions).

Levi-Setti R, *Trilobites: A Photographic Atlas*, Chicago, University of Chicago Press, 1975.

Kidd R, Oxnard CE, Patterns of morphological discrimination in the human talus: A consideration of the case for negative function, *Perspectives in Human Biology* **3**: 57–70, 1997.

Kupczik K, Dobson C, Fagan M, Crompton RH, Oxnard CE, O'Higgins P, Assessing mechanical function of the zygomatic region in macaques: Validation and sensitivity testing of finite element models, *J Anat* **210**: 41–53, 2007.

Lisowski FP, Van der Steldt A, Vis JH, Upright posture: An experimental investigation, *Act FRN Univ Comen* **5**: 127–136, 1961.

Matsuoka T, Ahlberg PE, Kessaris N, Ianarelli P, Neural crest origins of the neck and shoulder, *Nature* **436**: 347–355, 2005.

Moore WJ, Effects of dietary differences on the growth of the skull of the albino rat (Rattus norvegicus), Doctoral thesis, University of Birmingham, UK, 1958.

Moore WJ, Skull growth in the albino rat (*Rattus norvegicus*), *J Zool* **149**: 137–144, 1966.

Obendorf P, Oxnard CE, Homo floresiensis: A remarkable evolutionary paradox, *Proc Australas Soc Human Biology*, Melbourne, Victoria, 4 December, 2006. *Homo: J Comp Human Biol* **58**: 235–265, 2006.

Oxnard CE, Some variations in the amount of vitamin B^{12} in the serum of the rhesus monkey, *Nature* **201**: 1188–1191, 1963.

Oxnard CE, *The Order of Man: A Biomathematical Anatomy of the Primates*, Hong Kong University Press and Yale University Press, 1983/1984.

Oxnard CE, Biomechanics and architecture of cancellous bone, *Proc Australas Soc Hum Biol* **2**: 189–212, 1987.

Oxnard CE, *Fossils, Teeth and Sex: New Perspectives on Human Evolution*, Hong Kong University Press, Hong Kong University of Washington Press, Seattle, 1987.

Oxnard CE, From fossil giant ground sloths to human osteoporosis, *Proc Australas Soc Hum Biol* **3**: 75–96, 1990.

Oxnard CE, Mechanical stress and strain at a point: Implications for biomorphometric and biomechanical studies of bone form and architecture, *Proc Australas Soc Hum Biol* **4**: 57–109, 1991.

Oxnard CE, Ghosts of the past: The temporal fascia in some primates, *Proc Int Primate Conf*, Adelaide, Australia, Vol. 18, pp. 1–10, 2000.

Oxnard CE, Ghosts of the past II: Forelimb muscles and fasciae in some primates, *Am J Phys Anthropol* **S40**: 162, 2005.

Oxnard CE, Franklin D, Ghosts of the past I: Some muscles and fasciae in the head domain, *Folia Primatol*, 2008 (in review).

Oxnard CE, Franklin D, Ghosts of the past II: Muscles and fasciae in the primate forelimb domain, *Folia Primatolo*, 2008 (in review).

Oxnard CE, Wealthall R, Ghosts of the past: The temporal muscles, fasciae and bones in some primates, *Am J Phys Anthropol* **S36**:163, 2003.

Ruch TC, *Bibliographia Primatologica*, Springfield, Thomas, 1941.

Stedman HH, Kozyak BW, Nelson A, Thesier DM, Su LT, Myosin gene mutation correlates with anatomical changes in the human lineage, *Nature* **428**: 415–418, 2004.

Swindler DR, Wood CD, *An Atlas of Primate Gross Anatomy: Baboon, Chimpanzee and Man*, University of Washington Press, Seattle, 1973.

Testut L, Memoire sur la portion brachiale de nerf musculo-cutane avec le nerf median, *Int Monat Anat Histol* **1**: 305–341, 1884.

Wealthall R, Facial shape and growth: A Study of bone remodelling in *Macaca mulatta* and *Sus scrofa*, M.Sc. Thesis, University of Western Australia, 2000.

Wood J, On human muscular variations and their relation to comparative anatomy, *J Physiol Anat* **1**: 44–59, 1867.

Zuckerman S, Ashton EH, Flinn RM, Oxnard CE, Spence TF, Some locomotor features of the pelvic girdle in Primates, *Symp Zool Soc Lond* **33**: 71–163, 1973.

Chapter 4

Reversing Development: From Adult to Gene!

Looking Upwards from Genes and Downwards from Adults

The first two chapters in this book examine aspects of the mechanics of bone form and architecture, in particular their ontogenetic determinants. Such functional studies of bone and bones, carried out on exemplar anatomical units, look sideways, as it were, at the structures as they work. They demonstrate the degree to which the form and architecture of bones, muscles and connective tissues are related to the mechanical stresses and strain that are laid upon them during function by such loading factors as body weight and the actions of muscles, the motors of the body.

The third chapter looks at those motors, the muscles. It shows that, of course, many components of muscle are, inevitably, directly related to their functions, the postures and movements that they generate. It also shows, however, that others of their variations look downwards (as it were) towards the zygote reflecting what has happened during development of the individual, and look backwards at what has happened during development in the past: evolution.

This chapter returns to bones but applies the additional reverse developmental thinking derived from the muscle studies. Assuming that bones also reflect development and evolution as well as function, it asks whether the three different pieces of information can be disentangled.

Genes, molecules and cells, look upwards towards the adult showing the mechanisms and processes that lead to adult structures; this is developmental biology. The confluence of many developments over many generations, also looking forwards from the past to the present, i.e. from generations of genes and generations of adults to adults today, is the stuff of evolution. This chapter, in contrast, examines adult bones, to see whether, by looking in reverse directions, downward from the adult to the gene, as it were, we can identify the developmental mechanisms that underlie the production adult form. This chapter also examines whether, by looking backwards from the present structure, we can, likewise, identify the generations of developmental mechanisms that underlie the evolution of adult form.

Looking upwards from genes can only be performed in small numbers of experimental animals; it therefore gives general concepts but only elucidates individual cases. Looking downward from adults can be carried out on large numbers of specimens and species; it therefore can test specific ideas. Likewise, looking forwards from the fossils is based upon so little data, seeing how few fossils are known. Looking backwards from present day adults allows a more complete structure to be examined; again, therefore, it can test specific ideas.

Measuring Anatomies: Morphometrics

Thus, the questions being asked in studies of adult anatomical differences include the following. What can adult anatomical similarities and differences (using the comparative, observational methods of measuring shape: morphometrics) reveal about individual species? In particular, to what extent do these similarities and differences inform us about the determinants of adult anatomies, about functional adaptations, about developmental mechanisms, and about evolutionary relationships? Is it possible to partition the information content in such a way as to disentangle these different components?

Analysing anatomical or molecular characters (e.g. by cladistics, phylogenetic analysis) is a useful tool to discover groups among species, but seems to be less useful when characters are interrelated. This may be because such analyses often treat characters as though they were separate items of information. They do not easily allow for the fact that characters are not independent, that each character contains some information that is also contained in some parts of other characters. In other words, they do not easily recognise that characters are complex and correlated. Morphometric approaches (multivariate statistical analyses of quantitative anatomical data) in contrast, may more easily be able to partition biological determinants because they recognise that characters are not separate datum items, but are inter-related and correlated with one another. Morphometric approaches allow for these correlations and produce underlying descriptors that are, genuinely, uncorrelated. It may be these underlying descriptors, not the characters themselves that reflect the underlying biological complexity.

In other words, can developmental biology and evolution presenting views upwards from the genes and forwards from fossils, duly modified by the view sideways from mechanics into architecture, eventually meld with morphometric views downwards and backwards from the wide array of present adult differences that exist. Evidence is available on this problem, both from new data, and from reanalysis and reinterpretation of old data, that, working with many students and colleagues, I have garnered in a lifetime's work on the Order Primates.

Anatomies of Parts: Functional Adaptations

It has long been known that, in primates, morphometric studies of restricted anatomical regions (individual functional units such as, in the skull, the jaw and its teeth, and in the post-cranial skeleton, the joints of the trunk and limbs) result in clusters of species that seem to mirror functional convergences and divergences. Such species separations are usually irrespective of degrees of evolutionary relationship.

This was very clearly evident in the first investigations of the shoulder that we carried out (Ashton and Oxnard, 1964a, 1964b; Ashton *et al.*, 1966; Oxnard, 1967, 1968) in which shoulder features separated various primates according to convergences in shoulder function. As a result, for instance, these studies placed spider monkeys of the New World and gibbons of the Old World close together. Though they are completely different in an evolutionary sense (different grades, different

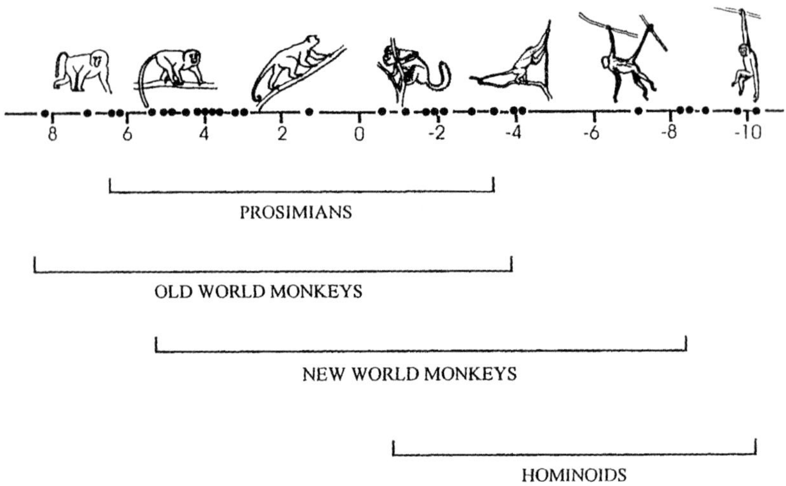

Fig. 1. The first canonical variate for the individual study of the shoulder in primates. The means for various primate species are represented by dots and they are scattered along the axis from right to left in a way that seems related to increasing degrees of arm-hanging and arm-swinging components of locomotion as shown by the cartoons. There are no relationships with the evolutionary groupings of these species shown by the brackets below. The scale is given in standard deviation units and thus indicates highly significant separations. For full identification of species see Oxnard (1973).

continents) they share some activities of the shoulder in locomotion in trees. Similar findings were evident throughout the primates (Fig. 1). This kind of result has now been replicated many times, in many other anatomical regions, both by ourselves (e.g. Oxnard, 1973, 1975, 1983/84) and by a host of other investigators using similar methods and asking similar questions (e.g. McHenry and Corrucini, 1975; Feldesman, 1976, 1979, 1982; Corrucini and Ciochon, 1978; Manaster, 1975, 1979; Senut, 1981; Tardieu, 1981; Stern and Susman, 1983; Larson, 1993). The concept is, therefore, not in doubt.

Nevertheless, I have many times attempted, over the years, to test the reality of the functional groupings of species in those studies, by examining the groupings of the anatomical features that produced them. How, in producing these apparently functionally adaptive clustering of species, are the anatomical features clustered?

Do the clusters of anatomical features actually fit with ideas about the mechanical adaptations of the particular anatomical regions?

In the earliest studies, we answered these questions using *ad hoc* methods that depended upon a degree of intuitive insight. For example, D'Arcy Thompsonian Cartesian coordinate transforms allowed us to visualise clusters of features in terms of overall form (Oxnard, 1983/84). The biomechanical relevancies of these aspects of form were further confirmed using mechanical stress analysis techniques. More recently, the transformation diagrams have been confirmed and extended (Fig. 2) using the more objective tools of geometric morphometrics such as thin plate splines (Albrecht, 1991). Such studies have almost always confirmed, if also usually extending, the earlier findings. That now, however, is old hat.

What is new is the information that comes from gradually adding together the various studies resulting in much larger data sets. In the old days this could not be done. I sometimes wonder if readers even know why we only ever looked at nine anatomical features in each of those early studies; it was, of course, because the computers and programmes of those days could only cope with matrix operations of that size. For many years now this has no longer been a constraint.

Anatomies of Wholes: Evolutionary Relationships

We therefore carried out much more extended investigations in which the original separate anatomical regions are aggregated into more complex combinations of body parts, eventually into the whole organism. Whenever we have examined these aggregations, and we have now carried out such studies in several independent ways, the functionally adaptive (convergent) clusters of species gradually disappear. Although the anatomical features that we analyse are the same, their examination in combination seems to clusters the species in a way that hides functional adaptation, but that, in contrast, reflects evolutionary relationships. Though evident in aggregated studies of cranial, jaw and teeth parts (e.g. Pan, 1998), this finding is most obvious when parts of the entire organism are compounded (Oxnard 1998, 2000).

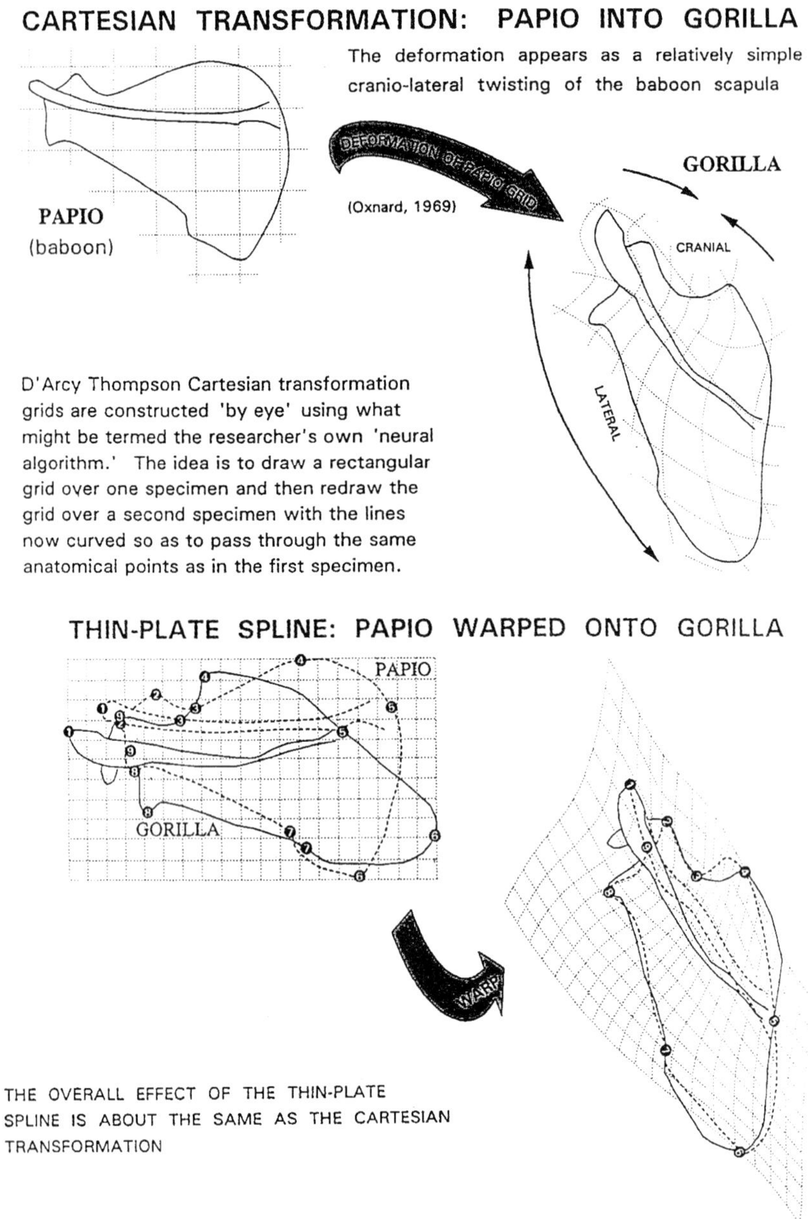

Fig. 2. Top frame shows the hand drawn Cartesian coordinate transformation between scapular shape in the baboon and gorilla. Bottom frame shows the computed thin plate spline transformation between baboon and gorilla (courtesy G. H. Albrecht).

A good example of this can be seen in the comparison between the studies of individual post-cranial anatomical parts and the coalescence of all post-cranial anatomical parts. In the former studies, for example, the apes are all clustered together (all using all four limbs in somewhat similar ways). Humans are quite separate (humans, being bipedal, use only their hind limbs for locomotion). In the latter studies, to choose the same species as the example, the lesser apes (various gibbons) are clustered with the Asian great ape, the orang-utan, whereas the African great apes (chimpanzees and gorillas) are clustered with humans. These new grouping seem most related to their evolutionary propinquity. How can these two different results be obtained from analyses of the same anatomical features?

In order to find out we have asked, again, the reverse question. How, in forming the (apparently) evolutionary clusters of the species are the anatomical features clustered? Are the clusters of anatomical features merely random assortments with no apparent biological meaning? They certainly are not the functional clusters found in the prior studies. Or, are there other ways in which anatomical features are clustered that might be of biological import? In order to understand, however, the meaning of these clusters, it is necessary first to understand the kinds of new information that have resulted from the experimental work of developmental biologists.

Separating Function, Development and Evolution: Jaws and Teeth

For example, studies on the development of lower jaw suggest that a series of developmental processes are involved. Homeobox genes start a cascade of developmental processes involving stages from cell populations through developmental components to morphological units. The cell populations initially derive from separate clusters of cells from the first segment of the neural crest. These neural crest cell clusters give rise to several sets of bone-producing cell populations responsible for, separately, the different portions of the jaw bone (Fig. 3; Atchley, 1993; Hanken and Hall, 1993). These include, separately, the parts of the jaw bone holding the incisors and the molars, together with

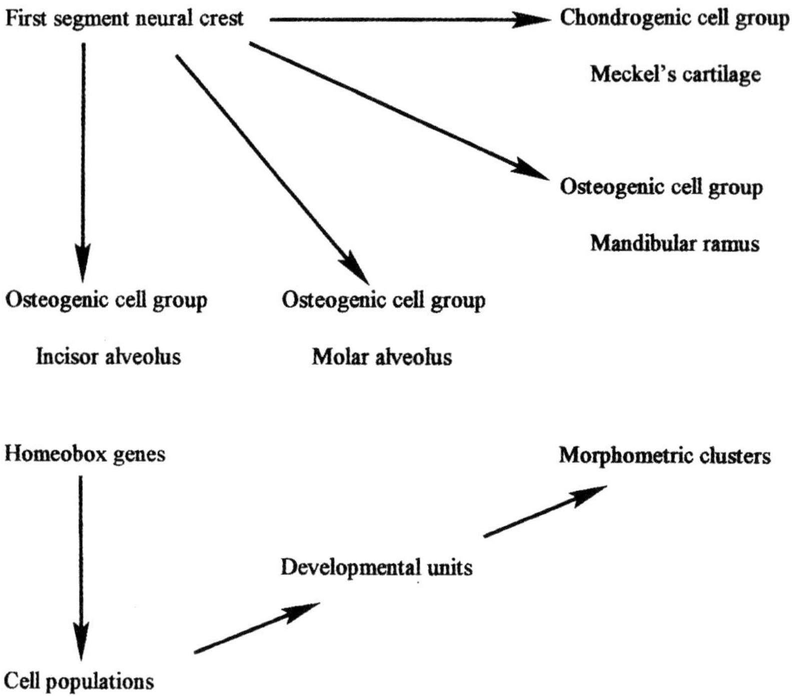

Fig. 3. Top shows the various neural crest cell populations that give rise eventually to the various components of the jaw. Bottom shows the cascade of processes controlled by homeobox genes giving rise to cell populations, developmental units and morphometric clusters of variables.

several other jaw parts: the body of the jaw bone, and its angle, coronoid process, and condyle (e.g. Fig. 4; Atchley, 1993; Hanken and Hall, 1993). These studies have generally been carried out in rodents and hence do not give information about those parts of the jaw bone that, in other species, bear other teeth such as canines and premolars.

In addition to this modular mechanism in jaw production, there are also at least two other developmental processes. One is a distinctly linear trend arraigned along the jaw from front to back: related, perhaps, to increasing molarisation of teeth. A second is another apparently linear process arraigned across the depth of the jaw from most cranial (closest to the mouth) to most caudal (the lower rim of the jaw closest to the neck).

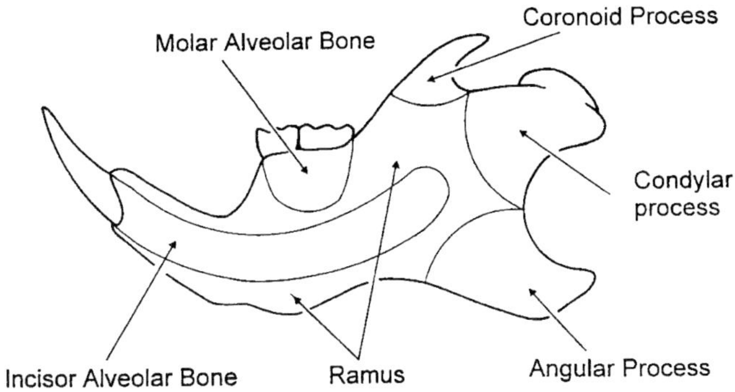

Fig. 4. The components of the rat mandible that arise from the different populations of neural crest cells in the embryo.

With these developmental ideas in mind, let us now look at the cluster of anatomical features in our two sets of analyses of these anatomical data. One set of analyses is of raw data and reflects merely the lengths and breaths of various teeths. As such it is not sensitive enough to recognise tooth patterning or complexity, a most important element of dental evolution. When these raw data were examined individually the species clusters included, for example, separations of herbivorous from omnivorous species, clearly related to functional adaptation. The clusters of features that provided these species separations included such combinations as aggregations of the lengths of incisor edges (food cutting), aggregations of molar areas (food crushing), and so on. Such clusters seem obviously relate to jaw function especially mastication, and this seems to show that the functional convergence of species is produced by functional adaptations of anatom ical features. Such an insight is nothing new; it is well-known not only from our quite restricted studies of primates, but to a much greater extent from innumerable studies of dental diversity by many investigators in a very wide array of animals over many, many years (an excellent review is available in Lucas, 2007).

When, however, the anatomical features are examined in combination, the species separations are not functional but evolutionary. For example, the species of African great apes are grouped with the human species; the species of Asian apes (that have similar functional

Table 1. Association between clusters of anatomical features, morphological descriptors and individual developmental processes.

Clusters	Anatomical feature	Morphologic descriptor	Developmental process
Cluster 1	All dimensions approximately equally	Overall size	Result of general growth processes (in both primates and rodents)
Cluster 2	Incisor dimensions	Incisor portion of jaw	Incisor alveolus (equivalent to incisor alveolus of rat)
Cluster 3	Molar dimensions	Molar portion of jaw	Molar alveolus (equivalent to incisor alveolus of rat)
Cluster 4	All molar and premolar versus canine and incisor dimensions	Relationships along jaw	Developmental gradient along the jaw

attributes as African apes) are completely separated from the African ape species and lie nearer to the lesser apes. How can such different clusters of species occur? This apparent paradox was resolved by seeing how the anatomical features are clustered in the second, combined, study (Table 1).

The clusters of features in this analysis do not link functional attributes like cutting lengths and crushing areas. Rather, they seem to be linked in ways that relate to the aforementioned developmental mechanisms that result in the development of the jaw. Can it be that these combined features of adult morphology which separate species in relation to evolution, do so because they are reflecting adult remnants left over, as it were, from the underlying developmental mechanisms? Unless, however, this can also be confirmed in other anatomical regions, it remains just an idea. Our data permit us to check this for the post-cranial skeleton.

Separating Function, Development and Evolution Again: Limbs

In the same way, however, as was necessary for the jaw studies just outlined, before we can interpret the limb studies, we first need to know a little about limb development. The development of the upper limb (the earliest such work was carried out a long-time ago, e.g. experimental grafting studies in bird embryos: Wolpert *et al.*, 1975) imply that limbs are the resultant of the actions of a series of homeobox genes initiating a cascade of events. These include first, a proximo-distal developmental gradient producing sequentially, from shoulder to finger tips, the bones of the limb; second, a cranio-caudal developmental array producing those parts of the limb that contain at least several cranio-caudal elements (e.g. thumb — cranial — and fingers — more and more caudal to the little finger — in the hand, big toe — cranial — and other toes to the little toe — caudal — in the foot); third, a dorso-ventral dichotomy separating the back of the limb from the front. The back of the limb, however, is not as readily identifiable as one might think because of the rotation that the limb undergoes during development. For example, the back of the forearm and hand, and the front of the shin and top of the foot, are all developmentally at the back (dorsal) even though the front of the shin and the top of the foot seem to be on the front, belly side (ventral), in the human adult. The initial developmental investigations were carried out on upper limbs, but it is not in doubt that the concepts also apply to lower limbs. This is a very short summary of a complex set of studies.

Let us now examine the analyses of the adult primate post-cranial skeleton. The situation is more complex than for the jaws and teeth. The anatomical features are measures of several different anatomical regions. And they were combined in several different investigations. One combines smaller anatomical units such as the shoulder, elbow, hip, thigh and foot (Oxnard, 1983/84, 1992, 1998). A second combines major parts such as upper limb, lower limb, abdomen and thorax (also Oxnard, 1983/84, 1992, 1998). A third combines some smaller dimensions and some larger body proportions (e.g. Oxnard, 2000).

In each case, as for the dental apparatus, when the individual anatomical parts themselves are examined, they seem to give information about functionally convergent clusters of species and functional clusters of features as described earlier. When, however, the anatomical parts are examined in combination, the species separations are no longer functional but seem to be evolutionary. A good example is the separation of New World monkeys from Old World primates (in spite of the aforementioned functional convergence between New World spider monkeys and Old World gibbons), and the separation of the Asian apes from African apes (in spite of considerable functional convergences among all apes) together with, get this, a close linkage between African apes and humans (e.g. Fig. 5 and Oxnard, 1983/84).

As before, we need to know the clusters of anatomical features that produce these (apparent) evolutionary groupings. These are

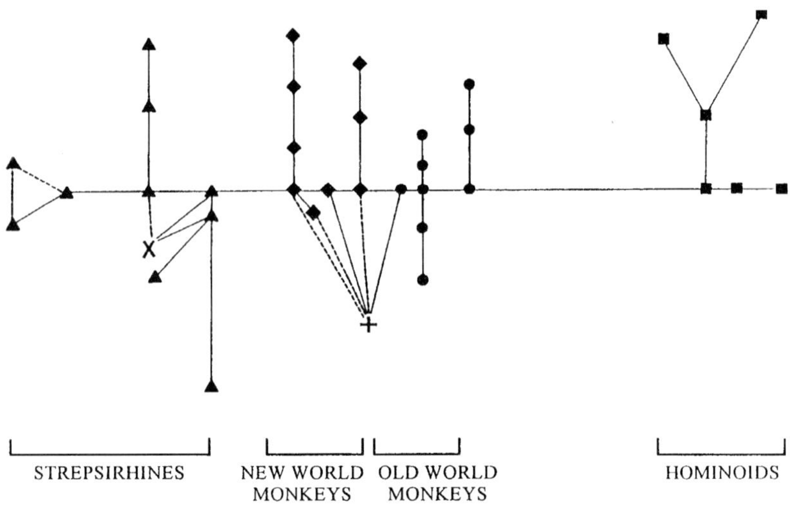

Fig. 5. Morphometric analysis of data on the combined bodily proportions of primates. The result is too complex for a simple two- or three-dimensional plot. The figure shows the minimum spanning tree of all species. The overall scale of the diagram is some 40 standard deviation units indicating very highly significant differences. The positions of species groups are as indicated. They are generally arranged according to the major evolutionary groups that they occupy. In particular, bush-babies, which are highly functionally convergent with tarsiers, are placed separately as indicated (×, position of bush-babies, strepsirrhines; +, position of tarsiers, haplorrhines).

Table 2. Relationships between clusters of limb variables and individual developmental processes.

Clusters	Anatomical feature	Morphologic descriptor	Developmental process
Cluster 1	Measures of major segments of both limbs	Measures of proximo-distally organised elements	Proximo-distal developmental gradient
Cluster 2	Measures of all bones of hand and foot	Cranio-caudal measures of hand and foot	Cranio-caudal developmental gradients

shown in Table 2. These two clusters of anatomical features are described by two of the morphometric descriptors, and seem to mirror, respectively, two of the developmental mechanisms. The first is a chain of proximo-distal anatomical features (the lengths, sequentially, of the long bones down each limb). The second is a cranio-caudal array of the lengths across the hand and foot. Is it possible that these anatomical features are clustered in this way because they are the adult remnants of the two underlying developmental processes? Is it then possible that this reflection of the developmental processes is why the clusters of species are evolutionary?

However, in both these examples, the jaws and the limbs, there are clusters of anatomical features that are not mentioned above. Let us now examine them.

The Power of Prediction

It would be amazing if there were a complete relationship between the developmental processes and clusters of anatomical features. In fact, there is not; there seem to be exceptions. It is therefore important to look at the exceptions. In the jaw studies, these exceptions include two clusters of anatomical features that are not reflected in the developmental biology of the rodent jaw. One relates to the premolar bearing portion of the jaw; a second relates to the canine bearing part of the jaw. Do these exceptions negate the general idea?

Table 3. Apparent disjunctions between clusters of jaw variables and individual developmental processes.

Clusters	Anatomical feature	Morphologic descriptor	Developmental process
Cluster 5	Premolar dimensions	Premolar portion of jaw	No process?[a]
Cluster 6	Canine dimensions	Canine portion of jaw	No process?[b]

[a] If developmental studies covered primates, would separate and additional premolar cell populations, developmental units and mandibular components exist?
[b] If developmental studies covered primates, would separate and additional canine cell populations, developmental units and mandibular components exist?

Surprisingly they do not! In the studies of the jaws and teeth (Table 3), the discovery of developmental factors is limited by the fact that most of the experiments have been carried out on experimental animals that have a reduced masticatory apparatus.

Rodents only have incisors and molars. As a result, jaw components in the adult, bone forming cell clusters during development, and the neural crest cell populations that are the originators, can only relate to incisor and molar bearing portions of the final adult bone. But primates have extra teeth and, therefore, extra tooth-bearing bone.

The situation in primates, thus, allows us to make two predictions for developmental biologist. The making and subsequently testing of predictions is one of the most important ways of taking research further. The two predictions relate, separately to canines and premolars. Thus, if developmental studies were carried out on an experimental animal that had canines and molars (rodents do not), would two other bone parts be found in the jaw, each of them underlain by their own population of bone cells, and each, in turn, having originated from two additional populations of neural crest cells.

Even this may not be the whole story — there are yet more missing teeth! Living primates, though having more teeth than rodents, still have teeth missing compared with ancestral primates (Fig. 6). Thus, studies of fossil primates indicate that originally there were perhaps as many as six incisors, four premolars, and at least four molars and possibly more on each side of each jaw. It is

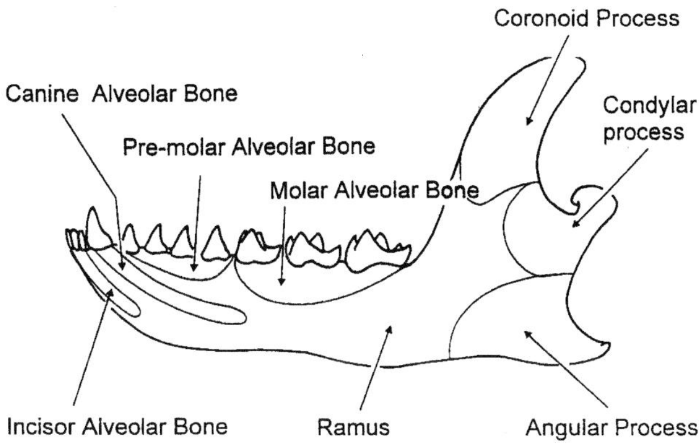

Fig. 6. The components of a hypothetical early primate mandible that might have arisen from the different populations of neural crest cells in the embryo if this could be studied (see text).

not impossible, therefore, that if the developmental experiments could be carried out on fossil primates (time machine needed!) the current findings might be further tested. For example, it is possible that there really were a larger number of neural cell populations, giving rise to a larger number of bone forming cell groups, and, therefore a larger number of jaw bone components. However, it is also at least possible that the process did not involve several groups of cells but just one long linear array of cells. The 'groupiness' of today's developmental biology (as it is generally accepted) may simply be an artefact of evolutionary missing components in an otherwise continuous process (Fig. 6 and Oxnard, 2007).

Some of these concepts could be tested by developmental biologists, others by anatomists. 'Time-machine' experiments cannot be done but indirect information about ancestral primates could be garnered by proxy, making inferences from fossils. None will be done by me (having not the materials, the expertise, nor, possibly, the time).

The Power of Prediction Again

As with the teeth and jaws, so for the limbs, it would be amazing if there were a complete relationship between the developmental processes and the clusters of anatomical features. Again, there are not; there are exceptions. It is therefore important to look at these exceptions. Thus, in the studies of the limbs there are some clusters of anatomical features that are not apparent in the developmental experiments (Table 4).

The first of these relates to cluster number 3 (Table 4). This consists of the dimensions of all elements of digit 4 in the hand! This would be most puzzling to anyone doing experiments on birds. But it is highly significant to any primatologist who would immediately recognise that the organisation of the hand in relation to the fourth digit (but not the fourth digit in the foot) separates all strepsirrhines from all anthropoids. We can therefore predict that if developmental biologists could do the grafting studies on primate exemplars, some mechanism might be found relating to the development of the hand axis in these different primates. Of course, primate conservation groups and ethical committees could not permit such studies, and in any case they would be technically very difficult. However, given the prediction, it might well be possible to test the idea in an inferential manner by examining developmental stages in embryos of exemplar strepsirrhines as compared with exemplar anthropoids. A few such materials are available in the museums of the world. Incidentally, the

Table 4. Apparent disjunctions between clusters of dental variables and individual developmental processes.

Clusters	Anatomical feature	Morphologic descriptor	Developmental process
Cluster 3	Dimensions of all elements of manual digit 4	Special feature of prosimians as compared with anthropoids	[a]
Cluster 4	Dimensions of alternate segments of upper and lower limbs	Serial elements of limbs	[b]

[a] If developmental studies included strepsirrhines, would there be a special developmental process relevant to manual digit 4?

[b] If developmental studies included upper and lower limbs together, would there be an alternating serial arrangement from the underlying homeobox genes.

fact that this digital axis was not also marked out in the foot is an especial strength of this prediction; this anatomical division is limited to the hand.

A second prediction arises from the limb studies. Thus, Table 4 lists a second cluster of anatomical features that comprises each alternate segment of upper and lower limbs. These measurements seem to define a morphological description of the limbs that may be termed 'alternating seriality'; a factor related to serial homologues between the limbs; this would be immediately obvious to the comparative anatomists of prior centuries. It seems to be relevant to limb structure. One of the primary patterns related to the homeobox genes throughout the body is the existence of seriality in the elements. If studies in birds or other exemplars were to include both upper and lower limbs, might such a serial relationship be recognised? Even experiments on developmentally produced 'third' limb pairs in experimental animals might provide information about this. Certainly there is a serial relationship of homeobox genes that need not preclude this result. The prediction remains to be tested (again a prediction for developmental biologists).

The Power of Negative Findings

In both the jaw and limb studies described above there are yet further matters to be considered. This relates to results that we might have expected but did not find, i.e. negative results (see Kidd and Oxnard, 1997).

In the jaw developmental story there is a developmental factor that relates to a trend from the tooth side of the jaw (cranial) to the neck side (caudal), a cranio-caudal gradient in the jaw. However, we could find no equivalent cluster of anatomical features (first row of the table below). If the above ideas are correct, we may ask, why did we not find a cluster of anatomical features that is related to this? The answer is clear; I was not clever enough to think that this would be important. I did not take the appropriate measurements. We therefore now have a prediction for me, the anatomist. With the insights of development, go back and take measurements that involve this aspect of the jaw; find out if they coalesce into a cluster.

Clusters	Anatomical feature	Morphologic descriptor	Developmental process
No cluster	No dimensions	No descriptor	Cranio-caudal developmental gradient in the jaw
No cluster	No dimensions	No descriptor	Dorso-ventral developmental gradient in the limbs

In the same way, in the limb studies described above there is a developmental factor that relates to a dorso ventral patterning of the limb (second row of the table above). Thus, we might have expected a morphometric cluster that reflected the dorso-ventral contrast evident to developmental biologists. Again, this cluster could not be found because, as an anatomist, it had not occurred to me to include any dorso-ventrally arranged dimensions of the limbs. Again, this yields a prediction for me, the anatomist. Go back, take a series of dorso-ventral measurements and see if they form a cluster in the multivariate analysis.

Again, therefore, speculative though the original suggestions may be, the coexistence of parallel suggestions in both jaws and limbs, together with the possibilities for testing, both positive and negative, add strength to the ideas.

Measuring Lifestyles: 'Niche-Metrics'

The above studies started with attempts to understand the relationships between the anatomies of anatomical regions and the mechanical functions of those regions. The data were the measurements of bones; the relationships were with the functions, largely due to animal locomotion, of the bones. However, it had always been an interest of mine to take the functional side of this equation into a much wider milieu: the overall lifestyles of animals. Could this be done?

Of course, detailed knowledge of primate anatomies has been available for many years; bones are easy to measure. The form and pattern of the primate skeleton overall has been explicated in a way more complete than ever before through the application of, originally morphometrics, and more recently, geometric morphometrics, to these measurements. Equivalent knowledge of primate lifestyles overall is considerably more sketchy, though of course, wonderful studies of individual primate groups in the field have been developed over almost a century now. Notwithstanding these pioneering investigation, primate lifestyles are still hard to study, primates are difficult enough even to see, large aliquots of time are required in the field, harsh conditions have to be endured, there may be much expensive travel, academic careers may be delayed, the importance of such work may not be academically recognised (compared to the recognition given to experimental laboratory work), and a whole new (new to me anyway) way of doing the science has to be learnt. Clearly these things are beyond an old-fashioned medical anatomist!

Once again, however, one of those serendipitous meetings arose that permitted a whole new attempt to look at animal lifestyles and anatomies. It began through my links with Late Professor Peter Lisowski originally one of the lecturers at the University of Birmingham, but later Professor at the University of Hong Kong and later still Emeritus Professor at the University of Tasmania, Hobart. Professor Lisowski led me to Dr Robin Crompton who was then a lecturer at the Chinese University (but is now a Professor at the University of Liverpool). One of Crompton's expertises in those days was in field studies of prosimians (as they were then termed) as well as in their functional anatomy. To this collaboration was later added the abilities of Dr Susan Lieberman, then my research and teaching associate the University of Southern California (but now involved in the World Wildlife Fund). Her academic background was in tropical ecology as well as quantitative biology.

Although my original aim was to look at anatomies and lifestyles in the whole Order primates, that was obviously too big a task for a first stab at the problems. A next best aim was to do this for one-half of the Order. What better collaboration could there be, anatomy, biomechanics, primatology, tropical ecology and statistics, with the various expertises being variously shared among the three of us. The original stimulus came from Crompton who wanted to apply my ways of analysing anatomies, to his data on lifestyles; but the investigation ended up totally shared.

The shapes, dimensions and forms, the anatomies, of our bodies are in large part related to the activities, the lifestyles that we undertake. These activities, lifestyles, include the current ways that we move, feed, and manipulate our internal and external worlds. Some characteristics of our anatomies cannot be easily related to our current lifestyles and may be an inheritance from the different anatomies and lifestyles of our ancestors. Our original aim, as a collaborative group, was to try to disentangle the various aspects of what is clearly a most complex relationship.

As we move from anatomies, to functions, from functions to lifestyles, we enter ever increasingly complex realms. It is hardly surprising that it is the first of these three that has received the greatest attention over the years and is furthest developed. The first component, dissection of cadavers and measurements of skeletons, though not necessarily easy in an absolute sense, seems easy in comparison. The second component, function, has been less well studied though in recent years it has yielded much to newer fields such as biomechanics and kinesiology. Such studies are inherently and technically more complex because biological materials are not uniform or simple in their mechanical characteristics, unlike the materials on which most engineering science is based. The third component, lifestyles, is, however, exceedingly complex. It includes not only what animals do, but the wider behavioural and adaptive roles that they take up, and even relevant physical characteristics of the habitats that they occupy. Such studies particularly share in greater ethical and practical problems because of the rarity, almost extinction, of subject materials, as the world's habitats steadily diminish. And yet it is axiomatic that all three of these elements must ultimately be involved in the study of the evolution and adaptation or organisms.

An attempt to propose a programme of research in this area began in discussions between Crompton and Oxnard among the mists, crags and rains of Lushan, one of the sacred mountains of China, while we were guests of the Chinese Anatomical Association at its 1982

Congress. We were foreign guests, i.e. we were two of six representatives from the University of Hong Kong, not then an integral part of China. Our discussions continued in the dry heat of a Southern Californian hacienda in Pasadena a year later. Later still they included Lieberman in my laboratories at the University of Southern California. These discussions centred upon commonalities, limitations, and prospects for study.

Because of my own prior work, the anatomical components of the study were well advanced. The morphometric methods that were being employed appeared a particularly powerful way of handling multiple species comparisons and complex combinations of anatomical features. The lifestyle elements were mainly verbal and cinematographic descriptions that were not yet in a useful form for comparison. Thus, Crompton had a great deal of data about prosimians (as they were called in those days) obtained from his own field studies, partially culled from his earlier mentors (Rose, 1973; Walker, 1974, 1979) and from fellow workers, and from the literature about the pattern and frequency of postural and locomotor activities. His data included also information about the structural and behavioural contexts of these activities. They included yet again the diets that were garnered by the activity performances in the local environments. But, in contrast to the anatomical information, there was no satisfactory summary in a way that allowed analysis for comparison with the anatomy.

Crompton wished to take these data further by transforming them into a quantitative base which could then be analysed using the same statistical techniques that I had already employed on the anatomical features. We would later call this 'niche-metrics' as a foil to 'morphometrics'.

Accordingly (mainly through Crompton's expertise), we changed the lifestyle descriptions into several groups of variables: first, of locomotor activities (such as leaping, scurrying, slow climbing), second, of environmental features (such as small branch, undergrowth, large branch, canopy) and third, of diets (such as leaves, fruits, nectar and animal products). Though scarcely functional in the sense of functional morphology, they are extremely 'functional' in the sense of ecological adaptation (Oxnard *et al.*, 1990).

At first it was important to view the data univariately. Because it rapidly becomes difficult to visualize large numbers of parameters for each species, we borrowed a multidimensional grouping method that Oxnard had previously obtained from Welsch (1976) a graphical data analyst. Figure 7 gives an example of how one particular species could be visualised using this technique. Figure 8 shows how similarities between species could be readily seen. Figure 9 shows that it is easy to recognise the one species (in our study) that was completely different

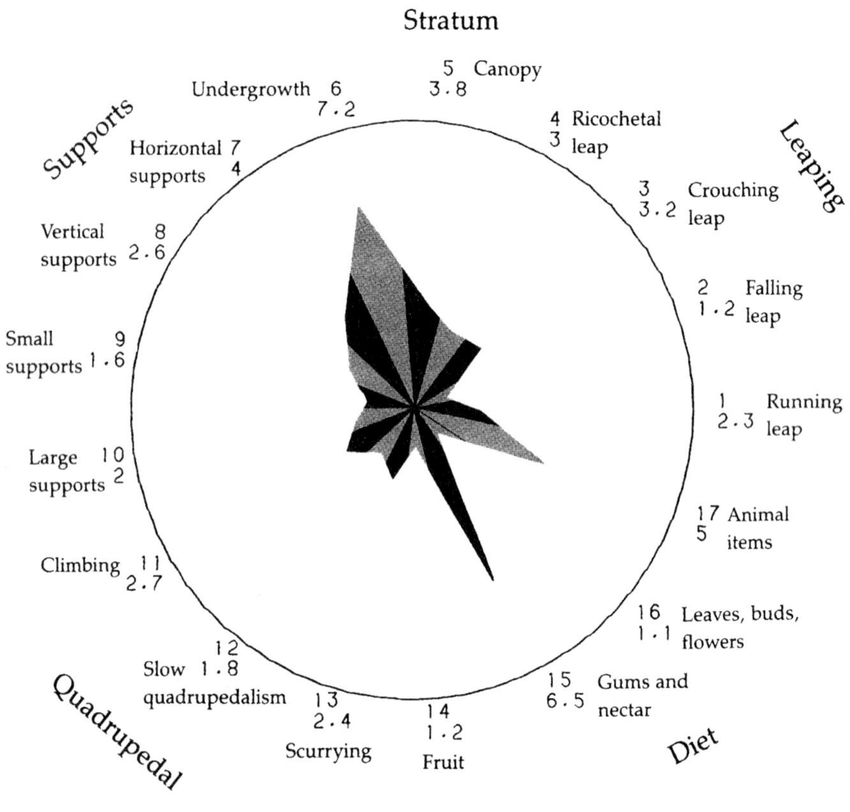

Fig. 7. Polar coordinate plot showing the niche variables for a single species of bush-baby. The centre of the diagram is the zero value for each variable/each ray of the star is the actual value for each variable (also indicated in numbers around the edge of the plot).

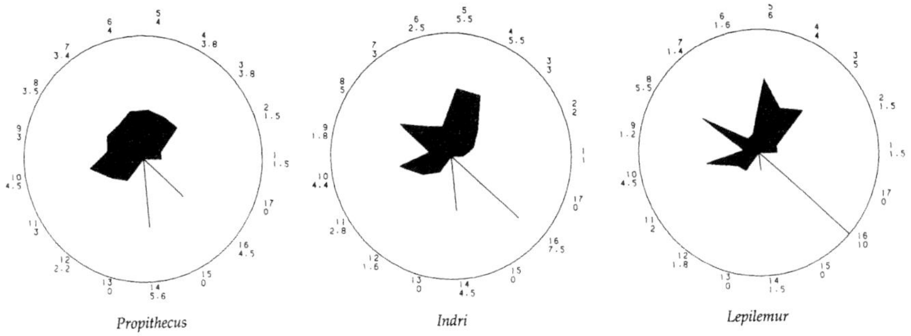

Fig. 8. Polar coordinate plots for a group of species that are all rather similar (two indriids: *Propithecus* and *Indri*, and *Lepilemur*).

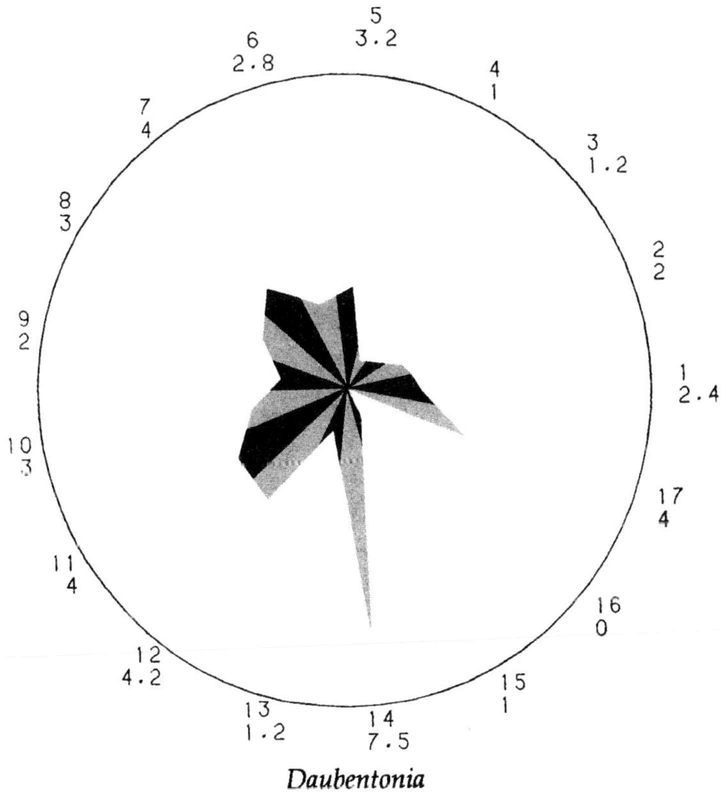

Fig. 9. The polar coordinate plot for one rather unique species: *Daubentonia*, the aye aye.

from all the others; not surprisingly, this is the Malagasy species, the aye aye (*Daubentonia*)!

We then applied multivariate statistical methods (specifically the same methods that we had used on the anatomical features) to these data (see Crompton *et al.*, 1987). When we asked questions about how individual species are clustered, the groups clearly seem to reflect information about ecological similarities, the equivalent, in morphology, of functional convergence (Fig. 10).

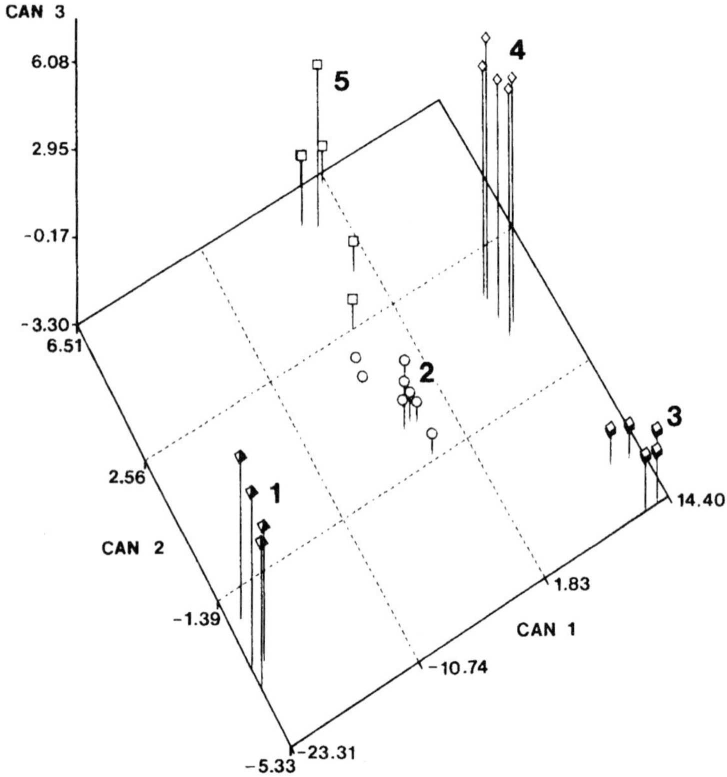

Fig. 10. Niche-metric results (three canonical axes as indicated). Analysis showing the niche groups in the data. 1 = Slow climbing lorisines, 2 = lesser leaping cheirogaleines plus bush-babies (but not the highly leaping *Galago senegalensis* and *G. alleni*, 3 = the highly richochetal leaping tarsiers plus *G. senegalensis* and *G. alleni*, 4 = the very highly leaping indriids plus *Lepilemur* and *Hapalemur*, 5 = the least leaping, most quadrupedal lemurs plus *Phaner* and *Daubentonia*.

For example, some of the most extreme leaping bush-babies are grouped with tarsiers (also extreme leapers). As another example, all of the lorisines are separated from all of the other species (the lorisines are slow climbers; the others are fast running, climbers and leapers). A third example is the confluence of the somewhat lesser leaping dwarf lemurs with some of the lesser leaping bush-babies. Although such species clusters relate to niche similarities, the niche in lifestyle is the equivalent of functional convergence in anatomy.

It is therefore not surprising that the results of the niche-metric studies of the individual species are incredibly concordant with the results of the morphometric studies on individual anatomical units of the species (Table 5).

But the application of the same statistical methods to ask questions relevant to evolution, i.e. to discover how species combined into their families are clustered, arranges the species in a different way. In this second analysis (Fig. 11), the three tarsiers, for example (tarsiids) are no longer linked with any bush-babies (galagines). Indeed the tarsiers are completely separated from all bush-babies. All lorisines (previously quite separate in a niche sense from all other species, are, in the combined analysis, inseparably linked with the bush-babies, as in the evolutionary grouping of the family of lorisids comprising bush-babies (galagines) and lorises (lorisines). Other separations of major evolutionary interest are also readily evident including the clear differentiation of the single species *Daubentonia*, as the family Daubentoniidae, from all other lemurs. Thus, information of evolutionary import seems also to be present in these data. Asking the right question of the data is necessary to allow it to appear.

In other words, information about the niche convergences in the study of individual species is over-ridden by the evolutionary relationships inherent in the analysis of species combined into families.

In a manner parallel to what we have done in the various morphometric investigations, we looked to see how the individual niche features are clustered.

This was first carried out for the analysis of individual species. The question being asked is: in the study of individual species, do the

Table 5. Comparison of morphometric and niche-metrics.

Morphometric species clusters	Niche-metric species clusters	Niche
Four species of lorisines *Perodicticus*, *Nycticebus*, *Loris*, *Arctocebus*	Four species Same	Slow climbing, insectivorous, small branches
Three cheirogaleines *Cheirogaleus major* and *medius*, *Microcebus*, Plus Two galagines *Galago demidovii*, *G. crassicaudatus*	Same species Plus *G. garnetti* and *G. elegantulus*	Leaping with body in quadrupedal posture, scurrying, insectivorous small branch milieu
Three tarsiiformes *Tarsius bancanus*, *T. spectrum*, *T. syrichta* Plus One galagine *Galago demidovii*	Same species Plus *Galago alleni*	Very strongly leaping with body curled up, richochetal leaping, insectivorous
Three indriids *Propithecus*, *Avahi* and *Indri* Plus Two lemurs *Hapalemur* and *Lepilemur*	Same species Plus *Phaner*	Leaping with body extended, largely vertical supports, Leaves, buds and fruit
Four *Lemur* species Plus *Daubentonia*	Same species	Leaping creatures but less so than those above, more quadrupedal, insectivorous and fruit

clusters of the variables make ecological sense? This is easily seen to be so. The variable clusters are not random or undecipherable. In contrast, the locomotor variables are all placed in one part of the multivariate space, and the environmental variables placed in a neighbouring

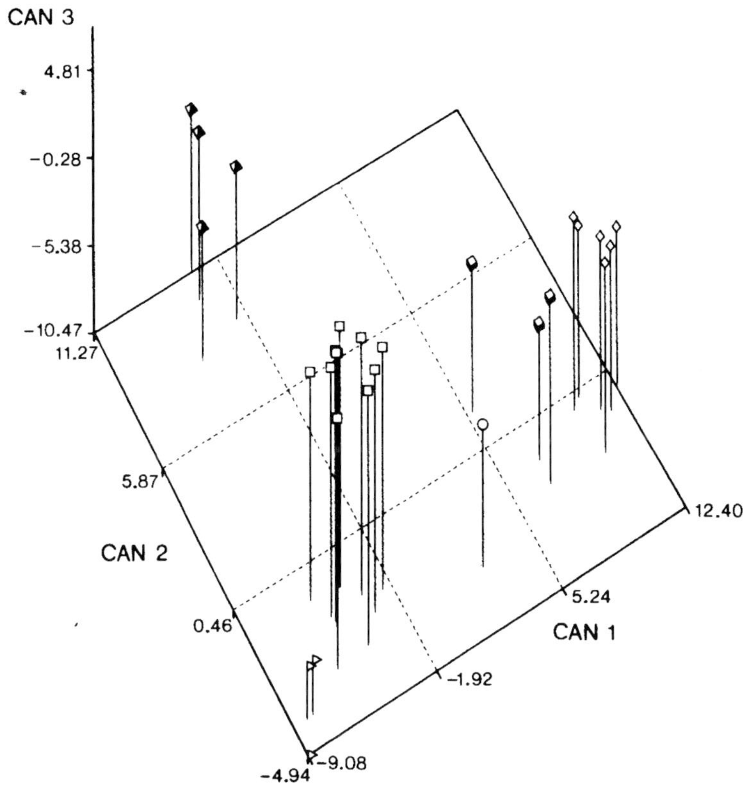

Fig. 11. Niche-metric results (three canonical axes as indicated). Analysis showing the species combined into families. All members of the family Cheirogaleidae — pyramids, all Indriidae — cubes, all Lemuridae — diamonds, all Lorisidae (lorisines and galagines) — squares and the family Daubentoniidae represented by a single genus — a uniquely separated circle.

part of the analytical space. Lying on the periphery of this combined locomotor/environmental variable region are each of the dietary variables. Each of these is, however, individually, about as far distant from each of the others as it could possibly be (Fig. 12, Oxnard *et al.*, 1990). Clearly diet is a critical differentiating factor when seen in the light of locomotion and environment.

Further, within this general picture, it is possible to isolate restricted variable neighbourhoods. In each case, the variable

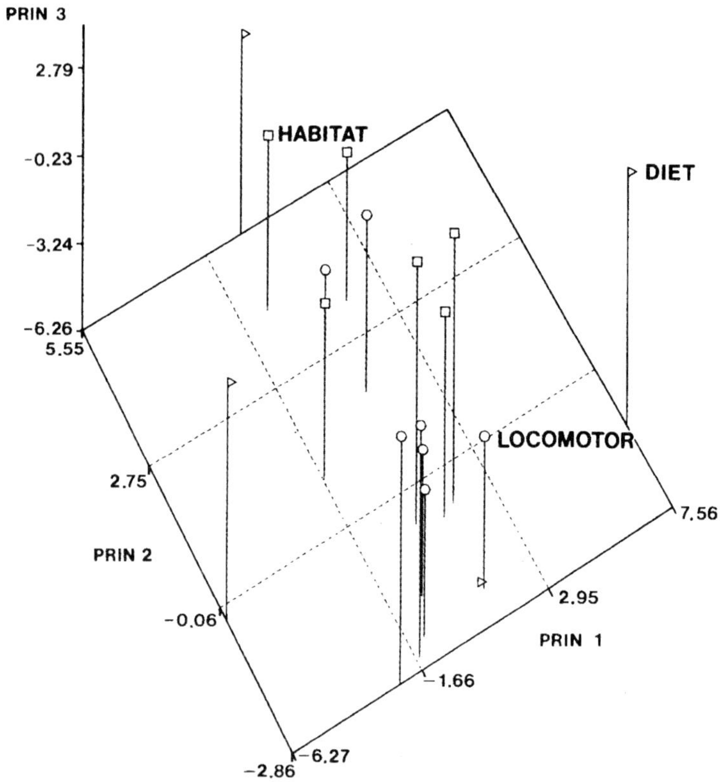

Fig. 12. The plot of niche variables (squares, locomotor variables; circles, environmental variables; triangles, dietary variables) responsible for the niche group separations in the individual niche-metric study of species. Two local neighbourhoods of specific variables are identified.

neighbourhood makes sense in relation to the ecological adaptations outlined by particular clusters of the species. Thus, one neighbourhood of niche variables includes fruit eating, using many horizontal supports, leaping, scurrying and climbing; this is the milieu which contains the convergence of some of the lesser leaping bush-babies and the dwarf lemurs. Another neighbourhood of niche variables includes leaf-eating, living on large supports, and leaping; this is the milieu which contains the convergence of the indriids, *Lepilemur* and *Hapalemur*. Such variable neighbourhoods make reasonable ecological sense in a study that groups the species ecologically. It is evident, then, that the information content of these data not only identify niche groups of species, but also niche groups of variables.

The second question that can now be asked is: what are the clusters of variables that produce the evolutionary separations in the combined analysis? Table 6 shows the list. One example is a cluster that contrasts variables with high positive loadings (slow quadrupedalism, large supports, and diets containing fruits, leaves, buds and flowers) to variables with high negative loadings (scurrying, undergrowth, small supports and animal dietary items). This contrast is responsible for clustering bush-babies with lorises (that is, into the family of all lorisids). This grouping of variables is not adaptive. It is related to the

Table 6. Relationships between clusters of niche variables and evolutionary groups.

Variable clusters	Variables involved	Evolutionary separations of families
Cluster 1		
High + loadings in four variables High – loadings in four variables	Slow quadrupedalism, large supports, fruit, leaves as contrasted with scurrying, undergrowth, small supports, animal diet	1. Separates **lemurids** from all others groups and, 2. Clusters galagines and lorisines in the **lorisids** and 3. Separates **galagines from tarsiids**
Cluster 2		
High + loadings in three variables and High – loadings in three others	Falling leaps, scurrying, fruit eating as contrasted with undergrowth, vertical supports, animal diet	4. Unifies **cheirogaleids** and 5. Separates them from **lorisids**
Cluster 3		
Loadings in four variables	Falling, crouching and richochetal leaps, animal diet	6. Separates **tarsiids** from all other groups

Note: Bold text indicates critical family groups as distinct from niche groups in the previous table.

evolutionary confluence of two differently adapted sets of species into a single higher evolutionary family grouping. Other similar variable clusters perform equivalent phylogenetic separations (also Table 6).

Of course, more 'distal' links to developmental phenomena are not evident. But then, at least at the present, we have no inkling of the developmental processes (if any) that might be involved in the behavioural/environmental/dietary side of the organism. Perhaps we need to know more about that.

Measuring Brains: 'Neurometrics'

In a precisely similar manner, we have extended these concepts to studies of the brain, the fount of behavioural flexibility. The main story of the brain is provided in another chapter. But there is one part of that story that is also relevant to the questions being asked here. Thus, simple studies of volumetric measures of various brain parts (kindly supplied to Dr de Winter and I by Professor Heinz Stephan) imply that the major information content of such data is primarily related to size: brain size and body size. Indeed, such studies usually imply that something like 98% of the variation in the data is correlated with overall body and brain size, respectively, and that very little other information is present (e.g. Finlay and Darlington, 1995). Such information suggests that the developmental constraints on the mammalian brain are enormously tight; that is: that the simplest way in which one particular brain region could enlarge in evolution (in relation, for example to some special behavioural adaptation) would be through evolutionary enlargement of the entire brain (or even, by extension of the argument, through enlargement of the entire body).

However, data pertaining to particular behavioural adaptations may more properly be about functional relationships between brain parts. For example, the proportional relationship between the cerebellum and the medulla, or between the neocortex and the palaeocortex, might relate more specifically to 'neurological talk' between these regions during function. When the data are reorganised to reflect such cross-talk between brain regions, analysis provides a totally new picture.

Thus, separations of species are no longer primarily along a single axis representing the overall sizes of the brains. The species separations now relate to major evolutionary differences (e.g. the total distinctions of insectivores, bats and primates along almost mutually orthogonal axes, together with lesser separations of a few high-level phylogenetic groups within each (see later chapter; de Winter, 1997; de Winter and Oxnard, 1997). These new studies are also sensitive enough that many functional parallels and convergences can also be detected (e.g. groupings, separately, of various semi-aquatic insectivores, various fish-eating bats and various acrobatic arm-swinging primates, irrespective, each, of lower evolutionary separations within these parallel groups). Indeed, the richness of this way of organising and analysing the data is far greater than that of size alone. It certainly does not present the picture of a highly constrained developmental process in the brain that is common to all mammals.

We may therefore now ask our, by now, customary question: what about the clusters of brain features? These studies are not yet complete and the full answer is not available. Nevertheless, we can already say that the axis that characterises differences among primates is associated with brain features that are the measures of proportional expansion of the highest levels of the motor hierarchy, as distributed over three main regions, the neocortex, striatum and cerebellum, relative to the amount of somato-sensory and motor information, through the medulla, that is exchanged with the body. These are features that are involved in the planning of complex motor behaviour and cognition which relate to abilities to strategically plan behavioural acts in advance of their execution. They are clearly strongly related to overall brain (and body) size. Similar functional findings are evident in each of the other major mammalian groups in this study.

We are only just starting to ask the developmental questions of brains. The best we can say at the moment is that, in primates, one sequence of brain features separates all primates in a linear manner consistent throughout the entire Order (including humans). Humans are at one extreme of this linear sequence and their next closest relative is the chimpanzee. But there are other clusters of brain features that separate humans (and only humans) from all other primates. This is a picture implying that new developmental changes have occurred (and possibly are still occurring) since the separation of prehumans from pre-apes (presumably, from pre-chimpanzees). This may well relate to special features of the development of the human brain that do not occur in non-human primates, not even in the chimpanzee. This is all described in more detail in a subsequent chapter.

The Information Content of Structure

All these investigations are attempts to recognise the interrelated and multifactorial nature of the information content of the features of adult organisms. They are also, however, attempts to discover the degree to which such integration can be disentangled. They assume that most observable features of organisms, the anatomical measures of evolutionary differences, are the resultant of the several underlying mechanisms and processes of both function and development. They therefore apply analytical methods to the features so as to allow the underlying elements, if they are truly present, to be revealed. As such, the studies require the features to be rendered quantitatively, and methods to have the capacity to partition the information content.

Thus, we can now summarise the results in all the investigations. When analyses are carried out at the individual unit level, species are grouped according to functional adaptation. There must be evolutionary information within the individual studies because the data are from related species. But in some way, the evolutionary information is hidden.

When, however, analyses combine several individual units, then species are grouped according to evolutionary relationship. In this case, we know that there must be functional information within the combined studies (we know that because they use the same data as the individual studies). In this case, however, it is the functional information that is largely hidden; the evolutionary information has become dominant.

How can this be? There are at least two possible explanations.

Applied Thinking: Partitioning Function, Development and Evolution

We can now present a summary of the information content in the various studies (Table 7). In terms of individual studies, this table

Table 7. Summary of information content of all studies.

Information content of teeth and jaws

- **Individual studies**
  - Separates species by masticatory function
  - Clusters features such as incisor edges for cutting, molar areas for mastication
- **Combined studies**
  - Separates sexes and species in relation to evolution
  - Clusters features reflecting genetic and developmental mechanisms

Information content of post-cranium

- **Individual studies**
  - Separates species in relation to locomotion (function)
  - Clusters features in relation to biomechanics
- **Combined studies**
  - Separates species in relation to evolution
  - Clusters features reflecting genetics and development

Information content of niche

- **Individual studies**
  - Separates species in relation to the niche convergences
  - Clusters features reflecting the niches
- **Combined studies**
  - Separates species into families
  - Clusters features in reflecting evolutionary separations

Information content of brain (studies incomplete — neurometrics)

- **Individual studies**
  - Separates species in relation mainly to overall size
  - Clusters features in relation mainly to overall size
- **Combined studies**
  - Separates species in relation to major evolutionary groups but functional convergences can still be discerned
  - Clusters features in relation to major evolutionary groups but relationships with function still clear

documents that in studies in which separations of species relate to the functional milieu within which the individual anatomical regions operate, clusters of variables relate to the adaptive factors pertinent to those functional milieux. In terms of various combined studies, in which the separations of species are most closely linked to what we know about their evolution, the clusters of variables are most closely allied with the genetic and developmental underlay of whole-organism diversity.

Perhaps a helpful way of describing the relationships between the morphometric studies of static differences in a diversity of adult forms, and the dynamic mechanisms and processes of development, growth and adaptation, can be provided by an analogy from Chap. 2. The patterns of trabeculae in a bone, static at any given time, are related to the many dynamic and constantly changing stresses and strains that have acted upon the bone as a result of the dynamic processes of function. It is almost as though the static pattern of trabeculae is a 'memory' of the resultant of past dynamic stresses and strains. In a somewhat similar way, the static adult differences as seen through morphometric analyses may be a 'memory' of the past continuously changing developmental, functional and evolutionary mechanisms and processes that produced them.

Theoretical Thinking: Partitioning Function, Development and Evolution

These investigations have also started me on some theoretical thinking that further indicates why individual studies might speak most closely to function, and combined studies to development and evolution.

Let us assume that any individual morphometric study of a particular functional unit contains information that is partly about functional adaptation and partly about evolutionary relatedness. This means we can think of the study as revealing mainly functional information 'f' say five parts, but nevertheless also containing (if not so obviously) a lesser degree of evolutionary information 'e' say three parts. (Of course, these phenomena are really continuous; I am describing them as discrete bits to simplify matters.) Let me now write the information down in a manner that looks like a mathematical formula.

But before you, the reader, read on, let me assure you that the following is not mathematics.

It is not something to be afraid of.
It very specifically is not something to be skipped over.
It is not difficult.
It is, in fact, enormously easy.
Just as one can express complex mathematics through a simple picture (and there are lots of these in this book), so, too, can one express complex ideas through simple statements. What appear to be mathematical equations are, in fact, just simple statements that, like pictures, are better than a thousand words.
I urge you to read on.

The information content '**I**', in one of the individual studies, might then be written as

$$\mathbf{I} = \mathbf{f1} + \mathbf{f2} + \mathbf{f3} + \mathbf{f4} + \mathbf{f5} + \mathrm{e1} + \mathrm{e2} + \mathrm{e3},$$

that is

I = a total of eight units (where **f**s are overt function and es covert evolution).

This implies that **f** (= biological function) appears to be the major part (5/8) of the total information and e (– evolution) the lesser part (only 3/8). Perhaps this lesser amount of e is partly why it is less easily recognised. I have used the bold format to designate the more easily recognisable portions.

Yet the above is too simplistic. It is likely that some of the evolutionary information will also be 'similar to' (i.e. correlated with) some of the functional information. After all, functional adaptation is not always convergence and, in any case, is also a major part of evolutionary diversity. This is the equivalent of saying that there will inevitably be at least some interactions between function and evolution. Let us describe this by showing when fs and es contain similar information. Let us assume that two of the fs and two of the es are the same. Then, the information in the above description might be rewritten as

$$\mathbf{I} = \mathbf{f1} + \mathbf{f2} + \mathbf{f3} + \mathbf{f4}(=e) + \mathbf{f5}(=e) + e1(=\mathbf{f}) + e2(=\mathbf{f}) + e3,$$

that is

I = still a total of eight units of information.

In this statement, however, the evolutionary information (e), although appearing in 5 bits out of 8 in our example has only 1 bit out of 8 (e3) in which it is clearly different from **f**. In contrast, function, **f**, appears to be much more identifiable because it is 7/8 of the information different from e. The portions where e and f are the same (f4 = e, f5 = e, e1 = f and e2 = f) will, thus, all be seen as **f**. Thus, in a study of an individual functional unit where there are interactions between function and evolution, it is easy to see why the smaller part — evolution — may be obscured, and why the larger portion — function — may appear extremely large indeed.

If, now, we had a series of such studies taken on different functional units (a, b, c, d and e) the following exposition shows how, though each individual unit might greatly emphasize function, a combined study of all units together might sum to something different.

Thus, for each individual unit, a, b, c, d and e, individual results might give descriptions like that above. As a result, the entire suite of descriptions might look like the following:

$$\text{Unit a} = \mathbf{f1a} + \mathbf{f2a} + \mathbf{f3a} + \mathbf{f4a}(=ea) + \mathbf{f5a}(=ea) + e1a(=\mathbf{fa}) + e2a(=\mathbf{fa}) + e3a,$$

$$\text{Unit b} = \mathbf{f1b} + \mathbf{f2b} + \mathbf{f3b} + \mathbf{f4b}(=eb) + \mathbf{f5b}(=eb) + e1b(=\mathbf{fb}) + e2b(=\mathbf{fb}) + e3b,$$

$$\text{Unit c} = \mathbf{f1c} + \mathbf{f2c} + \mathbf{f3c} + \mathbf{f4c}(=ec) + \mathbf{f5c}(=ec) + e1c(=\mathbf{fc}) + e2c(=\mathbf{fc}) + e3c,$$

Unit d = **f1d** + **f2d** + **f3d** + **f4d**(=ed) + **f5d**(=ed) + e1d(=**fd**)
+ e2d(=**fd**) + e3d,

Unit e = **f1e** + **f2e** + **f3e** + **f4e**(=ee) + **f5e**(=ee) + e1e(=**fe**)
+ e2e(=**fe**) + e3e.

Exactly as in our first example, for each of these descriptions, **f** appears to be 7/8 of the information even though f and e are actually split 5/8 and 3/8, respectively.

But when we add the data for each individual description into a single combined description, the totals could look very different. First, the various fs cannot be expected to add up beyond any single description because the functions in each unit are different. Second, in contrast, the various es (including those es that are related to fs) can be expected to add up across every unit because the information about evolution should be the same for each unit (after all, they are all parts of the same animals).

Accordingly then, the total **es** are 25 (5 for each equation and **e** can now be written bold) but no single f is any greater than 3 (e.g. 3 fas, 3fbs, 3fcs and so on). Now **e** (evolution) shines out strongly at 25/40; no single f (function) shines out more strongly than 3/40, even though different functions combined total 25/40. There is just as much functional information present in the combined study as in the total of individual studies but it is now hidden.

This theoretical exposition alone could be the reason why **function** is clearly evident in individual studies, but **evolution** in combined studies. I do hope that the apparent intrusion of what appear to be algebraic equations has not turned any reader off. This is **not mathematics**, it is **just thinking**.

The Next Step

These various investigations speak especially to the relationships of humans within the primates broadly, and within hominoids more specifically. But they also speak to what can be learnt about dynamic underlying processes from static adult anatomy.

In terms of the individual functional units humans are generally uniquely different from all other primates. This presumably relates rather simply to the totally new functional milieux that humans have come to inhabit.

But in terms of combinations of functional units, the uniqueness of humans becomes buried. We generally recognise, instead, a picture that is the same as that evident from evolutionary (especially molecular) investigations, that is, a close grouping of humans with African great apes, and a clear separation of humans from Asian apes. This may be for the same reason as above; i.e. because these latter results reflect those common genetic and developmental phenomena that apply as much to humans as they do to other primates.

Such a discussion provides a solution to the apparent paradox of the position of humans among the primates. It requires us to integrate two different views of humans at one and the same time. We no longer have to take sides in a controversy that, on the one hand, sees humans as only slightly lower than the angels, and on the other, as just another animal. Before assuming a conclusion, let us see, however, how other chapters speak to this paradox.

References

Albrecht G, Thin plate splines and the primate scapula, *Am J Phys Anthro* **28**: 125–126, 1991.

Ashton EH, Healey MJR, Oxnard CE, Spence TF, The combination of locomotor features of the primate shoulder girdle by canonical analysis, *J Zool Lond* **147**: 406–429, 1966.

Ashton EH, Oxnard CE, Locomotor patterns in primates, *Proc Zool Soc Lond* **142**: 1–28, 1964.

Ashton EH, Oxnard CE, Functional adaptations in the primate shoulder girdle, *Proc Zool Soc Lond* **142**: 49–66, 1964.

Atchley WR, Genetic and developmental aspects of variability in the mammalian mandible, in Hanken J, Hall BJ (eds.), *The Skull*, Vol. 1: *Development*, University of Chicago Press, Chicago, pp. 207–247, 1993.

Corrucini RS, Ciochon RL, Morphometric affinities of the human shoulder, *Am J Phys Anthrop* **45**: 19–38, 1978.

Crompton RH, Lieberman SS, Oxnard CE, Morphometrics and niche-metrics in prosimian locomotion. An approach to measuring locomotion, habitat and diet, *Am J Phys Anthrop* **73**: 149–177, 1987.

de Winter W, Perspectives on mammalian brain evolution: Theoretical and morphometric aspects of a controversial issue in current evolutionary thought, Doctoral Thesis, University of Western Australia, 1997.

de Winter W, Oxnard CE, The primate brain and its mammalian context: A morphometric study of volumetric measures of brain components, *Am J Phys Anthrop Suppl* **24**: 243–244, 1997.

de Winter and Oxnard, Evolutionary radiations and convergences in the structural organization of mammalian brains, *Nature* **409**: 710–714, 2001.

Feldesman M, The primate forelimb: A morphometric study of locomotor diversity, *Univ Oregon Anthr Papers* **10**: 1–154, 1976.

Feldesman M, Further morphometric studies of the ulna from the Omo Basin, Ethiopia, *Am J Phys Anthr* **51**: 409–416, 1979.

Feldesman M, Morphometric analysis of the distal humerus of some cenozoic catarrhines: The late divergence hypothesis revisited, *Am J Phys Anthr* **59**: 73–76, 1982.

Finlay BL, Darlington RB, Linked regularities in the development and evolution of mammalian brains, *Science* **268**: 1578–1583, 1995.

Hanken J, Hall BK, *The Skull*, Vol. 1, *Development*, University of Chicago Press, Chicago, 1993.

Kidd R, Oxnard CE, Patterns of morphological discrimination in the human talus: A consideration of the case for negative function, *Perspectives in Human Biology* **3**: 57–70, 1997.

Larson SG, Functional morphology of the shoulder in primates, in Gebo DL (ed), *Postcranial Adaptation in Nonhuman Primates*, Northern Illinois University Press, DeKalb, pp. 45–69, 1993.

Lucas P, *Dental Functional Morphology*, University of Cambridge Press, Cambridge, 2007.

McHenry HM, Corrucini RL, Multivariate analysis of early hominoid pelvic bones, *Am J Phys Anthr* **46**: 263–270, 1975.

Manaster BJM, Locomotor adaptations within the Cercopithecus, Cercocebus and Presbytis genera: A multivariate approach, Unpublished PhD Thesis, Chicago, The University of Chicago, 1975.

Manaster BJM, Locomotor adaptations within the Cercopithecus genus: A multivariate approach, *Am J Phys Anthr* **50**: 169–182, 1979.

Oxnard CE, The functional morphology of the primate shoulder as revealed by comparative anatomical, osteometric and discriminant function techniques, *Am J Phys Anthr* **26**: 219–240, 1967.

Oxnard CE, The architecture of the shoulder in some mammals, *J Morph* **126**: 249–290, 1968.

Oxnard CE, *Form and Pattern in Human Evolution: Some Mathematical, Physical and Engineering Approaches*, The University of Chicago Press, Chicago, 1973.

Oxnard CE, *Uniqueness and Diversity in Human Evolution: Morphometric Studies of Australopithecines*, The University of Chicago Press, Chicago, 1975.

Oxnard CE, *The Order of Man: A Biomathematical Anatomy of the Primates*, Hong Kong University Press, Hong Kong, 1983, Yale University Press, New Haven, 1984.

Oxnard CE, Developmental processes and evolutionary diversity: Some factors underlying form in primates, *Arch Oceania* **27**: 95–104, 1992.

Oxnard CE, The information content of morphometric data in primates, in Strasser E, Fleagle J, Rosenberger M, McHenry H (eds.), *Primate Locomotion: Recent Advances*, Plenum, London and New York, pp. 255–275, 1998.

Oxnard CE, Morphometrics of the primate skeleton and the functional and developmental underpinnings of species diversity, *Linnean Symp Ser* **20**: 235–264, 2000.

Oxnard C, Invited discussion, in Bock G, Goode J (eds.), *Tinkering: The Microevolution of Development*, *Novartis Foundation Symposium*, Vol. 284, Wiley, Chichester, 2007.

Oxnard CE, Crompton RH, Lieberman SS, *Animal Lifestyles and Anatomies: The Case of the Prosimian Primates*, Washington University Press, Seattle, 1990.

Pan R-L, A morphometric approach to the skull of macaques: Implications for *Macaca arctoides* and *M. thibetana*, Unpublished PhD Thesis, University of Western Australia, 1998.

Rose MD, Quadrupedalism in primates, *Primates* **4**: 337–357, 1973.

Senut B, Humeral outlines in some hominoid primates and in plio-pleistocene hominids, *Am J Phys Anthr* **56**: 275–284, 1981.

Stern JT Jr, Sussman RL, The locomotor anatomy of Australopithecus afarensis, *Am J Phys Anthr* **60**: 279–318, 1983.

Tardieu C, Morpho-functional analysis of the articular surface of the knee joint in primates, in Chiarelli AB, Corrucini RS (eds.), *Primate Evolutionary Biology*, Springer Verlag, Berlin, 1981.

Welsch RE, Graphics for data analysis. *Comput Graphics* **2**: 31–37, 1976.

Walker AC, Locomotor adaptations in past and present prosimian primates, in Jenkins F (ed.), *Primate Locomotion*, Academic Press, New York, 1974.

Walker AC, Prosimian locomotor behaviour, in Doyle GA, Martin RD (eds.), *The Study of Primate Behavior*, Duckworth, London, 1979.

Wolpert L, Lewis J, Summerbell D, Morphogenesis of the vertebrate limb, *Ciba Foundation Symp* **29**: 95–129, 1975.

Chapter 5

Now You See It, Now You Don't: Hidden Aspects of Form

A Brief History of Size and Shape

Questions raised by the consideration of biological size and shape are truly old. In the time of the Ancients, Pythagoras (about 582 B.C. cited in Koestler, 1968) wrote that:

> '... all things have form, all things are form, and all forms can be defined by numbers...'

Nearly two thousand years were to pass before Galileo (1638) recognised that bones in large animals were not simply scaled up versions of bones in small animals, but had also to be a different form.

A further three hundred years produced Thompson's seminal book: *On Growth and Form* (1917) that went further into the problems of size and shape. He wrote that the study of differences between diverse and similar organisms might constitute a proof that:

> A comprehensive 'law of growth' has pervaded the whole structure in its integrity, and that some more or less simple and recognisable system of forces has been at work.

This law of growth included the concept of different rates of growth of parts of the body relative to the body as a whole. He further wrote:

> The study of form may be descriptive merely, or it may become analytical. We begin by describing the shape of an object in the simple

> words of common speech; we end by defining it in the precise language of mathematics; and the one method tends to follow the other in strict scientific order and historical continuity.

Fifteen years after Thompson, Huxley's *Problems of Relative Growth* (1932) generalised the simple formula for such allometry (as it came to be called) and showed that it had a very much larger number of effects on animal (and plant) form than previously realised. He explained such phenomena through growth centres and growth gradients.

Since Huxley enormous number of studies of allometry have been carried out. Yet it was not until near the end of the century that the problem of relative growth was elucidated beyond the point at which Huxley had left it. This is evident in some of the developments of what has come to be called 'geometric morphometrics' in the seminal works of Bookstein (1978), Kendall (1984), Bookstein (1991), Kent (1994), O'Higgins and Cohn (2000), Slice (2005) and, by now, many other workers.

My first part in this (in the early 1960s) was to believe that I could not do better than Huxley and that, therefore, I should do other things. Which I did! But the matter has festered within me over the years. Recent advances in the concepts related to times and rates of growth (derived from analytical explorations of size and shape, the experiments of developmental biology, and the invention of geometric morphometrics) have allowed me new insights. Thus, I have been lead to examine what I have called 'the tyranny of size' as the attempts to understand 'shape' are muddied by the attempts to measure 'size'.

For a parallel concept, we can look at many studies that have examined heredity and environment. So frequently this is done by measuring heredity (e.g. through twin studies) and, assuming that heredity and environment are separate and additive, subtracting heredity from unity to obtain the remainder, assumed to be environment. This is so simplistic; there are surely many different positive and negative interactions between heredity and environment so that they are intermixed and do not sum to 1.

In a similar way, size and shape are so interlinked that attempts to examine them separately hide much that is important. Size contains a considerable portion of shape, but not all. Equally, shape contains a lot of size, but not all. The relationships between them, moreover, are far more complex than merely additive. Further, neither 'visual associations' nor 'calculated relationships' necessarily imply 'biological causation'. Our 'eyes' may 'see' one thing, our 'measurements' another.

Puzzles and Perplexity

I am, therefore, not blind to the fact that these methods, visual and mensurational, of studying biological structure have their weaknesses. One weakness stems from a similar limitation of ordinary human minds and standard statistical methods. Both human minds and statistical methods think, most easily, in terms of straight lines or simple curves (Fig. 1), or the discreteness of nicely spaced and rounded or elliptical groups (Fig. 2).

As we think, however, about the possible structures that data may take up matters are sometimes more puzzling? For instance, when data are two-dimensional, the eye often does better than simple statistical methods. Figure 3 shows four sets of data. Our eyes readily see that they are different. The measurements can mislead. The means and standard deviations of the measurements of each of these plots are identical. Yet so often, especially in these days of statistical packages, we look only at the output and not the input.

This difficulty in 'seeing' groups is further explicated in the following puzzles. Some of them appear to show that the 'eye-and-the-mind-behind-the-eye' can do very well and statistics sometimes very badly. But some seem to show that the 'eye-and-the-mind-behind-the-eye' may sometimes have difficulty seeing unusual situations that may be revealed by other approaches.

Figure 4 shows three plots of artificial data. What groups do you 'see'? I think the top frame shows what appear to be two separate roughly circular groups, the centre frame shows a single dumb-bell-shaped group, and the bottom frame shows two linear approximately

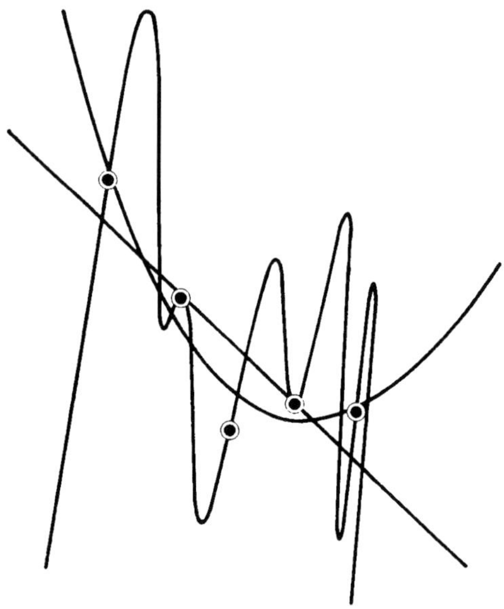

Fig. 1. Five points and three plots representing the relationships of the points. The straight line is fitted 'by eye', the simple curve is the statistical best fit, the complex curve is one of an infinite number of complex curves that go exactly through every point — 'perfect fits' (after Duda and Hart, 1973).

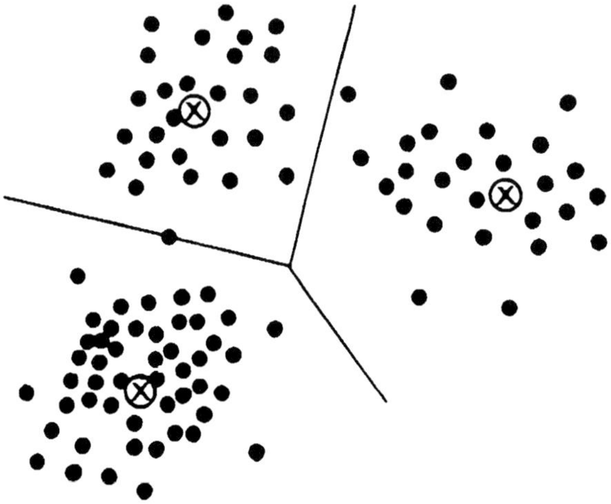

Fig. 2. A two-dimensional view of three round, clearly demarcated clusters. The 'eye' readily sees these clusters.

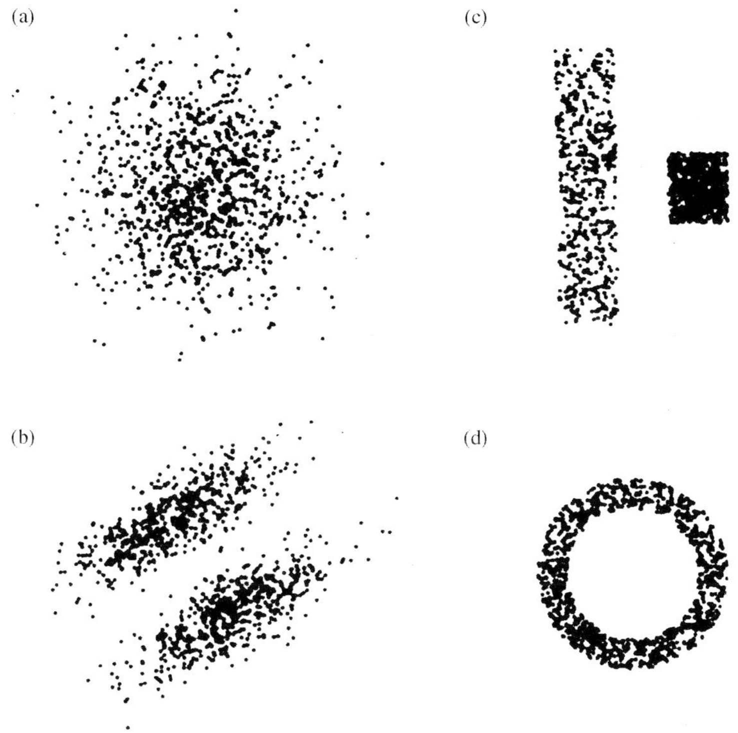

Fig. 3. Four data sets deliberately arranged so as to have identical mean and standard deviation. But the 'eye' readily distinguishes them. If one did not look, however, the statistics would make them appear the same (after Duda and Hart, 1973).

parallel groups. These are regularly identified as such by my students. Some students find, by eye, three groups in the central frame: two large circular ones and the small central linear one. There are, however, other possibilities.

Thus, Fig. 5 shows the same artificial data analysed using a neighbourhood-linking algorithm. The technique does as well as the 'eye' for the top data set; the two circular groups are readily found. The dumb-bell-shaped group in the centre is not recognised by the algorithm which only finds two slightly less circular groups. The two linear groups in the bottom frame are not found at all by the algorithm. Instead, it finds two approximately circular groups of different

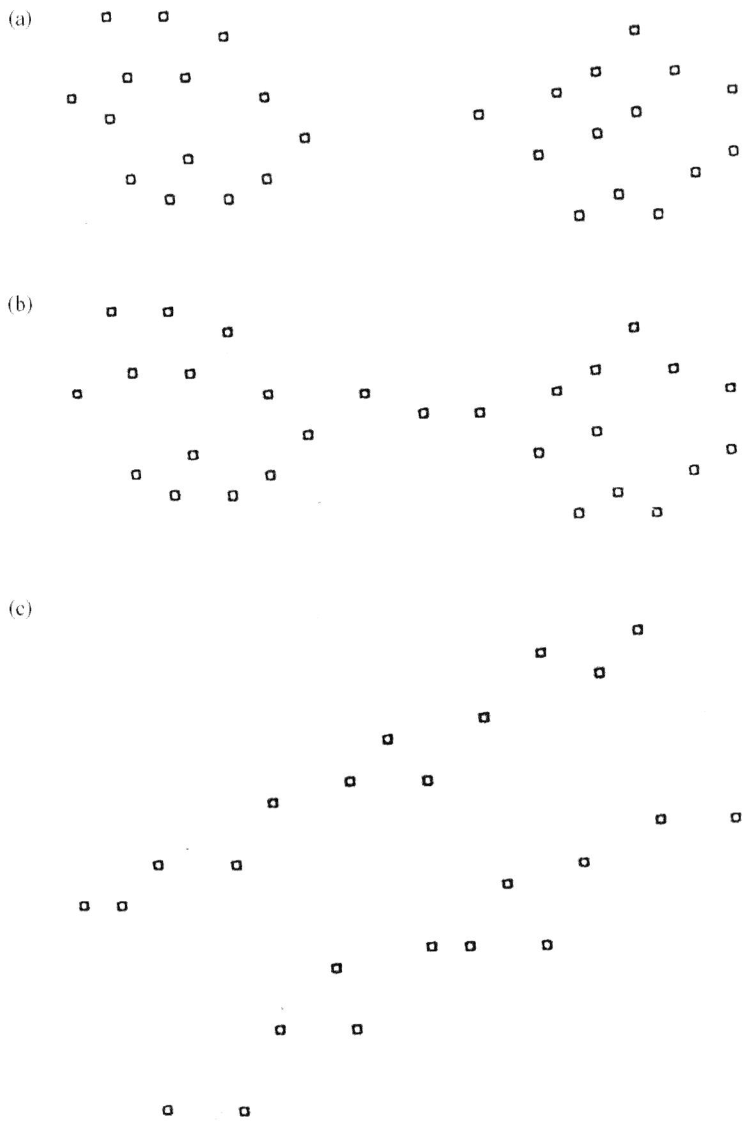

Fig. 4. Plots of artificial data. What groups do you see? The top frame shows what appear to be two roughly circular groups, the centre frame, a single dumb-bell shaped group, and the bottom frame, two linear approximately parallel groups. These are readily identified by students (after Duda and Hart, 1973).

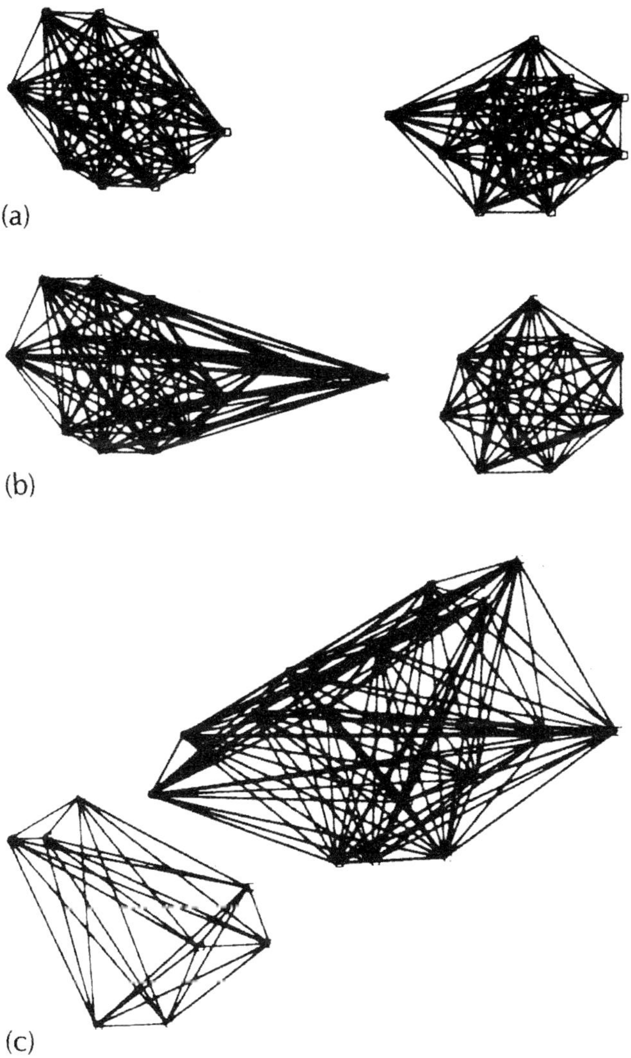

Fig. 5. The same artificial data analysed using a neighbourhood-linking algorithm. The two circular groups are readily found. The dumb-bell shaped group is not easily found, it looks rather more like two slightly less circular groups. The two linear groups are not found at all. Instead the algorithm finds two approximately circular groups of different size. Is this because such algorithms have difficulties with groups that depart markedly from circular or elliptical? (After Duda and Hart, 1973.)

size. Do we get this result because such algorithms have difficulties with groups that depart markedly from circular or elliptical; certainly such groups abound in biology? The 'eye' does not seem to have a problem with this. And who is right anyway, the 'eye' or the 'machine'? And does it matter?

Figure 6 shows yet another set of artificial data. The question might be: how many groups are there? One answer might be: two, a central small tight group superimposed upon a large much less dense group. Another answer might be: one, a single group with a high central peak of density and a low widely dispersed periphery. Other answers that have been given by my students include: many, depending upon which of the small subgroups in the plot the 'eye' happens to see.

A neighbourhood analysis of those data shows (Fig. 7) a possibility that my students never thought of: a complex star-shaped cluster. Could it be real? Do we know?

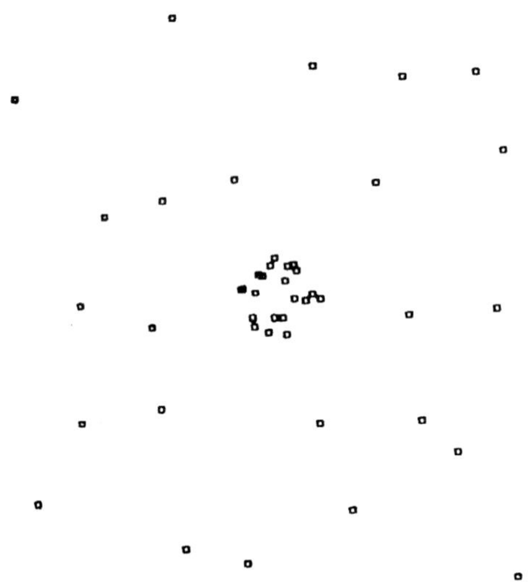

Fig. 6. Yet another set of artificial data. How many groups are there? One answer might be: a small tight group in the centre superimposed upon a large much less dense group. Another answer might be: a single group with a high central peak of density and a low widely dispersed periphery. Other answers that have been given by students are: a variable number of groups depending upon which subgroups in the plot their eyes happen to see.

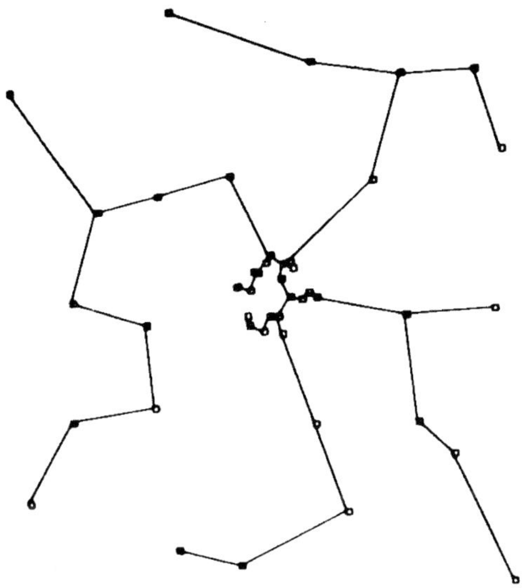

Fig. 7. A neighbourhood analysis of the previous figure showing a possibility that my students never notice: a complex star-shaped cluster. I have found just such an arrangement in real primate data (Oxnard, 1983/1984).

Continuous variation may be a further problem. The human mind tends to want to 'dichotomise' things, seeing the proverbial black or white, yes or no, good or bad, male or female. In spite of this, the human mind is well able to see 'continuous variation' under certain circumstances. Thus, in Fig. 8 the sixes grade imperceptibly into the fives. We readily identify this continuity by eye. Yet we may have some difficulties with some kinds of continuous variation. For instance, we have no names for the continuous shapes that lie in the continuity between threes and fives, or between fives and eights, in that same figure; we may, therefore, think that these numbers are uniquely different; we may miss the continuity.

Figure 9 shows data with several groups, some large and round, some small and round and one a sausage-shaped group. These groups can be easily 'seen' because this is a two-dimensional plot. If these data had been in several dimensions, however, recognising hypersausages might be more difficult! Analysis of these data using a group-finding procedure readily finds these groups (Fig. 10). The three

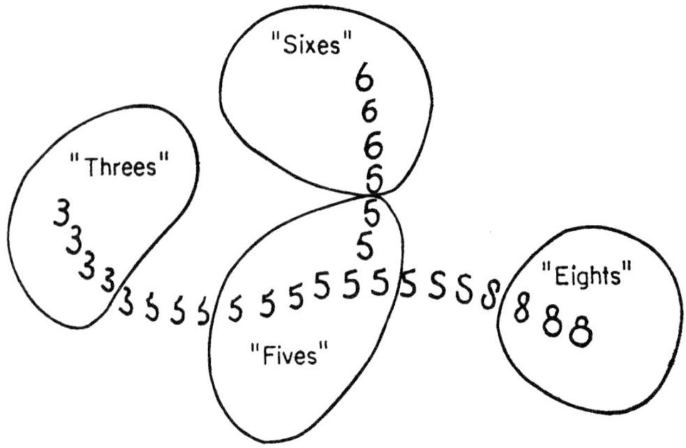

Fig. 8. Continuous variation between 'threes', 'fives', 'sixes' and 'eights' (redrawn after J. Cowan, personal communication).

Fig. 9. Plot showing groups of different size and shape including one long sausage-shaped group.

approximately circular groups of different sizes are readily recognised. So, too, however, is the long sausage-shaped group; and this would have been the case if the data were multidimensional not just two-dimensional.

What about the data in Fig. 11? Again my students find a variety of groups as they select different small localities within the entire plot. A group-finding procedure readily shows the answer (Fig. 12): the data are

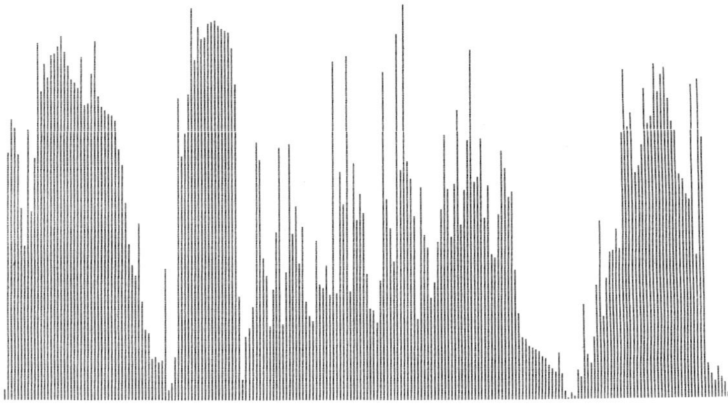

Fig. 10. Clustering programme designed to 'find' groups in the previous plot. All groups are found, even the sausage-shaped group.

Fig. 11. Plot of data. How many groups are there?

random. I know that because I used a random numbers table to produce them! This was done years before the days of modern computers.

Dumb-bells and Doughnuts, Sausages and Stars

In the preceding section, partly to make points, I have described artificial data in various curious forms. Thus, Fig. 3 shows a

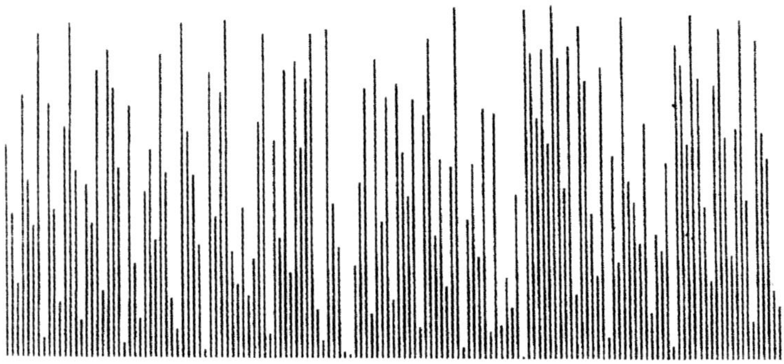

Fig. 12. The same clustering programme of Fig. 11 implies that it is likely that the dots are randomly arranged.

doughnut-shaped group, Fig. 5 shows a dumb-bell-shaped arrangement. Do we ever need to be able to recognise doughnuts and dumb-bells in real life? In fact, I have found arrangements of groups of real anatomical, functional and lifestyle data that are truly 'doughnut shaped' and 'dumb-bell shaped' in many dimensions (Fig. 13).

Sausage-shaped distributions or groups are common enough in biology; they are more usually called ellipses. Nevertheless a sausage-shaped group is presented in Fig. 9 and because its overall dimensions are so much greater than those of the round groups in the figure, this group might well be missed by standard methods. Even just having circular groups of very different size might be problematical (also Fig. 9). Of course, sausages (ellipses) abound in biology.

What about star-shaped arrangements? Fig. 7 shows the artificial data deliberately arranged in a complex star shape. Does this ever occur in real biology? It certainly does. I have found arrangement of groups of real anatomical, functional and niche data that truly were 'star shaped' in many dimensions (Fig. 14).

I aver that we should have all these curiosities (as my colleagues call them) in mind as we ask questions, analyse data, and solve, as we think we do, problems. Not to have them in mind is to miss them whenever, indeed, they occur.

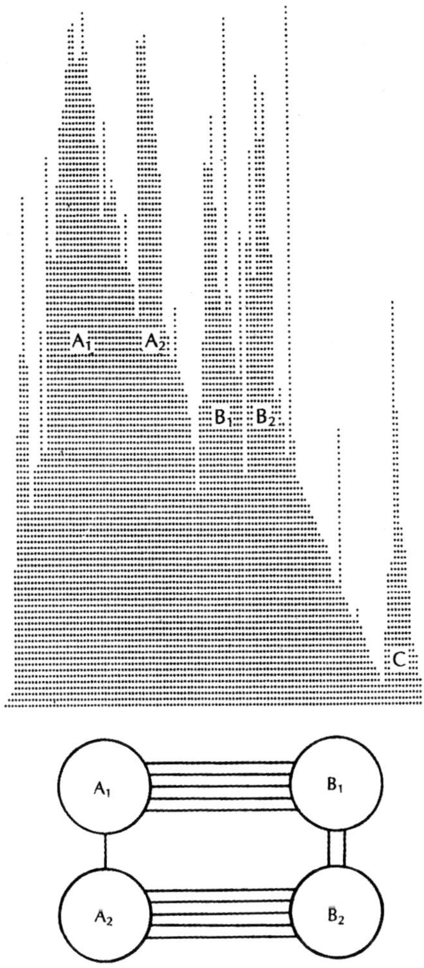

Fig. 13. Real data showing dumb-bell-shaped distributions.

Rainbows and Rocks

It is for reasons like these that I have always looked for methods that are better and able to show what our eyes have great difficulty seeing. This was how I became involved in what was called, in the old days, optical data analysis. Other chapters of this book provide examples within the context of the problems in those chapters. But a general survey of what

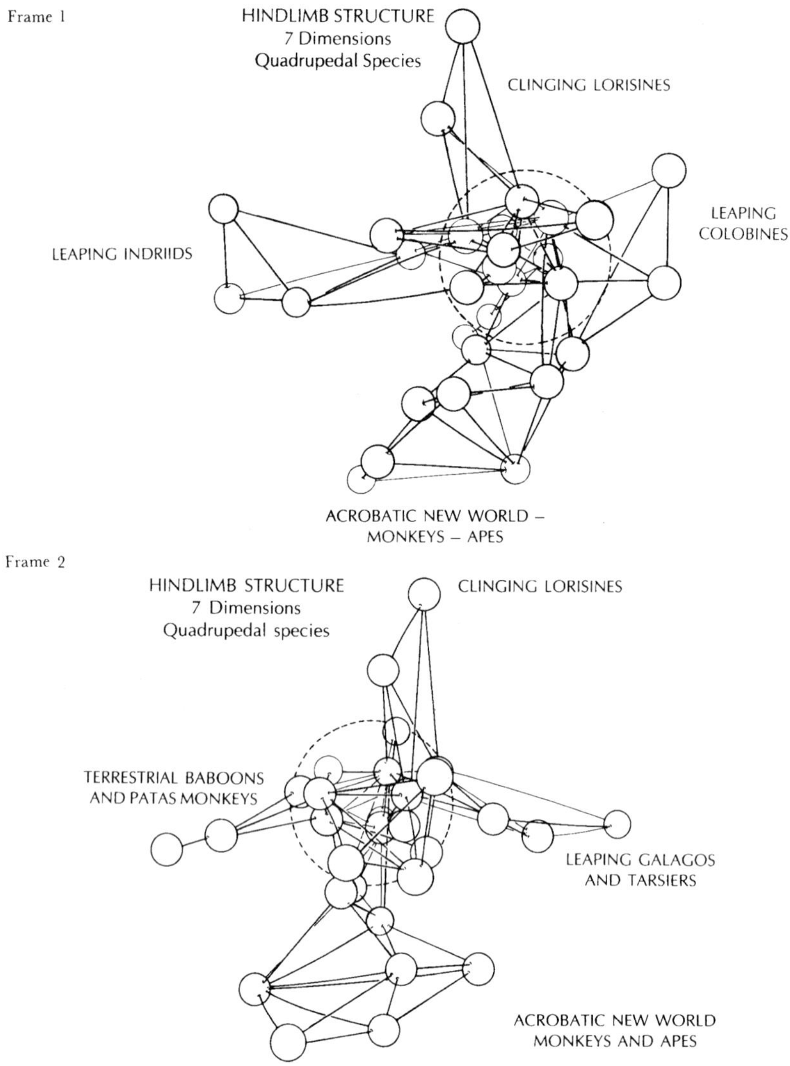

Fig. 14. More real data showing star-shaped distributions.

can be achieved by these methods is of heuristic value in any studies of biological (or indeed, any other) forms, patterns and textures.

Thus, complex forms, patterns and textures can be analysed using the method of 'Fourier transformation'. Though today this is always done computationally, when I was younger, in the days before powerful computers and programmes for calculating Fast Fourier Transforms, it could be done using an optical bench and a laser. One would never do it that way today. The method was technically quite hard. If a student dropped a book in the room, or a truck went past the building outside, the optical bench would be disturbed, and back we would all have to go to recalibrate it. It was also unconscionably long; at 22 feet the only way the Dean was about to give me space for it was to put a wall length-wise down a corridor. And because of the vibration problem, we had it mounted on automobile shock absorbers!

The optical way of describing the technique is much easier for a visual person (most biologists) to understand rather than trying to work one's way through Fourier mathematics with its double integral. Even just one integral was enough, in the old days, to frighten most people. For that reason I am giving the optical description here first.

One-dimensional Fourier analysis is what one sees in a rainbow. (In England, the rainbow comes after the rain; here in Perth the rainbow comes first; something to do with being in the antipodes, or Coriolis forces, I suppose.) A single beam of white light from a slit (one-dimensional, as it were) is broken down into its component coloured wave lengths (spectrum) using the refracting properties of a prism (Fig. 15). The equivalent process for a cone of light (rendered parallel by a lens system) which contains information about the two-dimensional area of an original pattern it has traversed is carried out by using a biconvex lens instead of a prism. The 'Fourier spectrum' is the intense spot of light at the 'focus' of the lens. Figure 16 shows the two-dimensional analogue of the prism. A camera at the focus gives a picture of the spectrum; after the second lens it gives a picture of the image of the original; after the third lens, the spectrum again; after the fourth lens, the image again; and so on.

Figure 17 shows how one can get information about a grating, its orientation and size, from the spectrum or transform. (The transform looks like the pictures from X-ray crystallography; indeed, this is the same process.) These two examples give us further information.

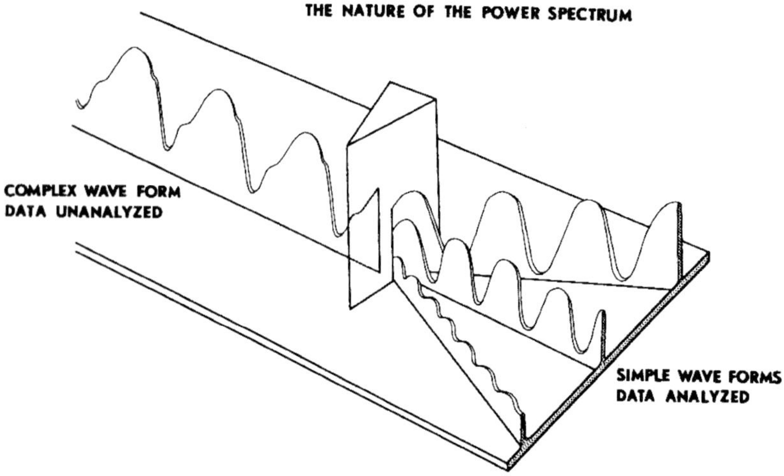

Fig. 15. The optical description of one-dimensional Fourier analysis — splitting a single beam of light.

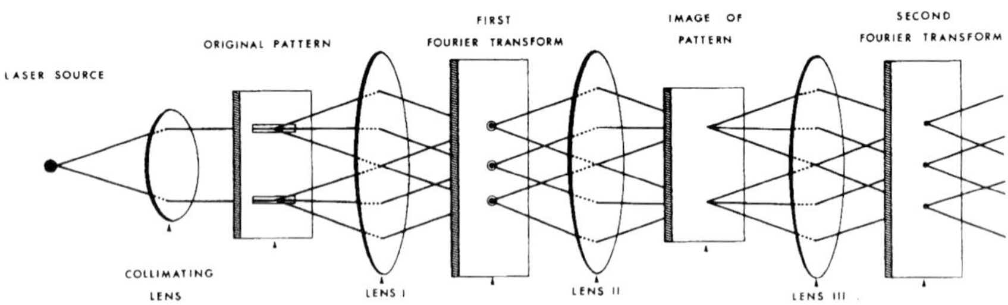

Fig. 16. The optical description of two-dimensional Fourier analysis — splitting the information in the area of a two-dimensional pattern. A second lens reconstitutes the original pattern, a third a second transform, and so on.

Orientations in the original picture are transformed by 90° in the spectrum (i.e. the lines of the grating at, say, 120° change to the line of the spectrum at 30°). Likewise, size information in the original is transformed through inversion in the spectrum (i.e. the small distance between the lines of the grating becomes a large distance between the central ray and the first harmonic — the first dot in the transform).

Figure 18 shows how the technique works in examining two more complicated, but still very simple patterns. The changes in size and orientation are readily apparent.

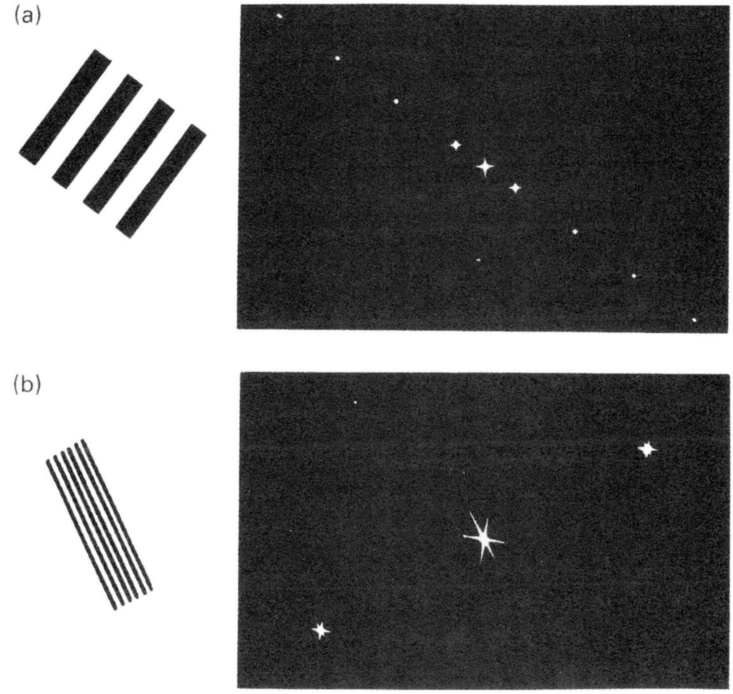

Fig. 17. Gratings and grids and their transforms (Lipson, 1972).

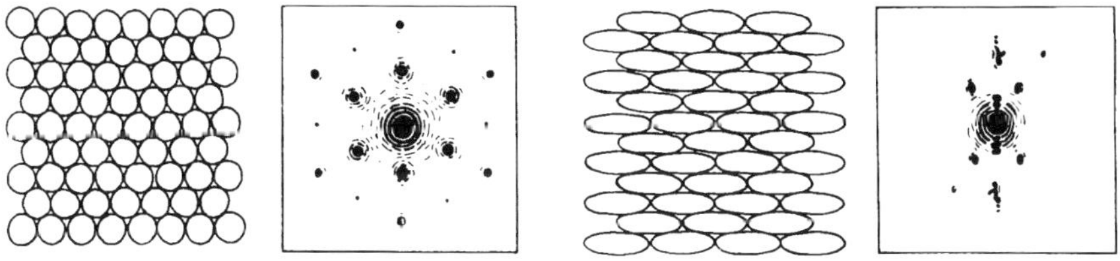

Fig. 18. A regular 'honey comb' and a squashed 'honey comb' and their transforms (Lipson, 1972).

I was first introduced to this technology by John Davis of the Kansas State Geological Survey. Of course, he was using it to characterise thin sections of rocks. Thus, Fig. 19 shows sections of rocks that are different. Figure 20 shows their transforms. It turns out that it is far easier to group different rocks by their transforms (a child could do it)

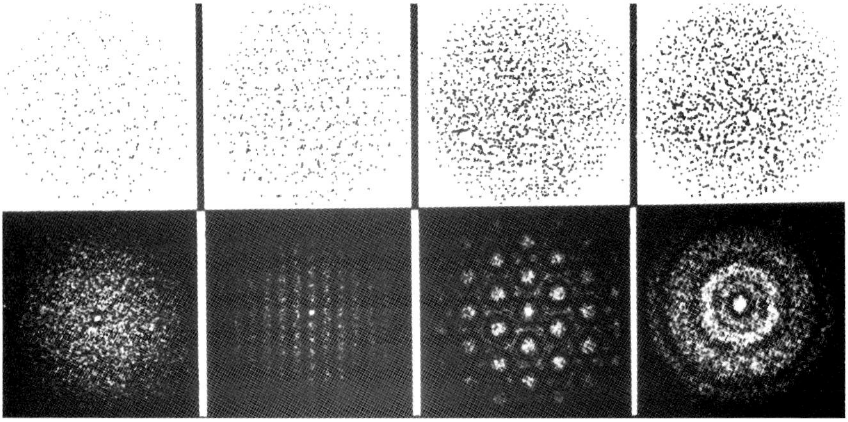

Fig. 22. The transforms of the dots in Fig. 21. The first transform is generally circular indicating that the pattern in the plot is essentially random. The second transform has vertical bars indicating that the second pattern of dots has a major horizontal trend within it. With the benefit of hindsight, this is evident to careful examination by eye. The third and fourth patterns of dots are obviously more complex. Their transforms, however, indicate completely different patterning unsuspected from the original plots (Lipson, 1972).

The technique turns out to be even more powerful in the study of textures. Thus, fabrics can be analysed in this way (Fig. 23). The transforms of the patterns of the first two fabrics show the quantitative differences between them though they are obvious enough without measurement. However, the transforms, given in the dots in the spectrum, are measures of the tightness of the weave, and can be used as a mask for quality control in a textile factory. It is much easier to use the transform as a mask rather than take continuous statistics of the fabrics themselves in order to control tightness of weave, etc. The third picture and its transform are from an extruded material, not a weave at all.

The technique was used in the early days in the recognition of cloud patterns. Thus, Fig. 24 shows pictures of clouds and their transforms. It is much easier to recognise the transforms rather than the original clouds.

All kinds of patterns can be so studied. Through the courtesy of anthropologist Joel Gunn, I was able to examine outline drawings of

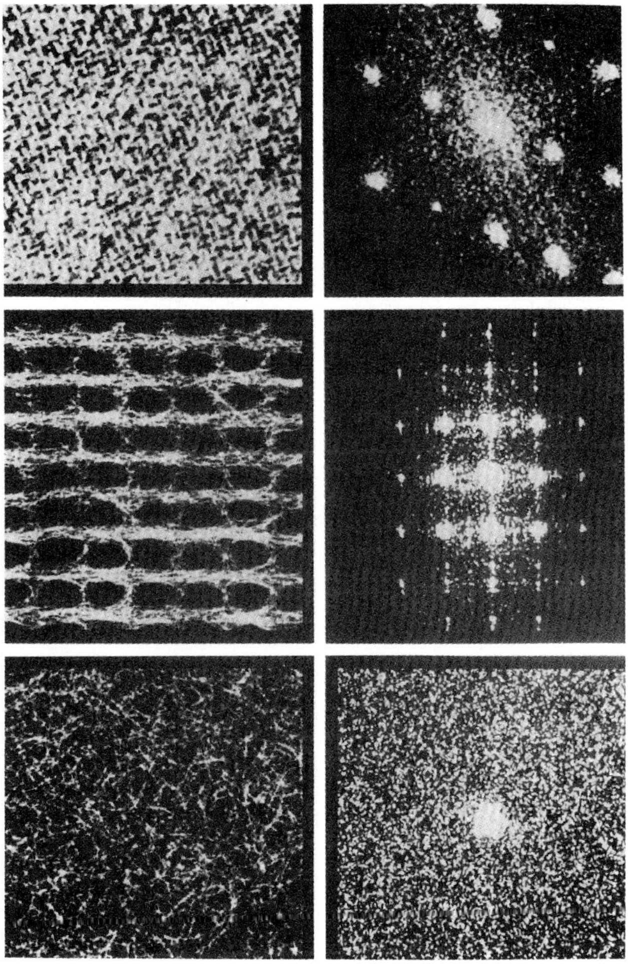

Fig. 23. Photographs of fabrics (left column) and their corresponding transforms (right column). The last fabric is not a weave but a man-made fabric the fibres of which are essentially random (Lipson, 1972).

micro-blade cores from Divostin, in the old Yugoslavia. It seems much easier to see information in the transforms rather than making many measurements of the originals (Fig. 25).

One very interesting example is the analysis of cellular nuclei. It can be difficult to recognise malignant as opposed to benign cells. Certainly

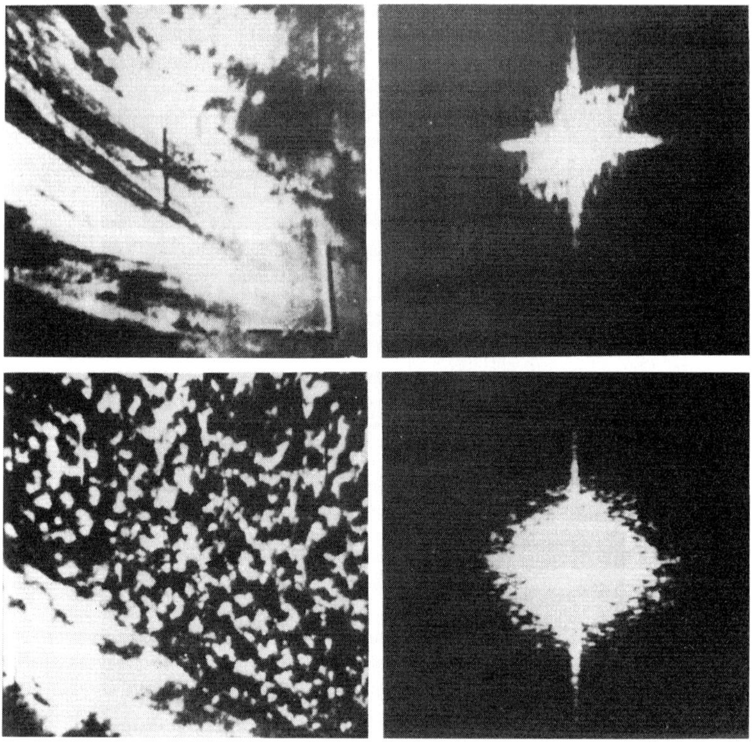

Fig. 24. Clouds patterns and their transforms (Rosenfeld, 1969).

technicians have to be trained to do it. Figure 26 shows the original and the transforms of two such nuclei, one benign and one malignant. The pictures are enigmatic. But the transforms are completely different. Further, the transform of a model malignant nucleus (based upon a specific hypothesis of the investigator) gives a transform not unlike the actual malignant nucleus. Again, technicians might find recognition of transforms much easier than recognition of the original nuclei.

Creation by Subtraction!

Normally one thinks of creating something by adding new information. How about 'creating' things by subtraction? For example, in

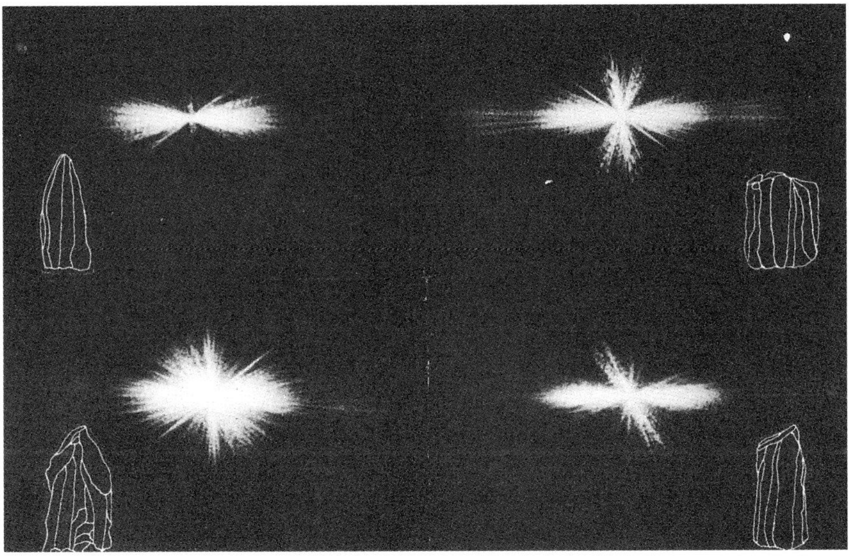

Fig. 25. Stone tools and their transforms (Courtesy, Joel Gunn).

(a)

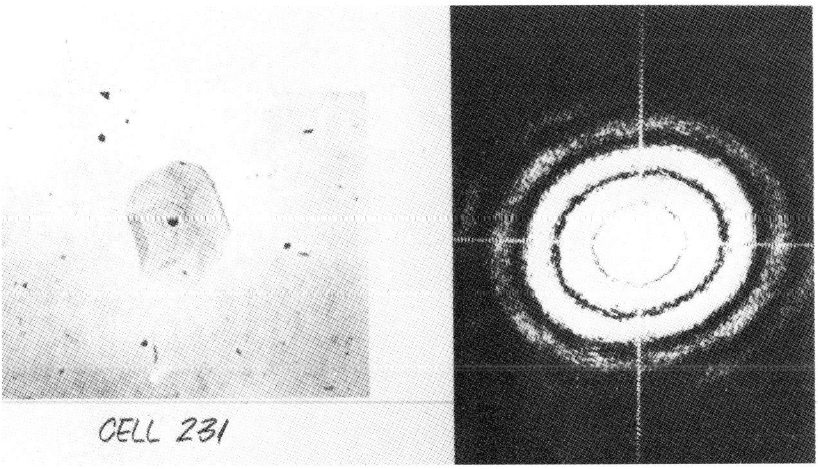

Fig. 26. The patterns of cell nuclei, (a) normal, (b) malignant, (c) theoretical, and their transforms.

(b)

(c)

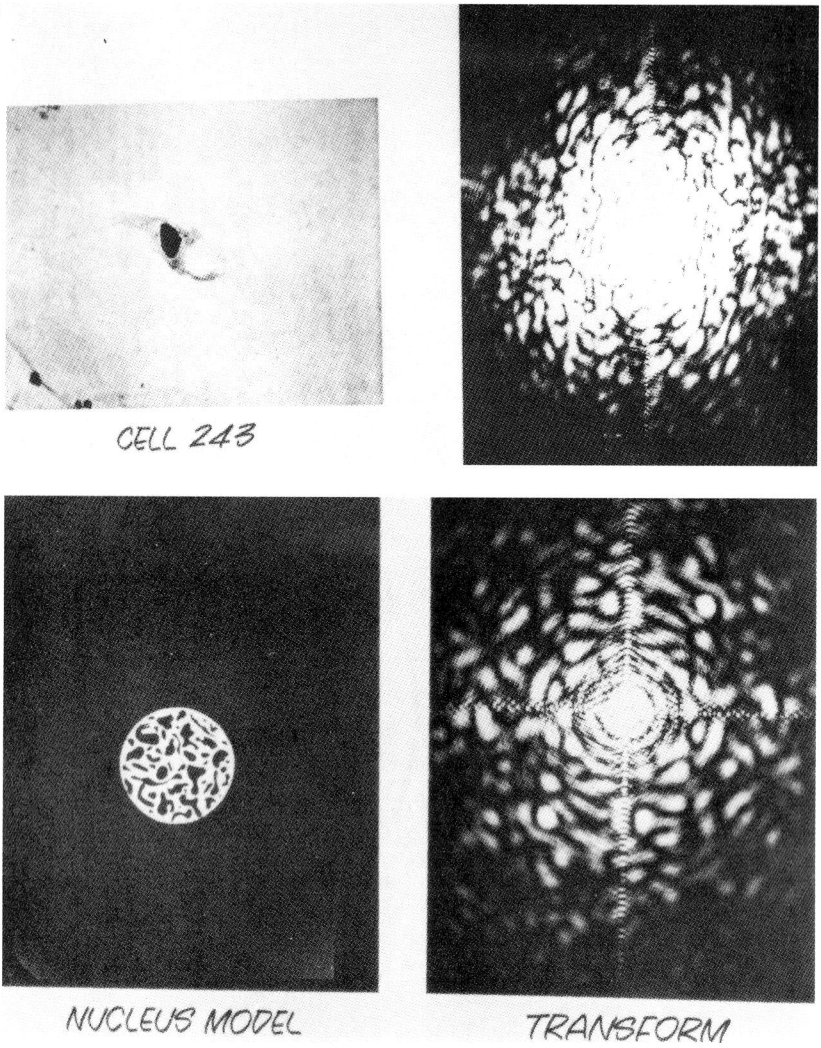

Fig. 26. (*Continued*)

those early days when Pincus was demonstrating his optical data analysis equipment, we were shown a curious ruffled-looking picture. No one knew what it was. Pincus explained. It was a military picture that had just been declassified; a picture of the tops of trees in a jungle as photographed from the air; a picture from near the Panama Canal (but presumably preparatory to jungle warfare somewhere else!). It was actually hard to see that these were the tops of trees. Pincus then used a slit as a filter to subtract all directions in the transform except one. In most directions of the slit nothing much changed in the reconstructed image. But at one particular angle (SW to NE) five lines could be seen traversing the picture. These represented damage done to the tops of the trees dozens of feet above the ground by the passage of five tanks through the undergrowth at ground level. What a powerful method!

This, thus, leads on to the matter of interfering with the optical process by subtraction. Given the nature of the optics it is clear that an image can be reconstructed after a filter has been put in place at the transform plane of the optical bench. (As will be explained later, this can be done in spades with computational Fast Fourier Transforms because one can use mathematical filters that could not be constructed as physical filters in the optical apparatus.) But again, the optical analogue is easier to understand.

Thus, Fig. 27 shows at top right, a picture of a hole in a plate illuminated by light from the back. The top-left frame shows the transform of that hole. Middle left shows the transform with the high-frequency elements (small features — remember the inversion!) removed; only that part of the transform representing the low-frequency elements, the central portions, are conserved. The middle-right picture shows the reconstruction that can be made from that subtracted transform; the large structure, the hole as a whole, is present; the small element, the edge of the hole, is absent. This is clearly a fuzzy hole; or if you like, a hole without an edge. Bottom left shows the original transform with a filter in place to block the low-frequency elements and allow through only the high-frequency elements (representing the smallest parts of the original hole). Bottom right shows the reconstructed image that can be made from this transform; an edge without a hole! If only ordinary dissection were so easy! One can see just how interesting and informative can be the process of Fourier subtraction.

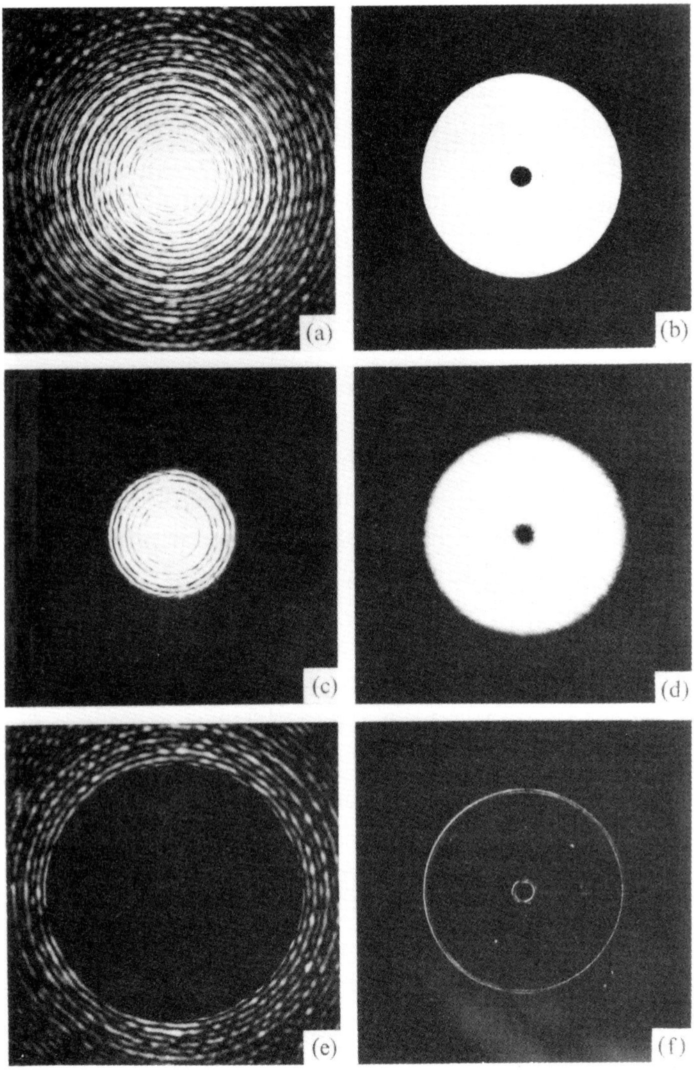

Fig. 27. A hole in a plate, its transform, and the effects of subtractions.

Figure 28 shows another example of Fourier subtraction. The top frame shows the original picture of a standard grating. The lower left frame shows a poor reproduction of that grating (achieved by adding noise to it, this was done by scattering black and white rug fibres over

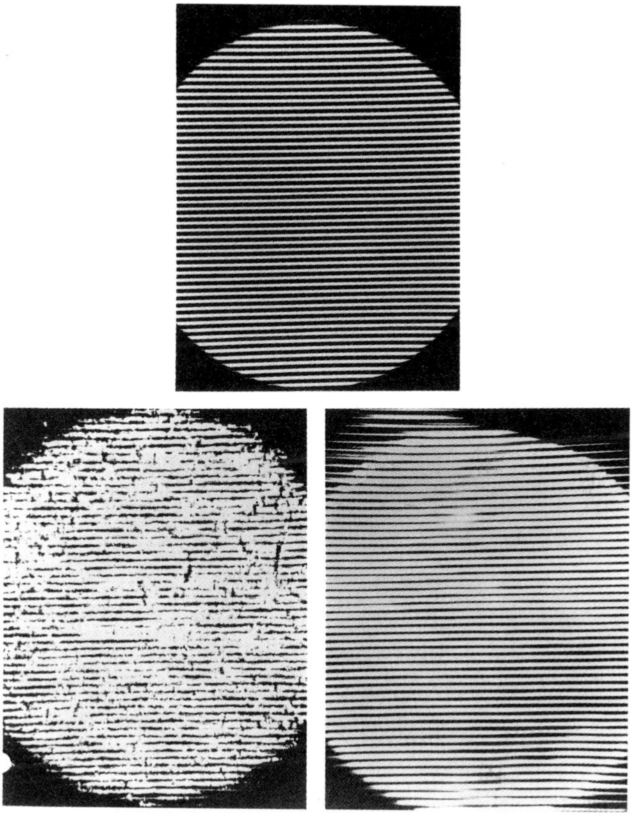

Fig. 28. A grating and the effect of modifying the original picture (Shulman, 1970).

the picture). The lower right frame shows the reconstruction that can be made from the left frame by subtracting from it the transform of scattered rug fibres alone. The lower right picture is a remarkable reconstruction of the upper. How wonderful it would be if one could subtract the transform of the matrix in which a fossil was embedded from the transform of the matrix plus fossil. Would one get a far better reconstruction of the fossil?

How about medical examples! Take a picture (say a mammogram) from a woman made last month and make a filter from its transform. Would using that filter to subtract from this month's mammogram, reveal only the difference between the two? The smaller the difference, i.e. the shorter the time between the mammograms, the more the difference would stand out because of the inverse factor in the Fourier transform. No doubt there would be technical difficulties. I am hoping that a new colleague (Ms Jo Firth, CEO of a breast imaging facility) may be able to take this up. Certainly she is excited about the possibility. And how brilliant a possibility it is! Such an idea can be applied to any clinical investigation where the result is a picture with texture, especially when that texture is a bit woolly.

Of course in my world, it is the texture of bone that is important. Figure 29 shows how sensitive the method can be in a radiograph of a small piece of a tibia. The enlarged view (upper left) shows mainly horizontal shadows. At upper right, is its transform; this shows no particular, easily recognisable orientational features, though there seems to be a slight horizontal predominance. In the bottom-left frame, however, the same transform is seen with a vertical filter in place to remove (subtract) these horizontal components. At bottom right, is the picture from the subtracted transform. The arrow indicates a small radio-opaque structure. It is present in the original picture but I challenge you to find it; it cannot be 'seen'. Subsequent histology of the specimen showed that it was a small ivory osteoma only 1 mm in diameter!

Figure 30 shows how the technique could be used to improve radiographs (not, perhaps, needed today, other techniques are now available). The original radiograph, left frame, shows quite clearly an abnormal ureter curling around in its passage from the kidney to the bladder. The right frame shows the reconstruction after optical subtraction to allow only the high-frequency elements to pass (the edges of structures). It shows the coiled ureter much more clearly; and also an enormous amount of detail in both kidneys, and in the vertebral column. Nothing is being created here. The details in the originals are just hidden by the low-frequency elements: real cases of 'now you see it, now you don't'.

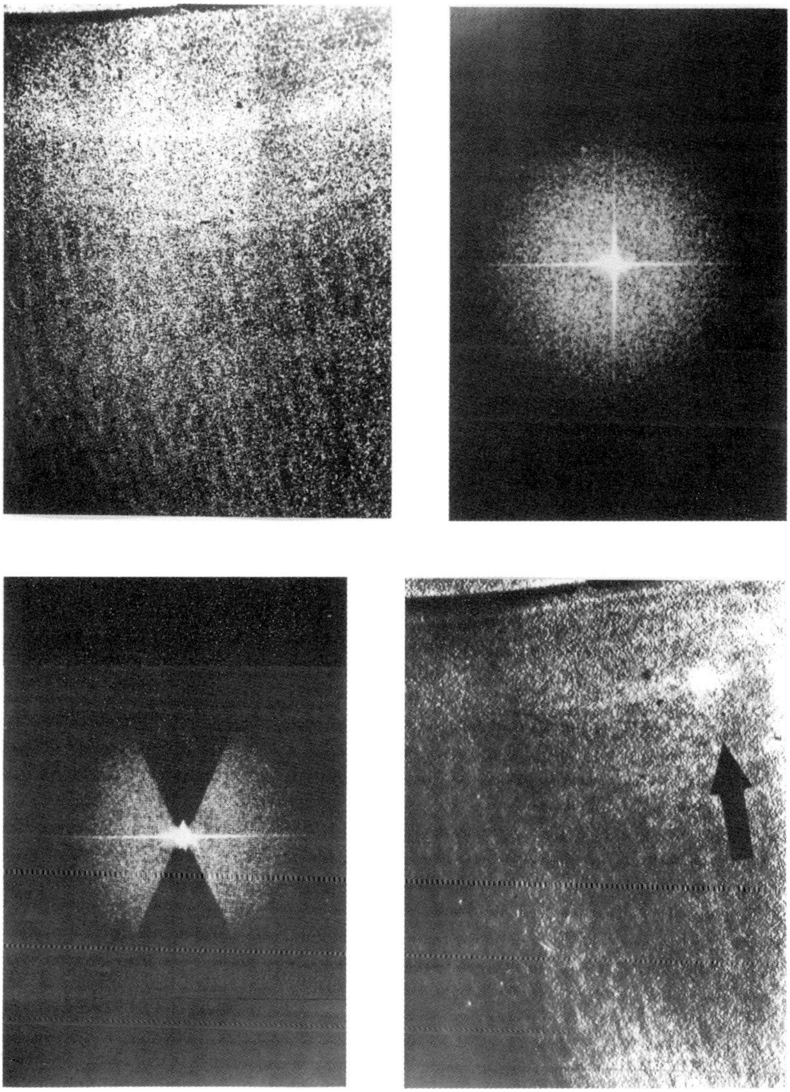

Fig. 29. An enlarged view of a radiograph of a small piece of tibia (upper left) shows mainly horizontal shadows. Upper right is its transform showing no particular orientated features. Bottom left is the same transform with a vertical filter in place to remove (subtract) the horizontal components of the original picture. Bottom right shows the reconstruction of the tibia from the subtracted transform. The arrow indicates a small feature. This is present in the original picture but cannot be 'seen' (Becker *et al.*, 1960).

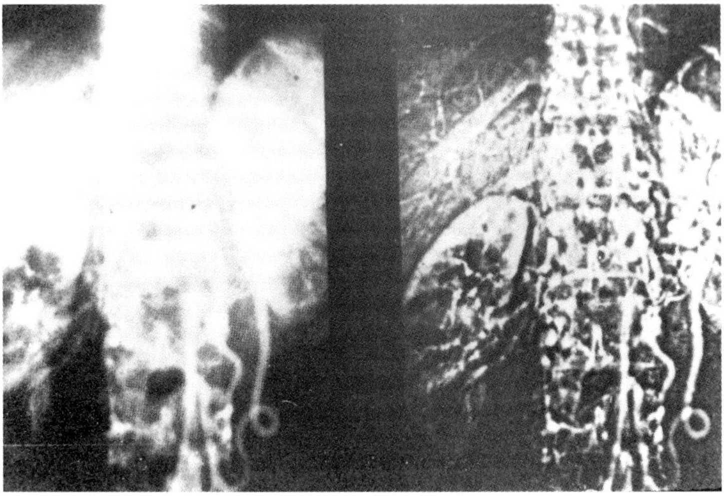

Fig. 30. Left frame shows a standard abdominal radiograph (from many years ago and therefore not of particularly high quality). Right frame shows the reconstructed radiographic when all the low frequency elements of the radiograph have been subtracted to reveal the edges of objects. A great deal in more detail can be seen in the reconstruction. Of course, nothing has been added; everything is present in the original radiograph; it is just revealed by the subtraction process (Pfeiler, 1969).

Apes and Humans

Using the original optical method on sections and later radiographs of bones (individual vertebrae) shows just how useful the method is for studying bone architecture. The optical Fourier transforms of lateral radiographs of vertebrae of various apes and humans show (work carried out by Dr Harold Yang, Oxnard and Yang, 1981) that most of the internal structures are orthogonally arranged (as we would expect, see earlier chapter and Fig. 31). But transforms show that the vertebra of the orang-utan is different. The radiograph looks similar, but the transform indicates that a different pattern of trabecular shadows exist. Again, this was confirmed by making sections (Fig. 32).

We suggested that the orthogonal arrangement in the African apes and in humans might be because, on average, most of the time, the principal stresses and strains generated by movement in these creatures is, most of the time, quite similar. It would be this

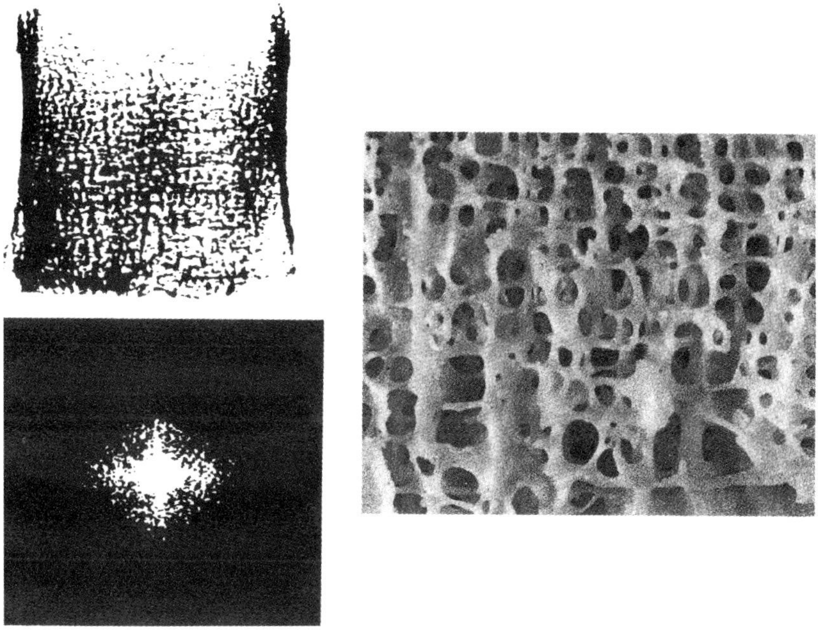

Fig. 31. Radiograph showing the transform and section of a chimpanzee lumbar vertebra. The picture is the same for gorillas and humans (Oxnard and Yang, 1981).

average to which the cancellous bone might be adapted. But in orang-utans, the mode of movement is very different. It does not involve similar postures and movements most of the time. The animal can hang by one arm, by two arms, by two arms and a leg, by two legs, by two legs and an arm, by all four limbs, or even just one leg during its acrobatics in the trees. In other words, the patterns of stresses in orang-utans would vary enormously more differently during locomotion and posture, than in African apes and humans. The best way to support such a wide array of stress/strain patterns might be a somewhat hexagonal arrangement as found here. This idea has yet to be tested.

One particular feature was especially interesting. The cancellous structure of the second human lumbar vertebrae (as shown in a sagittal section) could be very clearly distinguished from the fourth. Though not at all obvious in the vertebral radiograph, the

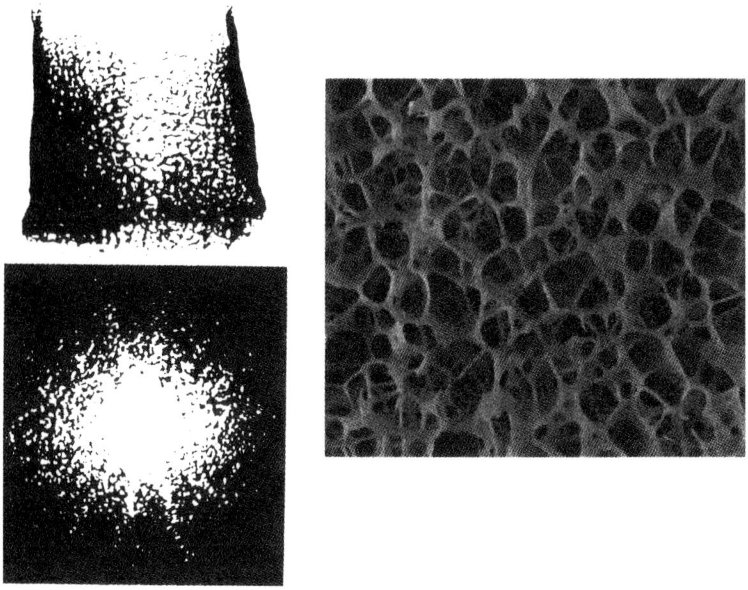

Fig. 32. Radiograph showing the transform and section of an orang-utan lumbar vertebra (Oxnard and Yang, 1981).

transform clearly shows that the second lumbar vertebra has specific items orientated at about 30° to the horizontal that are not present in the fourth. Again, we tested this by looking at scanning electron microscope picture of that vertebra; the off-orthogonal approximately 60° elements were quite easy to identify. They are not present in the fourth lumbar vertebrae (Fig. 33). We suggested that this might be due to the fact that the line of gravity passes through the body of the fourth lumbar vertebra in humans, hence most trabeculae orthogonal, but anterior to the second lumbar vertebra, hence the additional angled trabeculae because of the curvature of the lumbar column. This idea, too, has yet to be properly tested.

Normality and Disease

Later, we used computational Fourier transforms to examine radiographs of thick sections of bones of people of known ages (Fig. 34 and

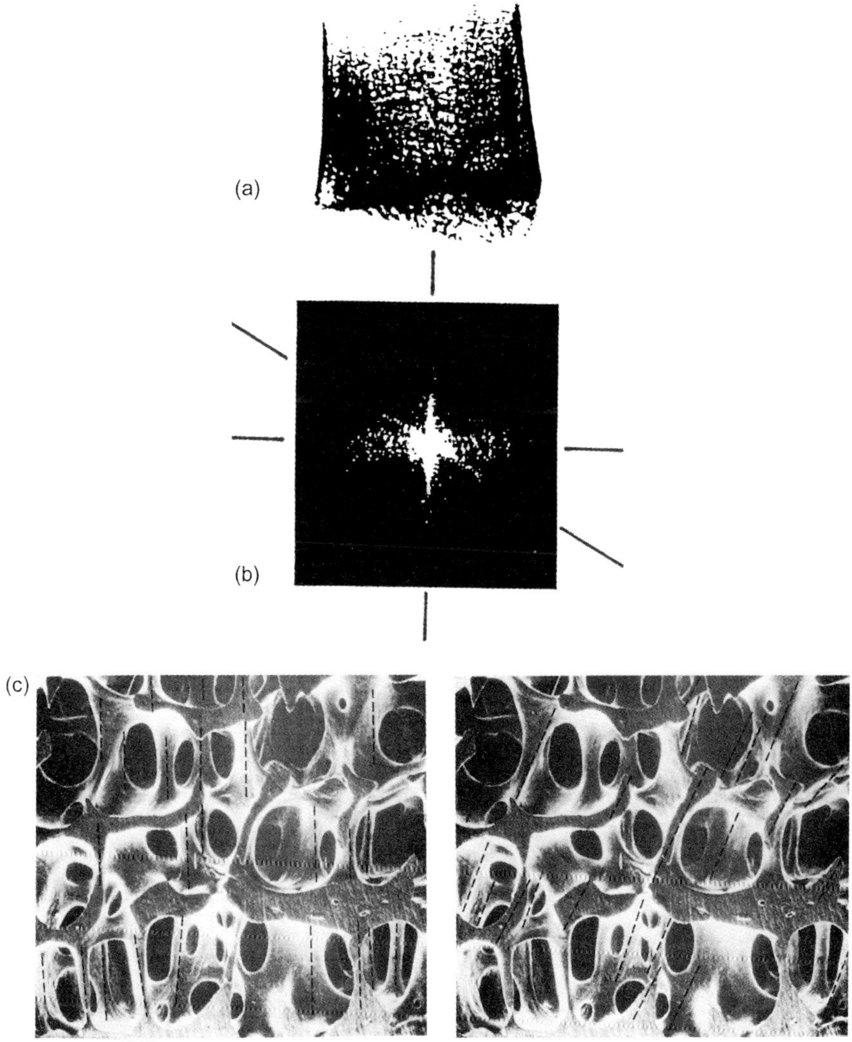

Fig. 33. Radiograph showing the transform and section of the second lumbar vertebra in a human.

work carried out by Dr Alanah Buck, Buck, 1998; Oxnard and Buck, 1996). These show the stigmata of osteoporosis, loss of the cruciate form of the transform, more evident in middle aged females than in middle aged males, though in old age, both are about equally affected.

As a final measure of the usefulness of the technique, we looked at samples of young individuals of each sex; a situation where we assumed the sexes would be approximately similar. To our surprise analysis of the transforms showed that there was a dichotomy between males and females (Fig. 35). Because we could not readily see this in the transforms of the whole vertebrae, we decided to

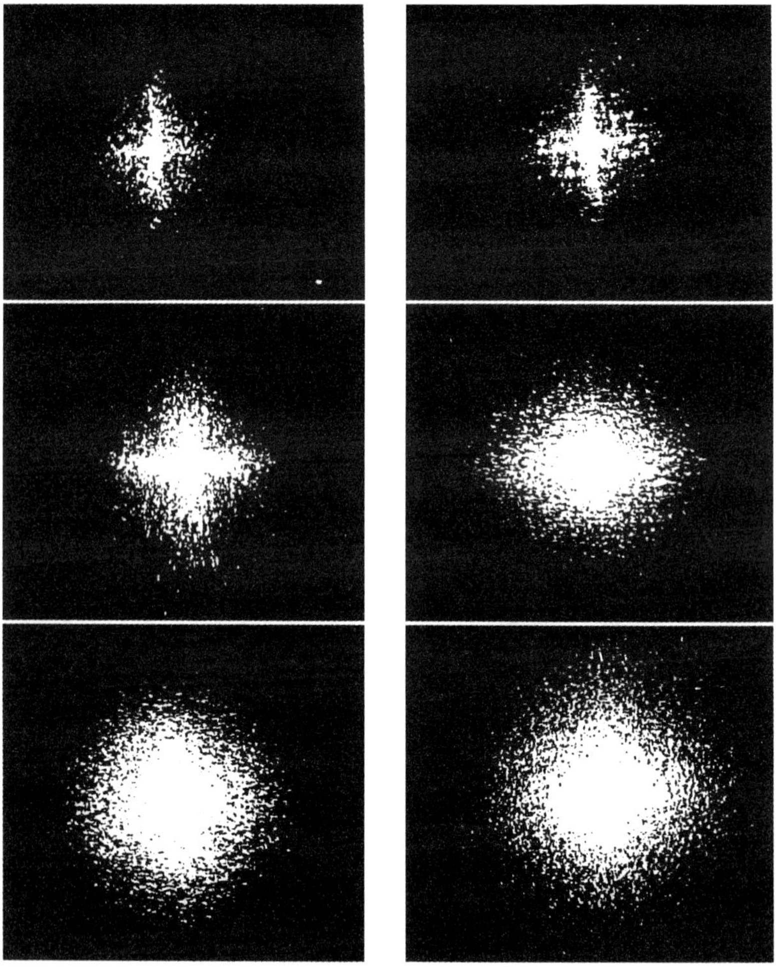

Fig. 34. The first column shows computational transforms of young, middle-aged and old male vertebral patterns, the second column for equivalent female vertebrae (Oxnard and Buck, 1996).

examine transforms of quadrants of vertebrae (Fig. 36). The reason for the dichotomy then became obvious. There was a difference between the third quadrant in males and females not evident in the other three (Fig. 36). A return to the transforms themselves showed the reason. All four quadrants in males (not figured) show transforms shaped like a Maltese cross. They are the same as three of the quadrants in females. But in females, the fourth quadrant (the anterior superior quadrant) showed a transform just like the transforms in the older women with osteoporosis (Fig. 37). This

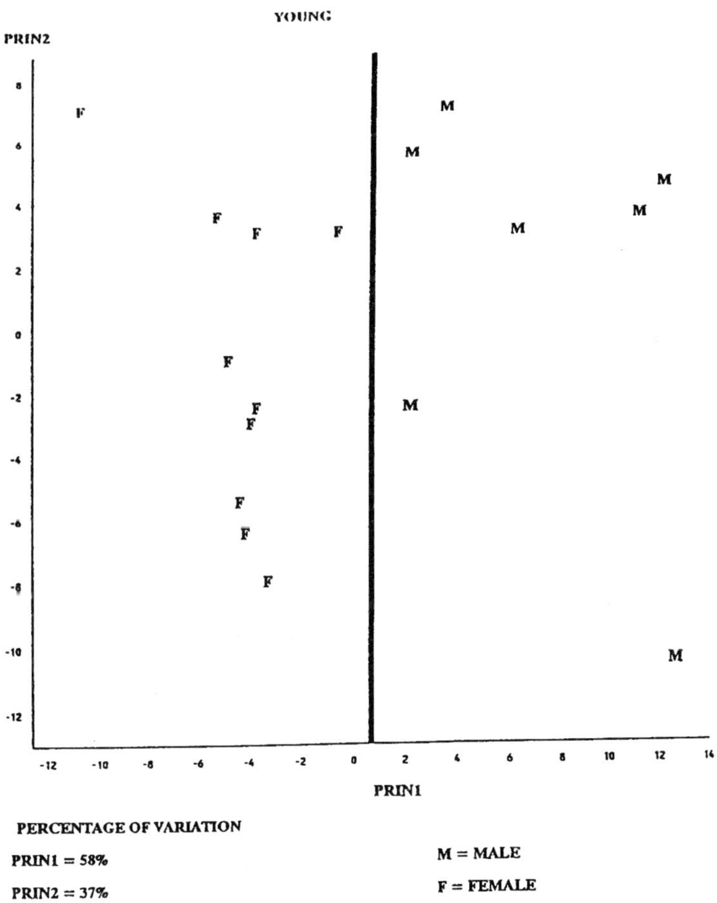

Fig. 35. Young males differ from young females (Oxnard and Buck, 1996).

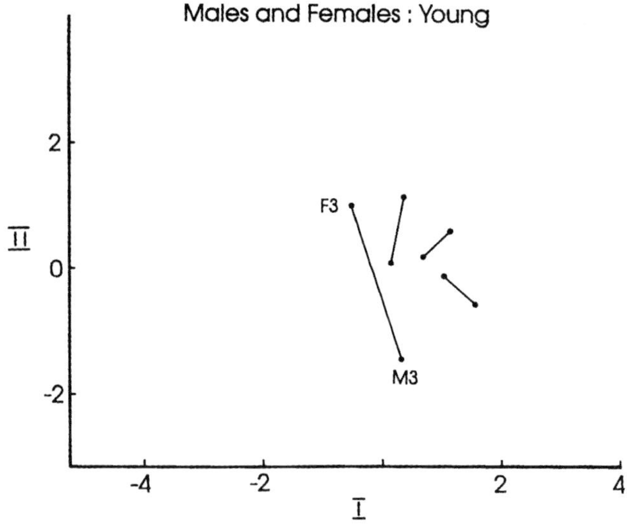

Fig. 36. The result of Fig. 35 is due to a difference in the anterior superior quadrant of some of the young females (Oxnard and Buck, 1996).

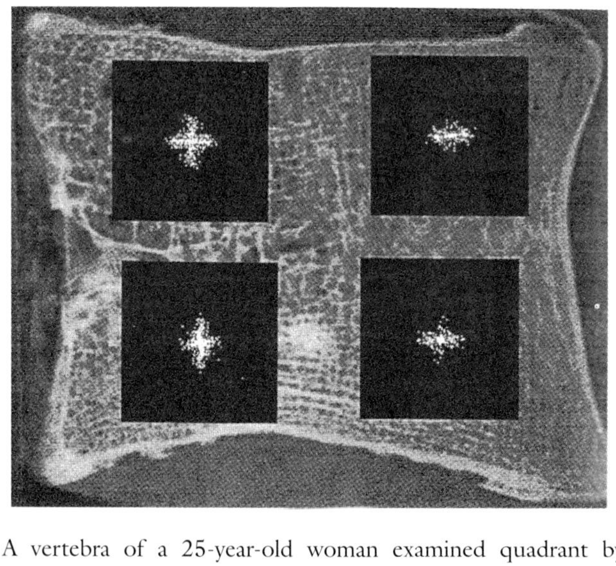

Fig. 37. A vertebra of a 25-year-old woman examined quadrant by quadrant. Three of the quadrants show a maltese cross just like in young males. But a fourth quadrant — upper right — shows a pattern where the cross has been largely eliminated — it looks just like an old (osteoporotic) female.

does not mean that the particular young females had osteoporosis. But it may mean that the beginnings of the process were evident in them. This possibility is supported by the fact that, in older women, the later complication of vertebral fracture very often occurs to greatest extent in the anterior segments of the vertebra thus giving the typical 'wedge' fractures. If this proposition is true then, in these young women, one would be able to give 'lifestyle advice' (e.g. exercise, take dairy products, stop smoking and consider HRT when older if necessary) and this might be a useful preventative.

Conclusion

The final conclusion from all these studies is a plea for us to keep our minds open. So frequently nowadays, one sees demands for 'world's best practice' to be followed, for the use of the 'gold standard' or nothing, for the one 'true' line. This standardisation, so frequently advocated, especially by those who are the supervisors in academia, not only prevents us from seeing what other methods may find, it also prevents us from asking new questions. I have always had a distaste for received scientific dogma; I have always wanted to test out the conventional wisdom; I have always had more fun in showing that something is wrong. I have had even more fun in finding something that was unexpected. Especially I have derived enormous pleasure from finding new complexity when everyone else saw nothing but simplicity, and new simplicity when everyone else saw only hopelessly complexity. These have been some of the greatest of my pleasures in the few pebbles on my beach.

References

Becker HC, Meyers PH, Nice CM, Laser light diffraction, spatial filtering, and reconstruction of medical radiographic images: Preliminary results, *Ann NY Acad Sci* **157**: 465–486, 1960.

Bookstein FL, The measurement of biological shape and shape change, *Lecture Notes in Biomathematics*, Vol. 24, Springer Verlag, New York, pp. 1–191, 1978.

Bookstein FL, *Morphometric Tools for Landmark Data*, Cambridge University Press, New York, 1991.

Buck AM, An investigation of vertebral cancellous architecture along the thoraco-lumbar column in humans using Fast Fourier Transforms: An anatomical study, Doctoral Thesis, University of Western Australia, 1998.

Duda RO, Hart PE, *Pattern Classification and Scene Analysis*, Wiley Interscience, New York, 1973.

Huxley JS, *Problems of Relative Growth*, Methuen, London, 1932.

Kendall DG, Shape manifolds, Procrustean metrics and complex projective spaces, *Bull Lond Math Soc* **16**: 81–121, 1984.

Koestler A, *The Sleepwalkers: A History of Man's Changing Vision of the Universe*, Penguin Books, London, 1968.

Lipson H, *Optical Transforms*, Academic Press, London, 1972.

O'Higgins P, Cohn MJ, *Development, Growth and Evolution*, *Linnean Soc Symp Series*, Vol. 20, Academic Press, London, 2000.

Oxnard CE, Functional morphology of primates: Some mathematical and physical methods, *Burg Wartenstein Symp* **48**: 1–42 (1970a). Also published in Tuttle RH (ed.), *The Functional and Evolutionary Biology of Primates*, Aldine Atherton, Chicago, pp. 305–336, 1972a.

Oxnard CE, The use of optical data analysis in functional morphology. Investigation of vertebral trabecular patterns, *Burg Wartenstein Symp* **48**: 1–30, 1970b (additional paper). Also published in Tuttle RH (ed.), *The Functional and Evolutionary Biology of Primates*, Aldine Atherton, Chicago, pp. 337–347, 1972b.

Oxnard CE, *Form and Pattern in Human Evolution: Some Mathematical, Physical and Engineering Approaches*, University of Chicago Press, Chicago, 1973.

Oxnard CE, *The Order of Man: A Biomathematical Anatomy of the Primates*, Hong Kong University Press, Hong Kong, 1983, Yale University Press, New Haven, 1984.

Oxnard C, Buck A, Bone, age, sex, and osteoporosis, *Persp Hum Biol* **2**: 63–82, 1996.

Oxnard CE, Yang H, Beyond biometrics: Studies of complex biological patterns, *Symp Zool Soc Lond* **46**: 127–167, 1981.

Pfeiler M, Image transmission and image processing in radiology, in Graselli A (ed.), *Automatic Interpretation and Classification of Images*, Academic Press, New York, pp. 399–415, 1969.

Power PC, Pincus HJ, Optical diffraction analysis of petrographic thin sections, *Science* **186**: 234–239, 1974.

Rosenfeld A, *Picture Processing by Computer*, Academic Press, New York, 1969.

Shulman AR, Optical Data Processing, Wiley, New York, 1970.

Slice DE, *Modern Morphometrics in Physical Anthropology*, Kluwer Academic/Plenum, New York, 2005.

Thompson D'AW, *On Growth and Form*, University Press, Cambridge, 1917.

Yang HC, The development of optical analysis for the investigation of radiographic patterns: Its application to the study of the lumbar vertebral column in man and the great apes, Doctoral Dissertation, University of Chicago, 1978.

Chapter 6

The Origins of Ancient Humans: 8,004,004 BC!

Fossils and Molecules and Time

It has always fascinated me that, as the evolutionary story unfolds, each new fossil find is hailed as the key fossil, often also the oldest fossil, certainly the most important.

Some of the robust australopithecines (*Australopithecus robustus*) were, when they were first discovered, postulated as early human ancestors around 2 million years ago (Broom, 1938; Broom and Robinson, 1952). Today they are generally recognised as lying on a side-branch of the mainline to humans. A particular hyper-robust australopithecine initially named *Zinjanthropus* was later called *Australopithecus boisei*, and was (at allegedly 1.0–2.5 million years) also posited close to human ancestry; so close, indeed, that a look of wisdom permeated the reconstructed portrait as drawn by Neave Parker for Louis Leakey in the pages of *Nature*. Today, it, too, is relegated as an extinct side issue. Many colleagues now prefer to return to an older name and call the robust forms *Paranthropus*.

A similar fate is starting to befall the gracile australopithecines, originally *Australopithecus africanus*, from South Africa dated at around 3 million years. Though long ago in 1977, I wrote an article in Nature with the title 'Australopithecines: Grounds for Doubt' it has not been until quite recent years that the South African australopithecines have been removed, by many workers, from a direct link with humans. Even so, these fossils are still generally recognised as within the bush of lineages related to humans though now usually placed on their own distinct and extinct twigs of that bush not leading to humans *per se*. The reason for this change of view, however, was not my 1977 study of the Sterkfontein fossils in their own right (that study has been largely ignored, see Cartmill, 2004), but rather the later discovery of even older fossils at about 3–4 million years, australopithecines from the Afar Valley in Ethiopia (*Australopithecus afarensis*). These are now often placed at the divergence between lineages leading to later australopithecines and paranthropines on the one hand, and more human species on the other.

Whether eventually even *Australopithecus afarensis* will be relegated from the direct human lineage is problematical but at least possible. It is, in fact, anatomically enormously similar to *Australopithecus africanus* though older, and this, together with other new discoveries that are even older still, strengthens the possibility of that relegation. Some of the other new discoveries are: *Kenyanthropus platyops,* at nearly 4 million years, *Australopithecus anamensis,* at beyond 4 million years, *Ardipithecus ramidus,* at 5 million years, *Orrorin tugenensis* at 6 million years and *Sahelanthropus tchadensis* at nearly 7 million years. *Kenyanthropus platyops* is at least as old as *A. afarensis* but has been suggested as being more like some *Homo* specimens (especially a younger specimen from East Rudolf now named *Homo rudolfensis*) than it is like *A. afarensis.* Most of these fossils, though not yet fully studied scientifically, have already been entered into the human tree in positions that now exclude all of the australopithecines from a direct line to humans (e.g. reviewed in Carroll, 2005). So the saga continues!

Many fossils more recent than the australopithecines are also treated as in the human lineage. These include the *Homo rudolfensis* mentioned above (but some, not wishing to completely give up the australopithecine link, call it *Australopithecus rudolfensis*!). They include also: *H. habilis* and *H. ergaster,* at around 2 million years, *H. erectus,* from 2.5 to less than 1 million years, and then *Homo antecessor, H. heidelbergensis,* at 1 million years and less, *H. neanderthalensis* at around 0.75 million years and up to 30,000 years, and finally modern *H. sapiens* within the last 200,000 years (reviewed in Carroll, 2005). There seems little doubt that these species, whatever their dates, are closer to modern humans than the australopithecines.

Yet other changes of opinion have occurred. The Neanderthals were once thought to be so like us as to be named *H. sapiens neanderthalensis*. Indeed, we were urged at one time to believe that we might find ourselves sitting next to one in the bus (it seems, so often, that we are). They tend today to be relegated as much more distant relatives, even a different species: *H. neanderthalensis*.

All these views imply, however, that these fossils lie on evolutionary branches located within the human part of the tree. Even today when it has become fashionable (and welcome it is) to see, in the human evolutionary story, the fine twigs of a 'bush' rather than the single direct 'branch' leading to humans, it is only the human lineage that is bushy. Thus, there are thousands of fossils in the human part of the evolutionary tree but only one fossil chimpanzee has been reported on the basis of three teeth, two incisors and a molar, probably from the same individual (McBrearty and Jablonski, 2005). Most curiously of all, there are no species in the bushes that lead neither to humans nor apes.

Is all this because prior fossil hunters have thrown out the fossil pre-chimpanzees, or even the non-chimpanzee/non-humans that they found? Because they are mainly interested in the more important pre-human fossils! Of course not! It is perhaps because fossil pre-chimpanzees when found are not recognised as such (though, as cited above, McBrearty and Jablonski have noted one). Is it also perhaps because we tend to emphasise those data in the fossils that place them in the human part of the tree? We seem less able to recognise the data that could deny that hypothesis.

One explanation often given for the lack of pre-chimpanzee fossils is that remains of woodland or forest creatures are not well preserved. Yet this does not seem to apply to many other mammals including other primates such as monkeys. Another explanation may stem from our treatment of characters. Thus, it is also often assumed that many of the features of modern African apes are primitive characteristics from which humans have diverged. Could it be, instead, that some of the adaptations of modern chimpanzees (and also of course of modern bonobos and modern gorillas) are new features, not present in their progenitors? Instead of being assessed as old characteristics (e.g. knuckle-walking, polygyny, and unbalanced adult sex ratios) to be looked for in fossils near the root of the links with the pre-human species, perhaps these are relatively new features in chimpanzees and gorillas, evolved since the departure of these apes from their progenitors. Could this be why ape ancestors are not easily recognised? Could it be that at least some of the allegedly pre-human fossils that we already have should actually be placed in a chimpanzee-bonobo-gorilla bush? This is an alternative hypothesis that we rarely seek to test.

Finally, it has always surprised me that, whether on branches or in bushes, so few of the fossils (so far, none) are placed on branches (or in bushes) that are totally separate from both modern humans and modern African apes. It is true that the robust australopithecines are generally now regarded as lineages with no extant descendants. Even they, however, seem to be solidly ensconced in the human part of the story. Indeed, a few workers still look upon them as direct human ancestors. Yet in most animal groups extinct branches are the stuff of evolution. Why are there no fossils that are both non-human and non-chimpanzee? Could it be that some of the fossils that we already have should be placed as both non-human and non-chimpanzee in relationship? This is another hypothesis that we rarely seek to test.

With these irreverent thoughts go others. Why has it always been possible to produce attractive descriptions so easily latched onto by the popular media? Phrases like 'missing links', 'killer apes', 'aquatic origins', 'people of the lake', 'children of the ice' and most recently 'hobbits' all flourish. Yet there is no such thing as a missing link; or, to put it another way, there are thousands of missing links. Equally, there is no evidence for killer apes, though it is true that today's apes kill, even their own, hence the unbalanced adult sex ratios in apes. The aquatic origins story seemingly implying that they were some kind of primate seal is a total furphy that continues to exist; yet the localised influences of water in relation to river-sides, lake-sides and ocean-sides may well have been a part of the human evolutionary story that we seem to be unwilling to publish. 'People of the lake' and 'children of the ice' are literary allusions that make the telling of the stories so much more attractive. These descriptive phrases then merge truly into total fiction such as 'the mammoth hunters', although undoubtedly mammoths were hunted. The 'hobbits', allegedly *Homo floresiensis*, the most recent of these fanciful descriptors (and it looks as though they did have disproportionately large feet compared to ordinary humans, though with no evidence of hair on the tops!) may yet prove to be something completely unexpected (see a later chapter).

Studying the fossils is so fraught with subjectivity. The problems are enormous. The fossil hunters undertake a mixture of very, very hard work, in difficult conditions (rain, heat, mosquitoes, leeches, deserts, droughts), through prohibitions (not only local and even governmental interference, but often damaging, dangerous, and in a few cases lethal effects of poachers), and of unrewarding times (often years without recognition or proper support, even years without finding anything). They depend upon huge aliquots of serendipity mixed with unquestioned and quite remarkable expertise. Notwithstanding all this, there may, sometimes, be an 'interest' in the fossils that may get in the way of objective assessment.

Thus, the times of common ancestors provided in many molecular evolutionary studies depend, at some point or other, upon times estimated from studies of fossils. For instance, though molecular investigations sometimes use internal information for calculating times (information about possible rates of molecular change, for instance) most often such studies are calibrated using external measures derived from the study of fossils. Molecular evidence about the human/chimpanzee divergence time is often estimated using the much earlier times from studies of the fossils thought to be close to the common ancestor of all great apes and humans. Molecular assessments of the origins of modern humans are often calibrated using times only a little earlier than the oldest australopithecine fossils as an estimate of the common ancestor of humans and chimpanzees. Of course, in both cases the actual fossil common ancestors themselves are not known.

For example, the common ancestor of all hominoids is usually set at *circa* 25 million years ago (Ma), that is, slightly earlier than the earliest known hominoid fossils (e.g. *Motopithecus* at 20 Ma, Gebo *et al.*, 1997, though this is controversial). Likewise, the common ancestor of humans and chimpanzees is usually set at 5 Ma; that is, slightly earlier than the earliest known pre-human ancestors, particular quasi-australopithecines (e.g. *Ardipithecus ramidus* at 4.4 Ma, White *et al.*, 1994, 1995). The latest findings from Kenya (with a morphology claimed as being more bipedal than the australopithecines) dated unequivocally at 6 million years ago makes the 8 million year date an even more likely possibility. Yet pre-chimpanzee (but non-human) fossils at about these times have never been recognised!

With the advent of what seem much more scientific methods of investigating evolutionary relationships through collaborations involving palaeontological methods: geological dating, palaeo-ecology, palaeo-climatology and palaeo-environmental assessments, the data of the molecules, and the methods of cladistics, problems still arise. It cannot be just coincidence that, some six decades ago, when the australopithecines from South Africa were thought to be about 2.5 million years old, the molecular studies consistently came up with 3 million years for the putative human/chimpanzee common ancestor (Sarich and Wilson, 1976; but see Dene *et al.*, 1976). It cannot, equally, be just coincidence that, about four decades ago, when the australopithecines of the Afar (3.5 million years old) had been discovered and described, the molecular assessments started to provide dates of 4.0–5.0 million years for the putative human/chimpanzee common ancestor (Goodman, 1983; Sibley and Ahlquist, 1984; Hasegawa and Horai, 1991). These estimations were provided before the discoveries of *Australopithecus anamensis*, *Ardipithecus ramidus* and *Ororin tugenensis* implied that the date must be well beyond 5.0 million years. One aberration as late as 1984 postulated 1 million years as the date of the human/chimpanzee common ancestor (Hasegawa *et al.*, 1984); but aberration is surely just what that was! Now, in this last two decades even newer fossils (e.g. *Sahelanthropus tchadensis*) have been dated at up to 7.0 million years ago. As a result, some workers are starting to hint at as much as 8.0 million years for the human/chimpanzee common ancestor (e.g. Wood and Collard, 1999; Wood and Constantino, 2004). Some even more recent studies are willing to envisage 10.0 millions of years for that elusive ancestor (Arnason *et al.*, 2000). Whether this will be followed by further re-evaluations of the molecular data along the same lines is yet to be certainly seen, but 13 million years has also been suggested (again, Arnason *et al.*, 2000)!

Why should we be so bothered about the age of the chimpanzee/human common ancestor? After all, no common ancestor has actually been found (and it is highly unlikely, on statistical grounds that it ever will be found). However, because chimpanzees are our closest living relatives, estimates of the time of our common ancestry with them is used in the calibration of molecular studies of the variety of modern humans. For instance, if we use the 3.0 million year estimate of the 1960s and 1970s for the time of chimpanzee/human common ancestor, then mitochondrial DNA studies (mitochondrial DNA is largely inherited from mothers), had they been achievable at that time, would have resulted in a date for the maternal origins of modern humans of about 25,000 years ago. That would have been clearly denied by the fossils of those days, never mind of today.

If, however, we apply the 5 million year date of the chimpanzee/human ancestor as the calibrating date, then both mitochondrial DNA and Y chromosome analyses (Y chromosomes are inherited by males only from fathers) give around 150,000 years ago as the date of modern human origins. We rarely remember, however, the enormous limits on each side of that figure for both mtDNA and for Y chromosomes (Ayala, 1995; Fu and Li, 1995). Using times based upon the newest fossils (i.e. extending the chimpanzee/human common ancestor to 8 million years) more than quadruples (to 600,000 years) the maternal (mitochondrial DNA) estimate of the origin of modern humans (Ayala, 1995).

Likewise, if, in the Y chromosome investigations we take account of the size of the male breeding population (Fu and Li, 1995) then the paternal origin of modern humans can be pushed out to 700,000 years with a population of only 5000 individuals. The limits on such an estimate are quite wild: 114,000 to 721,000 years! This would make the new distant time for the origins of Neandertals (Ovchinnikov, 2001) quite close to the range of modern humans, rather than far outside it, as is generally assumed today.

There is something going on here that we need to know more about.

Unexpected Invitation, New Ideas

These conflicting ideas seem to make it worth while trying to tease out some of the factors that might be involved in species (of whatever kind) evolution. Some years ago I, therefore, started a series of simplistic attempts to try to see what kinds of factors might affect divergences of species and individuals over time. These activities were initiated by an invitation I received to speak at the International Institute for Advanced Studies in Kyoto in 1993. After being invited, after immediately accepting, and after the acceptance was confirmed, I then asked why I had been invited. It turned out that the institute was attempting to gather together a mixture of geneticists, linguistics, human biologists, molecular biologists and paleontologists, all with views, often differing, on modern human origins.

> "But why me?" I asked.
>
> "You are a mathematical modeller. Perhaps you could model mitochondrial DNA evolution" they said.
>
> "Fat chance", I thought.

For though I have used a variety of mathematical methods over the years, they have mostly been applied in the analysis of real data. Except for theoretical stress analyses in bone biomechanics, a totally different context, I had never tried theoretical modelling. But the idea was intriguing and so I agreed to participate in what turned out for me to be a fascinating workshop.

I quickly realised that modelling mitochondrial DNA evolution, a matter of change of molecular materials in lineages of interbreeding and migrating individuals, was beyond me. I almost equally quickly realised that even the evolution of populations of individuals sometimes capable of interbreeding and migrating (sub-species, in biology) was also too difficult. I, therefore, ended up trying to study the evolution of groups that do not (should not) interbreed (species, in biological terms) and where migrations were not initially at issue.

I first did this using pencil and paper, a random numbers table, and a dart, mainly on the grounds that unless I have actually seen, first, every step of the way myself, I do not easily understand computational answers where the intermediate steps may be hidden from me. It turned out that I was not modelling, in the sense of using some mathematical function to model what was going on. I was in fact trying to mimic evolutionary processes by taking account of factors like those that might actually exist. Thus, I mimicked the difference between relationships of species determined only from 'fossil' and 'living' species, with the relationships based upon the totality of species (fossil, living, and 'un-fossilised', i.e. never discovered, species). Fossilisation, together with the subsequent finding of fossils, is a fairly rare event. Most species are not, and probably never will be known as fossils. This mimicry took account, therefore, of the implications of data that are missing because the species never became fossils or, if fossilised, were never found.

The studies were simplistic in the extreme. Furthermore, because I was doing it manually, the number of runs that I could carry out was very small. Notwithstanding that, the mimicry actually suggested something. It suggested that, if I were to do it computationally, with large numbers of runs to get statistical parameters, and with greater degrees of complexities in the mimicry, I might get interesting insights. For even this simplistic example suggested two things. Real common ancestors might be much older than generally estimated from fossil and extant data alone. Fossils might not often lie on lineages leading to living species (Fig. 1, and Oxnard, 1995).

By the time the book resulting from the Kyoto workshop was reaching publication stage, I had taken the mimicry further. I had moved on to the question of modelling groups that could 'interbreed' and 'migrate' between 'continents', as it were. In other words, I had incorporated sub-species and geographic isolation into my simulations. Again, the pencil and paper nature of these models precluded any real conclusion. Yet the result, again stemming from only a small number of runs implied that only the very oldest migrations were identified (Fig. 2, and Oxnard, 1995). All recent migrations were hidden! This, too, encouraged me to continue.

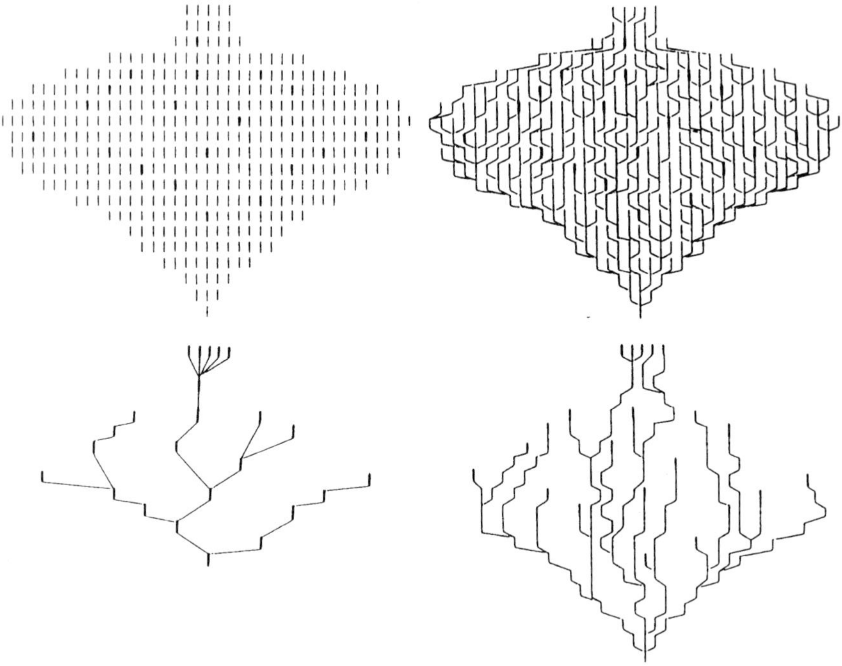

Fig. 1. My earliest attempt at evolving species. Top left shows the species that were evolved in the form of an amphora. Those that were chosen to be fossils are bold. Top right shows the links that actually existed among all the species. Bottom left shows the relationships between the species based only upon the bold (fossil) species of the first frame. Bottom right shows the true relationships among the bold (fossil) species. These last two frames provide completely different pictures of the relationship of the fossils.

Many studies (e.g. earlier — Raup *et al.*, 1973, and later — Sepkoski and Kendrick, 1993) had attempted to model the evolution of species. Most of these used computer techniques that modelled evolutionary branching by using mathematical functions (for instance, random processes) and then examining the groups (clades) that they produced. Such studies were mainly aimed at understanding the patterns of taxonomies and the evolution of higher groups, and were, I believe, highly successful. A recent study aimed at mathematically modelling the precision of timing of common ancestry implies,

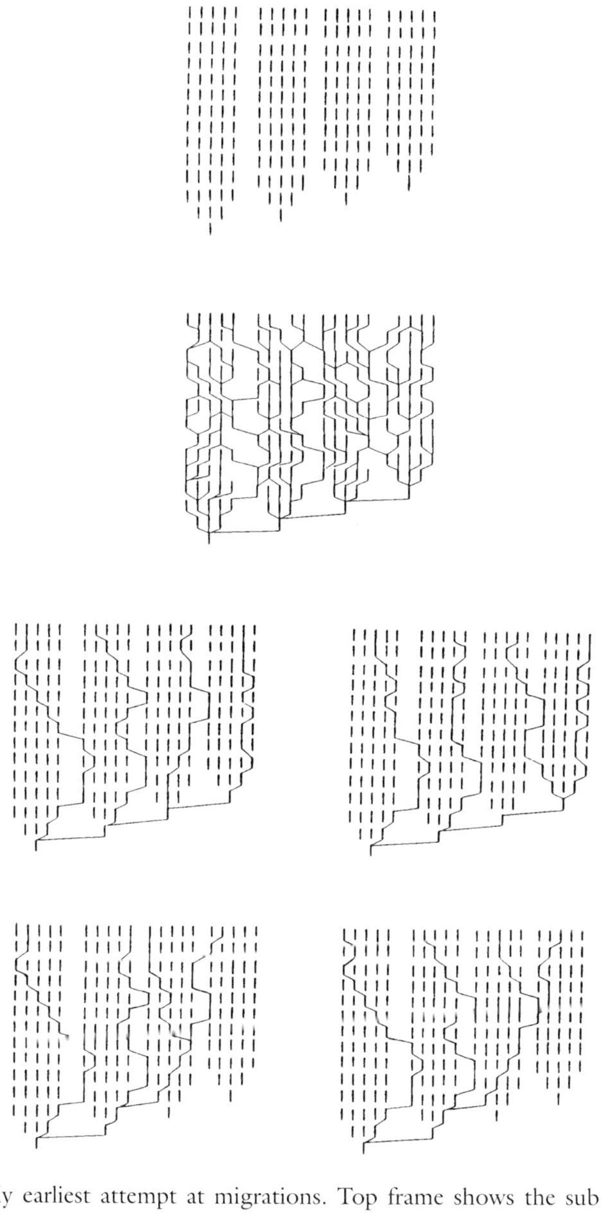

Fig. 2. My earliest attempt at migrations. Top frame shows the sub-species that have evolved on four 'continents'. Middle frame shows the true links among these sub-species. Bottom four frames shows four different sets of links in four different runs of this system showing that in each case, the key migrations were all ancient (near the bottom of the diagrams).

as do our investigations, that almost always, real times are likely to be much older than paleontologically estimated times (Tavare *et al.*, 2002). All of these studies differ, however, from the work that I was attempting because they do not copy specific biological mechanisms but they use mathematical functions. In contrast, my attempts are aimed at actually mimicking some of the mechanisms that occur, and they involve examining the implications not only of fossil species that we know but also of all those species that have existed but we have never discovered.

Avise (2000) has produced an extended introduction and review of these matters using the term: phylogeography. Avise outlines some of the principles and processes that govern the geographic distributions of genealogical lineages. The work described in this chapter and the papers leading to it (Oxnard, 1995, 1997) are attempts that actually mimic phenomena like those described in the later explanatory diagrams of Avise (2000).

New Colleague, Theoretical Physics, and Dr Ken Wessen

At this very point, Dr Ken Wessen, came into my orbit. Dr Wessen, holding a doctorate in theoretical physics, and with major skills in computation, had decided that he would like to apply his skills in a totally different field. He chose evolutionary biology for a new intellectual stimulus. He decided to do it through a research Masters Degree with me. The end result of this, however, was the conversion of his Masters thesis into a second Doctorate (Wessen, 2002), one good enough to earn him a Distinction, and leading to a book *Simulating Human Origins and Evolution* with Cambridge University Press (Wessen, 2005).

Ken Wessen has now developed mathematical mimicry of the divergences of lineages of species and sub-species (Oxnard and Wessen, 2001; Wessen, 2005) that are far more complex than the ones I had originally been playing with. He is able to vary factors such as extinction and fossilisation rates. He can change splitting of lineages (i.e. of species) and allow union of lineages (i.e. possibility of interbreeding of sub-species). He can allow (or prevent) migrations between continents and because, if migrations can occur then back-migrations can also, he can allow back-migrations. He can take account both of 'characters' that determine the next species generation and those that do not.

Thus, in Wessen's programmes, species 'evolve' through changes in 'characters'. There are two packets of characters for each species. The first includes the characters that are used to decide the make-up of the next species-generation (therefore 'hereditary' characters). The second comprises characters that can also change, but whose changes do not contribute to the determination of the species in the next generation ('non-hereditary' characters). The 'hereditary' characters of all species determine the 'real' pattern of evolutionary relationships between species, and they are descriptive of all species, living, fossil and non-fossil (Fig. 3, and Wessen, 2005). The totality of all characters introduces noise like that in the real world where we do not know which characters are hereditary and which not. It is also possible to use all the characters from fossils and living forms alone to 'reconstruct' evolutionary trees. This is what we do as paleoanthropologists. After all, as paleoanthropologists, we cannot certainly tell the differences between the two types of characters. And we certainly do not know anything about the characters of species that never appear as fossils. As a result it is possible to compare the true evolutionary relationships of all species with reconstructed relationships based upon fossil and living species alone.

Bowls, Vases and Amphoras

Wessen and I first tried out the effects of varying the rates of species generation, extinction and fossilisation. By varying these rates during the 'evolving process' we could produce different shapes of the distribution of species over time. This is what I had previously attempted with the paper-and-pencil. We repeated this idea making runs with generation, extinction and fossilisation rates that produced distributions of species that were 'bowl-shaped', 'vase-shaped' and 'amphora-shaped'. Figure 1 (earlier) showed an example of one of my original bowl-shaped distributions.

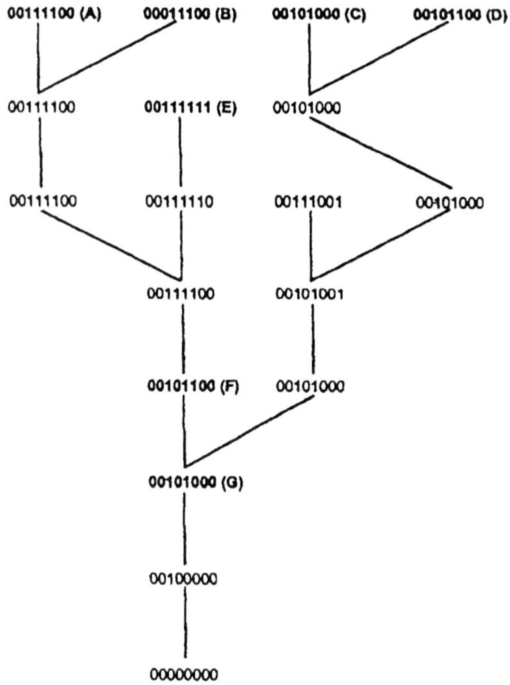

Fig. 3. Simplified diagram of species evolution as presented by Ken Wessen's method. The evolutionary tree starts with a single species at the root of the tree with 64 characters (but in this diagram only eight are shown, all zero). The characters can change: thus, the second species has a single change, 1, at the third character. Characters can change again so that the third species has two characters different from the first (in the third and fifth characters). Species can split to form two new species in the next species generation. Species can go extinct as indicated by lineages that terminate before the top of the diagram (which is the present day). Fossilised and extant species, the only ones whose characters are known to the palaeontologist, are indicated with letters A to G. The entire tree shows how the extant species actually arise.

The bowl-shaped species distributions were produced, starting from a single species, by allowing the numbers of species to gradually increase over time. This mimics what might happen early in an evolutionary radiation. The new niches that the species enter are empty, and, as a result, many different species rapidly evolve to fill them. The vase-shaped distributions produced by different values of generation, extinction and fossilisation rates, contained species distributions that show a gradual increase at earlier times and then a gradual decrease later. This mimics what might happen when a growing radiation has stabilised and then started to reduce, perhaps through increased competition. Both bowl-shaped and vase-shaped species distributions are well known in the fossil record (e.g. Simpson, 1953).

However, we were especially interested in runs where generation, extinction and fossilisation rates were modified in such a way as to produce amphora-shaped distributions (Fig. 4). In these, there is an initial increase in species numbers, as at the beginning of a radiation. This is then followed by a gradual decrease in species numbers, as the

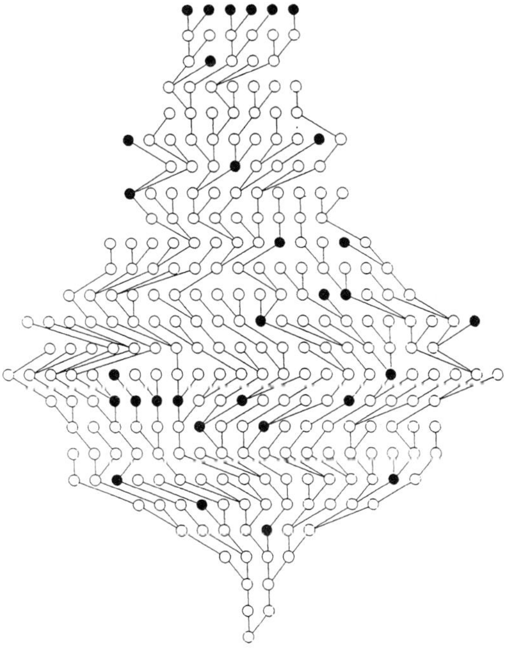

Fig. 4. A single run where splitting and extinction parameters have been adjusted to produce an amphora shaped distribution of species. The open circles are the unfossilised species. The closed circles are both fossilised species (throughout the body of the tree) and extant species (at the top, the present day). The links indicate which species are related to which. Species splits and species extinctions are also indicated by the pattern of the links.

radiation stabilises, the niches fill, and competition reduces the numbers of species. This in turn is followed by a long period of continued decreased species numbers (the 'neck of the amphora') leading to a marked reduction of living species. This describes the situation in many animal groups where there are only a few representatives living today of groups that had larger numbers of species in times past. It is especially relevant to the ape/human clade. At the present time, there are only a few ape species and only one human species. Yet the fossils indicate that there have previously been much larger numbers of species. As a result, there must have been much extinction, hence the neck of the amphora.

Mimicking Species Distributions as Amphoras

One thousand amphoras were mimicked in a first series of runs so that statistical information could be obtained. Figure 4 shows the complete evolutionary tree for one run starting with a single species at the base of the diagram. Species connections are indicated by the links between species at one time level and the next. Species continuity is indicated where a species continues into the next species-generation without a sibling species. Species splitting (divergence) is indicated where one species gives rise to two species in the next species-generation. Extinctions are indicated where a species has no onward connection into the next species-generation. Species not chosen for fossilisation are shown as open circles. Species chosen as fossils together with all extant species (at the top of the plot) are shown as closed circles.

Figure 5 shows the same phylogeny but using lines joining only fossils and living species and excluding the many stepped lines through the intermediate non-fossils in Fig. 4. That these direct lines summarise the stepped connections correctly can be checked by superimposing Fig. 5 on Fig. 4. In this particular run, the true common ancestor of all the living species is the particular fossil to which they converge in the diagram. This ancestor is 16 species-generations back in time (16 million years if we assume that 1 million years is approximately how long a species lasts). The entire run is 25 species-generations

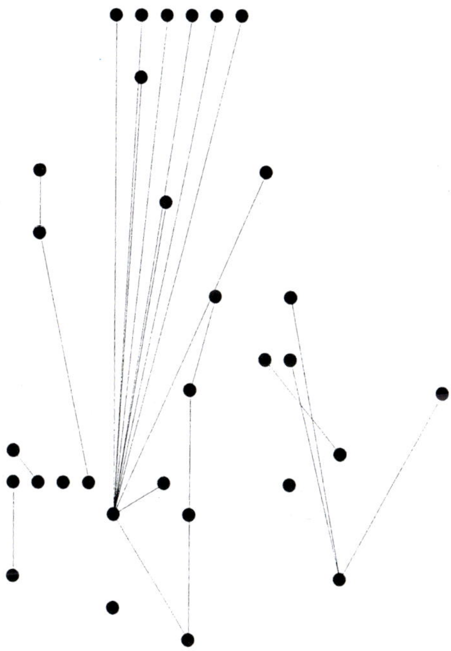

Fig. 5. The same runs as Fig. 4 showing the links of the fossils with each other together with the position of the true common fossil ancestor of all the living forms. Superimposition of this figure over Fig. 4 confirms that this is correct. The true common ancestor in this run is 18 species generations back from the present.

in time depth; 25 million years with the same assumption. Twenty five million years was chosen to mimic what is generally posited about the total length of time since the origins of all apes and humans.

In contrast to Fig. 5, Fig. 6 provides the evolutionary relationships based upon analysis of all characters (both 'hereditary' and 'non-hereditary' — i.e. including the 'noise') but for the living and fossil species alone, that is, excluding those species that are not known because they were not 'fossilised'. This is, of course, all the palaeontologist can do. Palaeontologists cannot be certain which characters are hereditary and which are not, though cladistics is an attempt to make such a separation. Further, palaeontologists cannot take any account of the characters of species that were never discovered. This figure shows that the common ancestor of all living species appears to

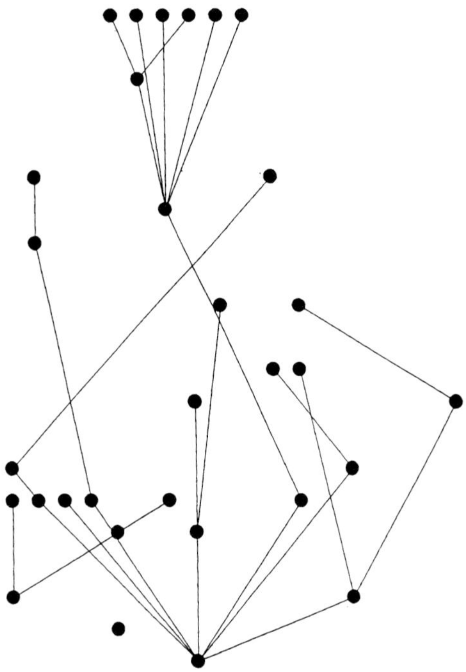

Fig. 6. The same run as Figs. 4 and 5, but with the fossil common ancestor as reconstructed from the characters of the living and fossils species alone. This ancestor is only six species generations back in time and this is, of course, incorrect by a factor of 3 times.

be only six species generations back in time (that is, 6 million years with the above assumptions). This is, by chance, similar to the time generally given for the common ancestor of humans and chimpanzees. Again, this can be checked by superimposition of Fig. 6 on Fig. 4.

Tables 1 and 2 provide a few statistics from these runs. Table 1 compares the average generations of true common ancestors with the average generation of reconstructed ancestors. It shows that reconstructions place the time of common ancestors between 2 to 8 times too recently. Table 2 shows that fossils are far more likely to be lying on lineages leading to living species than is actually the case (again by a factor of 2 times or more).

Table 1. Times of common ancestors of living species.

No. of runs	Generations back to true common ancestor	Generations back to apparent common ancestor
52	18 (range 15–23)	8 (range 3–11)
28	0	10 (range 6–16)
20	0	0

Table 2. Number of fossils lying on lineages to living species.

No. of runs	True number of fossils on lineages to extant species	Apparent number of fossils on lineages leading to extant species
1,000	11 out of 34	25 out of 34

If there is any verisimilitude to results like these, it really does mean that we should take the palaeontological assessments with a large pinch of salt.

Mimicking Sub-Species Distributions as Bowls

The parameters were next modified to permit not only splitting, continuation and extinction of species, but also 'union' of 'species', that is, mimicking the evolution of sub-species which, by definition, are capable of successful interbreeding. These runs were yet further modified to permit migrations (and therefore also back migrations) among as many as four different 'continents'. Migration rates and times can be varied, and migrations can be limited to specific time periods thus mimicking the opening and closing of migration routes produced by such phenomena as changes in sea level, movements of glaciers, tectonic events, grand climactic changes, and so on.

We carried out 1,000 runs in this mode. We initially had four 'continents', and migrations and back-migrations were permitted among them. However, these are such complex simulations with many sub-species and many migrations that any individual run is difficult to present visually. Let us, therefore, look first at a simpler example

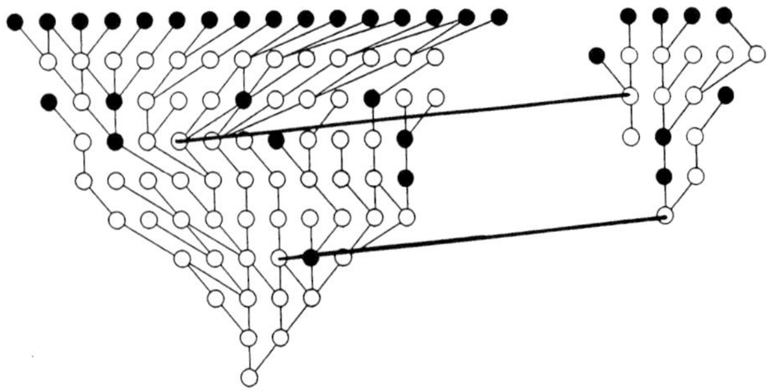

Fig. 7. A small run in which sub-species were permitted (i.e. union as well as splitting was allowed) and where there were two 'continents' with migrations allowed. The two bold lines indicate the two migrations.

(Fig. 7) where there are only 11 sub-species-generations, only two continents and only two migrations.

With the same conventions as before, Fig. 7 shows the complete evolutionary tree for all species and sub-species, living, fossilised and non-fossilised. The pattern of links among them indicates when union (interbreeding) as well as divergence (splitting) has occurred. In this particular run there were only two migrations and these happened to involve species or sub-species that were not fossilised (i.e. open circles). The number of fossils (shown, as before, by closed circles) was set again at 10%. However, in this case, the other parameters were set so as to produce a bowl-shaped distribution. This probably more closely relates to the situation in modern humans where there has been much migration, much interbreeding, and much expansion of numbers.

Figure 8 shows the tree reconstructed from all the characters ('hereditary' and 'non-hereditary' alike) but for fossil and living forms alone. This tree 'finds' a fossil common ancestor for all the living forms on each continent and a third, oldest-of-all, common ancestor for both continents. It even 'finds' only one migration (and that spans several time levels!).

This can now be compared with Fig. 9 showing the true tree. The reconstructed fossil ancestors are totally spurious. There is, in actuality, no ancestor that happened to be a fossil for each continent

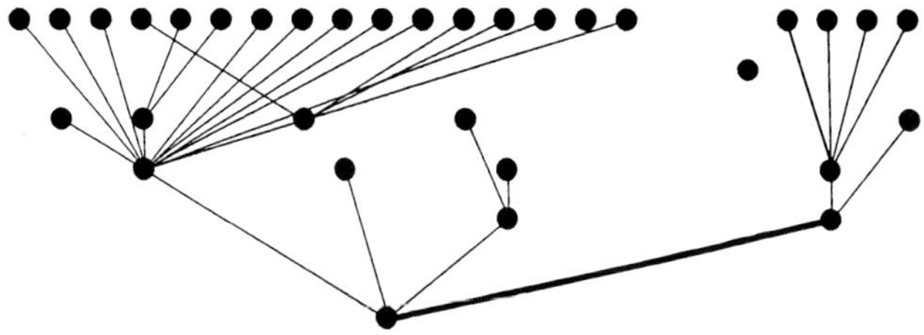

Fig. 8. The same run as Fig. 7, but with the relationships being based upon only the characters of the living species and the fossils. The links indicate that there were fossils that were the common ancestors of the living forms on each continent separately and the overall common ancestor for both continents. The bold line indicates the single migration that was 'found' and comparison with Fig. 7 shows that it is wrong.

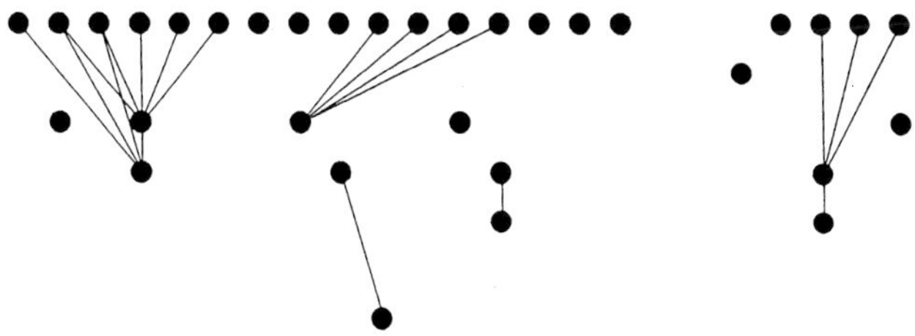

Fig. 9. The same run as Figs. 7 and 8, but showing only the living species and the fossils, and the true links among them. No fossil is a common ancestor. Perusal of Fig. 7 indicates that the common ancestors were all non-fossilised species

separately, nor for the entire suite of forms on both continents. There are common ancestors, of course, but they were all species that were never fossilised. Furthermore, the single reconstructed migration is spurious. The two real migrations in Fig. 7 are at quite different levels to that in Fig. 8.

At this point, we can now look at some results for bigger runs involving 25 sub-species-generations, four continents, A, B, C and D, and many migrations and back-migrations among the continents. The diagram (Fig. 10) displays the total number of sub-species at each time slice on each continent as the relative width of the continent. It also shows the migrations between the continents as connecting lines.

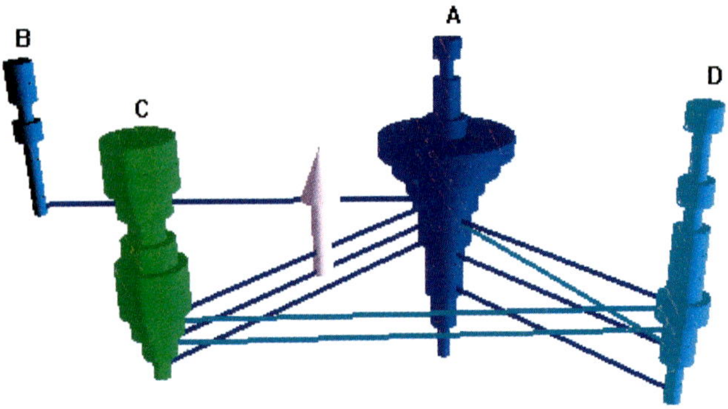

Fig. 10. A large run involving many species (indicated by the thickness of the diagram at each species level), many migrations (indicated by the joining lines) and the four continents themselves. This figure is too condensed to show each individual species in each species generation and the many links between them. The entire information, however, including the characters for every species is given in the data report for the run.

Table 3. Number and times of migrations.

No. of runs	Average no. of real migrations	Average generation of real migrations	Average no. of apparent migrations	Average generation of apparent migrations
50	23	16	12	9

There are so many sub-species that the diagram cannot show each individual as in the previous three figures. Even this is simpler than many runs that can be made. In all cases however, the full information is available from the printed computer output.

This example is only one from a number of such runs that provided the data in Table 3. Thus, when a series of 50 such runs were summarised, the reconstructed evolutionary trees were always unable to find the correct migrations. They usually found fewer migrations, usually migrations that were too recent, and always migrations that skipped several sub-species-generations. To this degree, the 50-run analysis confirms the information in the single run of Fig. 7.

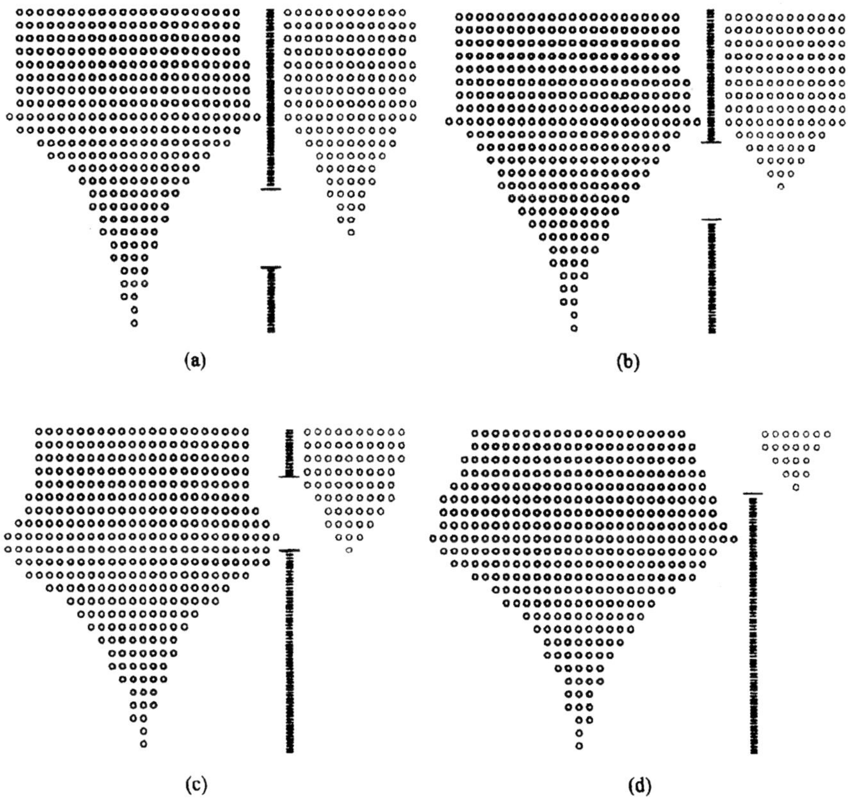

Fig. 11. A small run involving only two continents but repeated four times with migrations permitted at different locations in time. (a) Migrations were only permitted relatively early as shown by the gap in the barrier between the continents. (b)–(d) The migrations were permitted in windows that are later and later in time (Wessen, 2005).

All this is shown in Ken Wessen's results of the effects of changing the times of the migration periods (Figs. 11 and 12). These show the difference between the true migrations and the apparent (reconstructed) migrations. Reconstruction does not easily separate the migrations, and it fixes them at much later (and wrong) times.

This programme can be used for testing many different ideas. For example, how likely are modern human origins due to a recent set of

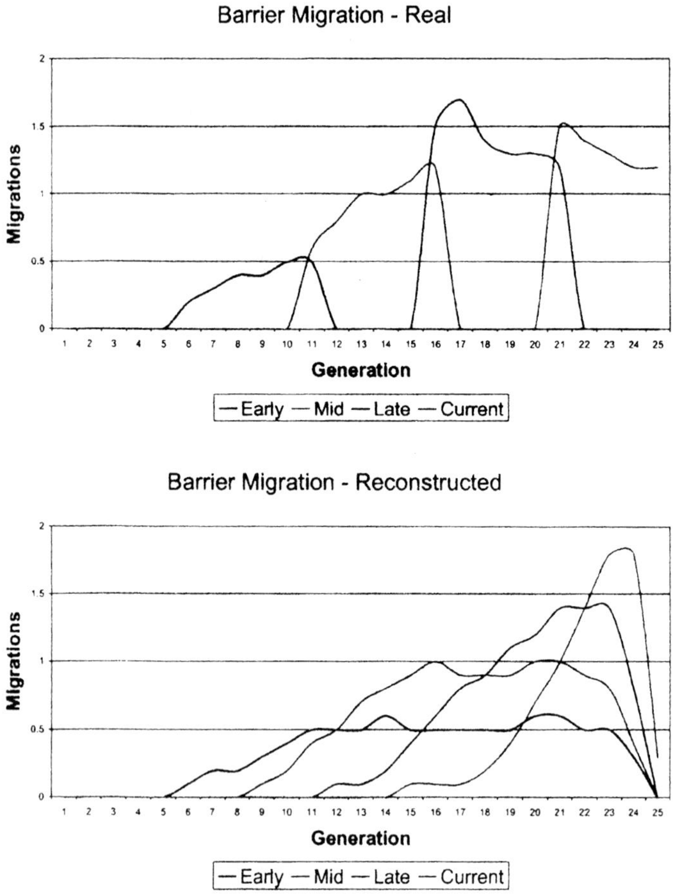

Fig. 12. Plots of the spread of real set of migrations for each of many runs for the four continents in Fig. 11 (upper graph) compared with the reconstructed sets of migrations (lower graph). Reconstruction almost completely fails to make correct identifications of migrations and implies them to be far more recent than in fact they are (Wessen, 2005).

migrations that completely replaced all prior human groups on other continents (the Out of Africa or the Noah's Ark hypothesis)? How likely are human origins due to evolution of ancient migrations to each continent (the Candelabra hypothesis)? How likely are human origins due to a combination of migrations at many different times between continents (the Multiregional hypothesis)?

We can also test a number of other evolutionary hypotheses. How complete are mass extinctions in evolution? How important are subspecies radiations in evolution?

Testing Evolutionary Ideas: The Noah's Ark Hypothesis

Thus, one view of modern human origins, the Noah's Ark Hypothesis, postulates that modern humans arose in Africa about 150,000 years ago and spread throughout the world, perhaps eliminating, certainly replacing those earlier humans that were already present elsewhere. This idea stems from morphological examination of fossils and from mitochondrial DNA and Y chromosome analyses.

In order to examine how well this idea would be supported by mimicry, I adjusted various factors (especially the migration parameters) so as to give a long period of evolution on a first continent followed by migrations to a second continent (other continents were ignored at this point in order to confine attention to one set of migrations). The facility for producing groups that could nevertheless 'interbreed' was turned on so that it is evolution of groups that sometimes interbreed that are being examined. Other factors (especially the extinction parameters) were adjusted to produce more interbreeding groups, representing both fossils and non-fossils, at later times. This resulted, on each continent, in group distributions that were bowl-shaped.

Figure 13 shows one run in which there are a small number of migrations between the two continents at about the half-way time level along the lineages on the first continent. In this particular example, the final living groups on the second continent are in two clades (red and blue), each with its own common ancestor also on that continent at levels following the migrations. There is no common ancestor on the second continent of all the living groups on that continent.

The figure also shows, further, that although the common ancestors on the second continent are derived from migrations, they are not derived from the same migration. A common ancestor of each clade in second continent lies, in fact, on the first continent. These

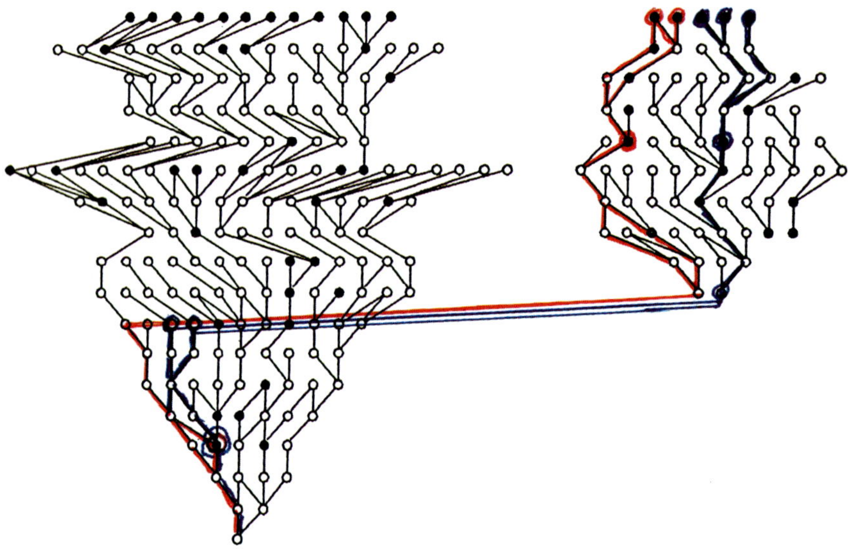

Fig. 13. An example of a run involving two continents where migrations to the second continent where not permitted until relatively late (about halfway through the simulation). The migrations are indicated by the lines between the continents. There are two related groups of species on the second continent. They each have their own fossil common ancestor on that continent. Their overall fossil common ancestor, however, is located very early on the first continent. Out of 50 runs, it was never the case that the species on the second continent all stemmed from one common ancestor on that continent. This mimics but tests the Out-of-Africa hypothesis.

common ancestors are not, however, close to the time of migration, but are much further back. The final single common ancestor of both clades on the second continent is also, of course, also on the first continent and lies very early.

Data from 25 such runs imply that common ancestors are only very rarely fossils, and that only extremely rarely are common ancestors found close to the time level of the initial migrations. A single migration never occurred with a set-up like this. The Noah's Ark view of migrations of individuals from a single population yielding all interbreeding groups on the new continent was not replicated in any of the 25 runs made with this model. These runs take a long time to analyse, however, and 25 runs are not enough to be certain statistically. In addition, these runs should be repeated with migrations to the full series of four continents.

Nevertheless, it already appears unlikely that the standard Noah's Ark idea could have occurred. It is too simple.

Testing Evolutionary Ideas: The Candelabra Hypothesis

The second model of modern human origins, the Candelabra hypothesis, is that modern humans have arisen on each continent from the interbreeding groups previously present on each continent. This is also easily mimicked by starting related groups on each continent (related because they, in turn resulted from the earliest migration) and then allowing migrations between them starting at about the same species generation as in the prior example. Again, to keep the picture simple, I show only two continents (Fig. 14) so that the links can be clearly seen.

This demonstrates, among others, four migrating generations between the two continents about half-way through. The final result of this process for the second continent is that there were three clades (red, green and blue) leading back to three different common ancestors. One of these common ancestors is actually on the second continent near its base. However, the pathway to that clade involved further migrations from the second to the first and then back again in the intervening period. The other two common ancestors are on the first continent yet are quite different. All common ancestors lie at very much earlier times than the migrations. It follows that there is no common ancestor to all living forms on the second continent.

Again 25 runs were carried out. The results were generally similar to those in the example figured above. Common ancestors are only rarely fossils. Only very rarely are these ancestors close to the time level of the initial migration. These ancestors were never from a single migration. The candelabra view of migrations influencing all species on the new continent was not replicated in any of the 25 runs made with this model. However, again, these runs take a long time and 25 are not really enough to be certain statistically. And again, as

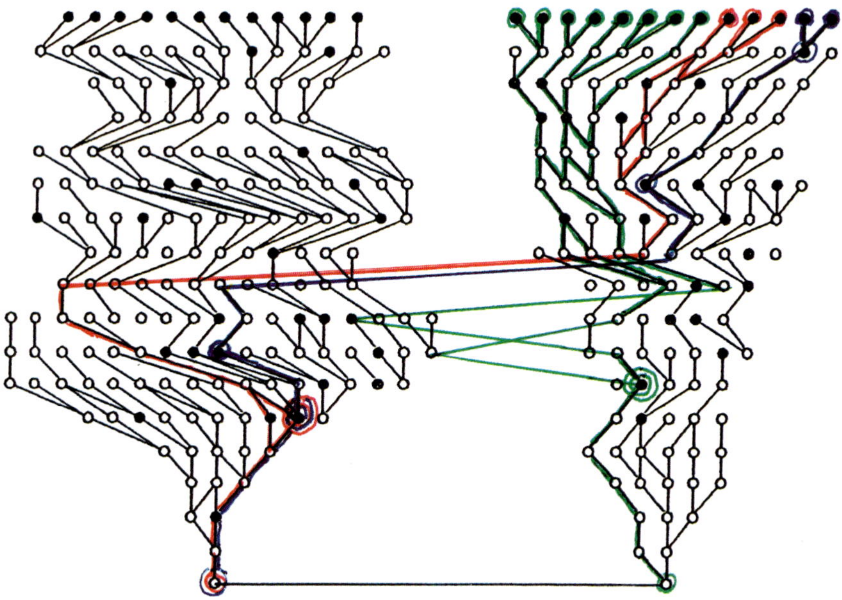

Fig. 14. An example of a run involving two continents where migrations to the second continent where permitted both early and late. The migrations are indicated by the lines between the continents. There are three groups of species on the second continent. One of them has a fossil common ancestor about the middle of the tree on that continent. The other two have their separate fossil common ancestors early on the first continent. There is no single fossil common ancestor for all species on the second continent. Out of 50 runs, it was never the case that migrations from the first to the second continent 'took over' completely the second continent. This mimics and tests the Candelabra hypothesis.

before, these runs should be repeated with migrations to the full series of four continents.

It again appears highly unlikely that the original uncomplicated Candelabra idea could have occurred.

If Not Noah's Ark, Nor Candelabra, Then What?

Of course these runs imply that by far the most likely arrangement would be where multiple migrations and multiple interbreeding of groups occurred on all continents and at several times. This might be thought of as simulating the multi-regional hypothesis (a later modification of the Candelabra idea). This seems to be by far the most likely evolutionary possibility. Even so, however, times and places of common ancestors, and times and numbers of migrations are rarely what they appear to be.

In addition to these ideas resulting from current controversies about modern human origins, it is also possible to mimic other situations that are common throughout the whole of evolution. One of these is the question of mass extinctions. Another is the role of subspecies.

How 'Mass' are Mass Extinctions?

The idea of the mass extinction in evolution is very old and based upon observed long-known marked discontinuities in the distributions of species over time together with new evidences as to causal agents. For example, there seem to have been a number of really major crises in the history of life where as many as 47–84% of species are estimated to have disappeared. One such example is the Cretaceous/Tertiary extinction, 65 million years ago, marking the end of the dinosaurs. However, there have been a number of other mass extinctions (though perhaps a little less 'mass' and confined to fewer species). One studied recently is the Cenomanian Turonian (C-T) mass extinction of echinoderms 91 million years ago. About 71% of fossil echinoderm species were lost.

Recently, however, this echinoderm mass extinction has come under new scrutiny. Thus, Smith *et al.* (2001), looking more deeply at distributions of sea urchin fossils, believe that the appearance of a major extinction event involving these fossils at the C-T extinction boundary is deceptive. Many of the fossils seem to have reappeared as much as 30 million years after the event. By their count, the C-T extinction of echinoderms shrinks from a catastrophic 71% to a rather meagre 17%. They suggest, not that there was no mass extinction, but rather that the magnitude of the mass extinction is greatly overestimated. An opposite example is provided by Long and Buffetaut's (2001) biogeographical comparison of dinosaurs and associated vertebrate faunas from the early Cretaceous. Their investigation suggests that dinosaur ancestors may have existed very much earlier than generally thought.

By varying fossilisation, extinction and selective advantage parameters, it is possible to produce an evolutionary distribution of species with a major catastrophic event at a particular time slice. The reconstructed relationships of the species obtained through comparisons of all characters for fossil and living form alone can then be compared with the real relationships based upon the entire set of lineages. The full suite of relationships for living species, fossils and all non-fossilised forms are given in Fig. 15. The lineages reconstructed from living and fossil characters alone imply that the living species mostly have common ancestors that are just at or subsequent to the extinction, i.e. the extinction was almost complete (Fig. 16).

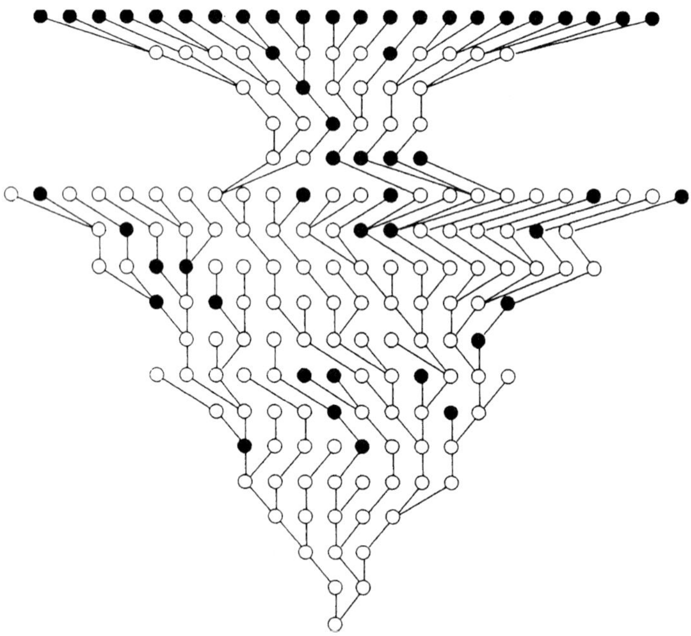

Fig. 15. A run in which the fossilisation and extinction parameters were arranged so as to produce an incomplete 'mass extinction' near the top of the distribution of species.

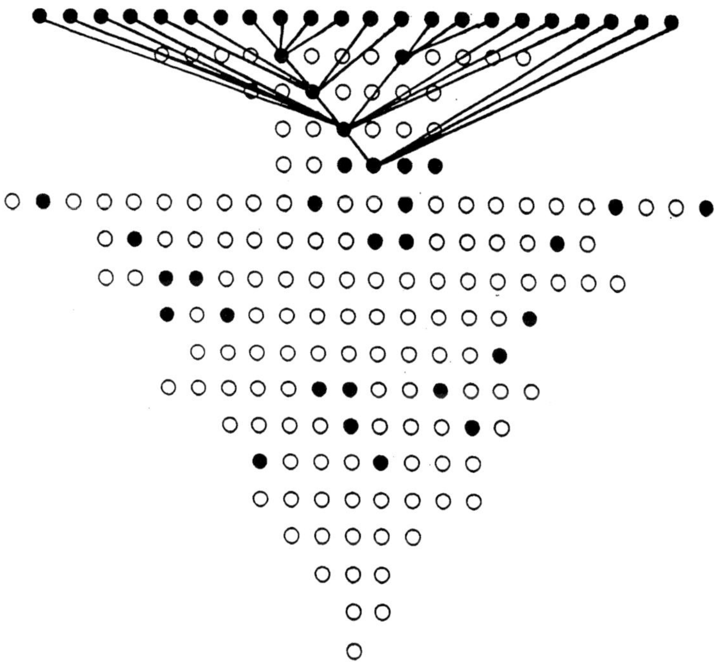

Fig. 16. The same run as Fig. 15 showing the reconstructed tree of extant and fossil species alone. It appears to show that the mass extinction produced a common ancestor of all the living forms that was at, or very close to, the extinction event.

In contrast, the true lineages show that the common ancestors of many of the living species existed long before the extinction event (Fig. 17). In particular, there are many lineages that pass the extinction boundary and then go on to produce an increased number of related species at subsequent time levels (Fig. 18). This is precisely what Smith *et al.* (2001) found in their examination of echinoderm species before and after the C-T extinction.

The reasons why these things happen in our runs, however, are not those presented by Smith *et al.* (e.g. sea level changes, sampling problems, etc.) because these matters have not been mimicked here. The reasons, here, are simply due to the fact that lineages predicted from fossils alone do not give good estimations of phylogeny. Again, this has only been run 20 times. Many more simulations must be

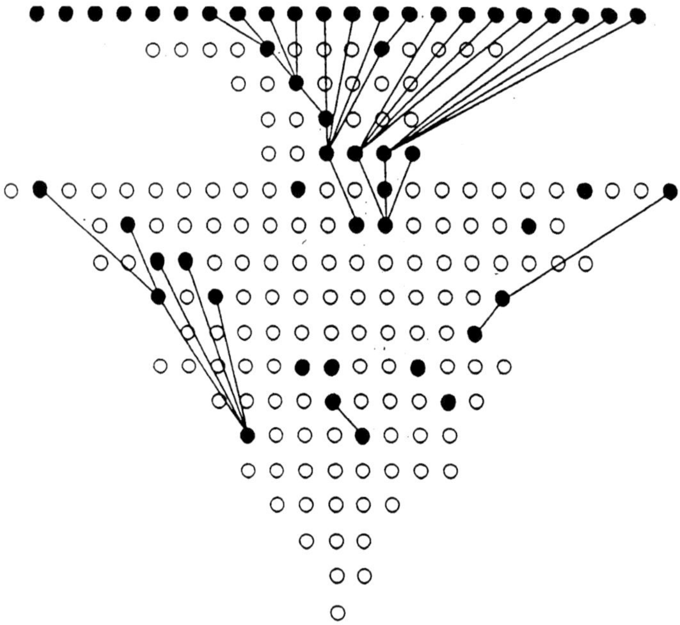

Fig. 17. The same run as Fig. 15, again, showing the true relationships of living and fossil species. Many of the living species have no ancestors that were fossilised. Many of the fossils do not lead to any living species.

made to confirm or deny these initial results. But the preliminary findings are most tantalising.

How Real are Sub-Species?

There is also the question that lies around the existence of groups that are capable of interbreeding (i.e. below the species level). These groups vary: local variations, hybrid zones, clinal differences, pan-populations and true sub-species. Some evolutionary biologists have concentrated their attention on species and their relationships within genera and higher groups. Such investigators see the realities of species and genera. They usually recognise fewer species and pay less attention to sub-species. Other evolutionary biologists also note the differences and overlaps among sub-species but recognise many more

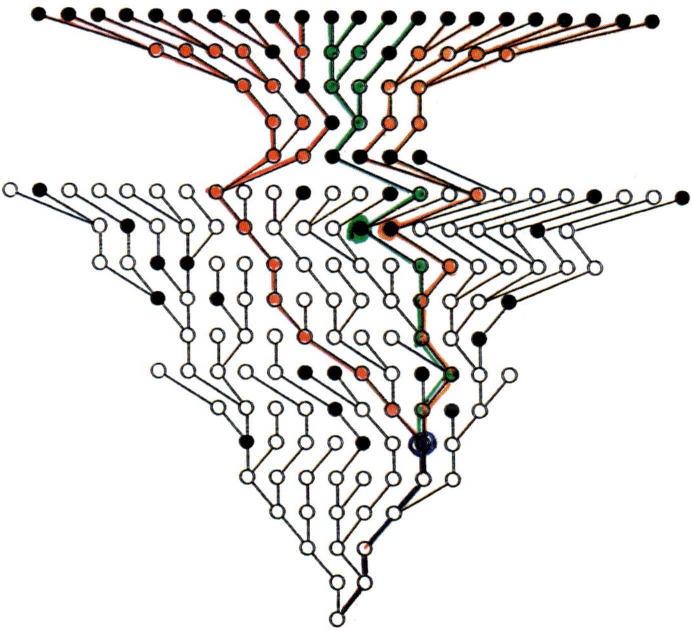

Fig. 18. The same run as Fig. 15, again, showing the true links and the trees leading to all living species. This shows that there are, in fact, three different groups of living species; their lineages, red, green and orange, all extend far back beyond the extinction event, and the final common ancestor is only six generations from the bottom of the entire tree.

of them, and recognise also, therefore, larger numbers of species. These various workers have been called 'lumpers' and 'splitters', respectively. Though some might see these terms as disrespectful, this need not be really so. They speak, rather, to matters of emphasis, emphasis on relationships between species in the designation of the genus, or emphasis upon local populations and sub-species in the designation of species. Both viewpoints are important in different contexts.

For example, Groves (2000), in studies of differences in many mammals, especially many primates, determines that a far larger number of sub-species exist than previously recognised. Groves' data are clear and his taxonomic distinctions are carefully made and replicable. The question then arises: what are the evolutionary implications of such diversity of sub-species? Is it possible that models like those described here might give insights into such problems? How stable or how ephemeral are sub-species radiations? Could they be detected through studies of fossils? What effects do they have upon subsequent evolutionary patterns?

Again, therefore, I decided to use Ken Wessen's facility to mimic such situations: both species and sub-species. In effect, my mimicry attempted to study the effects, on long-term phylogenetic assessments, of short-term sub-species explosions. On the surface it would seem that such extensive radiations would favour elaboration of species lineages. Figure 19 shows the relationship of all species within a setup producing a bowl-shaped distribution of species with no sub-species allowed. This can be compared with Fig. 20 where all parameters are maintained except that splits of some species into pairs of

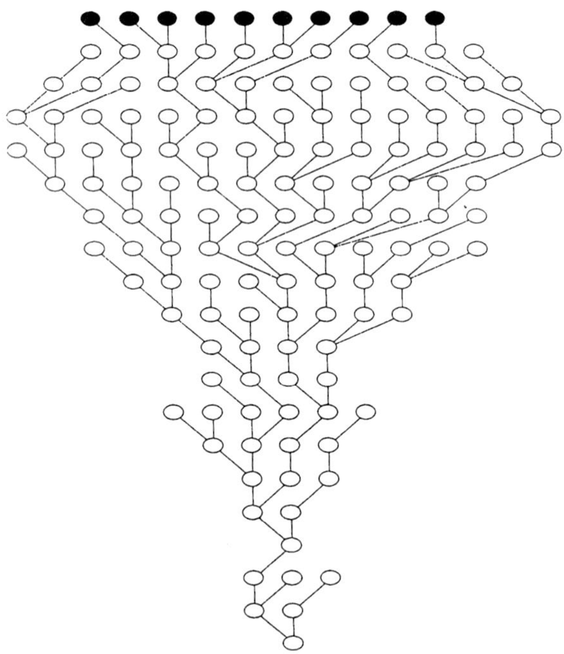

Fig. 19. A run, showing a deep bowl shaped distribution of species in which sub-speciation was not permitted.

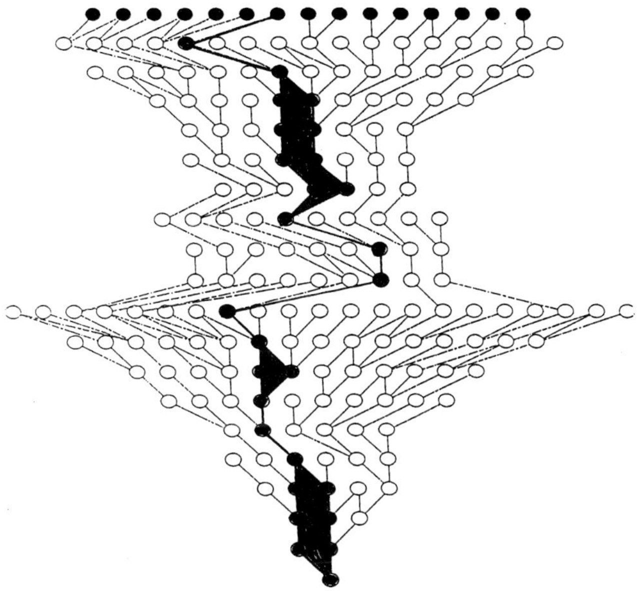

Fig. 20. A run, using the same parameters as Fig. 19 except that splits were allowed so that some species produced sub-species, and unions were allowed so that some sub-species could unite to form species. There are many such sets of sub-species in the various lineages of the model, but only one lineage of these has been darkened to make it more easily visible. The effect of this on the overall form of the distribution is marked. It produces a bottle neck in the species distribution. The effect upon the final distribution of species is, however, not at all marked; indeed, the various sub-species, though evident at different time levels, have a quite ephemeral effect upon the final distribution of the living species themselves.

sub-species are allowed. The shaded portions of Fig. 20 indicate the distribution of sub-species in one lineage from the base to the present day. This alters the shape of the distribution with a mini-bottle neck and two maxi-spreads. This can be compared again with Fig. 21 in which the complexity of sub-species is increased by allowing individual species to split into a maximum of three sub-species. In this case, there are three mini-bottlenecks and four maxi-spreads. This has been repeated with four and five sub-species allowed (but not figured here) and the results follow exactly the series outlined above, with further mini-bottlenecks and further maxi-spreads.

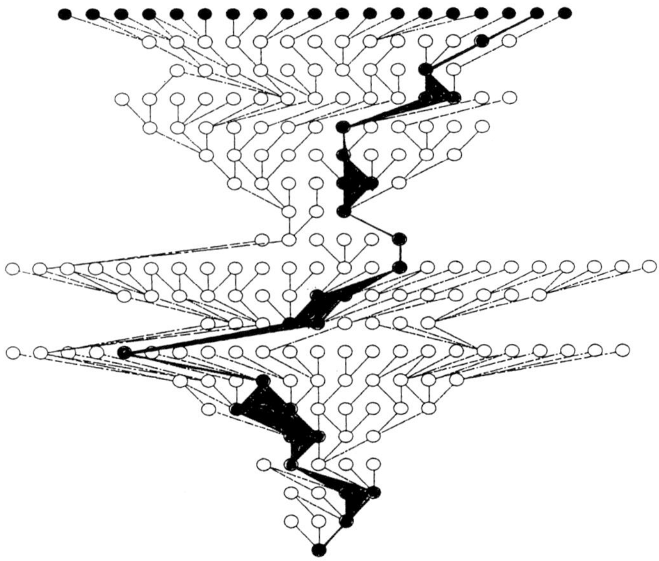

Fig. 21. A run, with the same general parameters as Figs. 19 and 20 but where the species were allowed to make both double and triple splits, and allowed to reunite into single species. Again, the effect upon the species distribution is marked, there being two broad bottle necks. Again, however, the final effect on the living species is negligible. Entirely similar effects are noted in runs (not figured) where species could divide into as many as four and even five sub-species at any given time slice.

The overall results show, however, that, though the distributions of sub-species are increased, and though the fossils have different distributions, the effects on the eventual overall lineages are quite ephemeral. Sub-species do occur, and they can be identified. There are a lot of them. But they seem to appear and then disappear in the phylogenetic record. They have effects upon the final species distributions at the times that they occur. But they have little effect upon subsequent species distributions. This does not mean that they are of no importance, and certainly not that they do not exist. However, it would be extremely difficult to identify them in the fossil record (though in these particular runs, I did not bother with fossils). It presumably means, however, that their effects can be negated by the very feature, interbreeding, that they permit. Is this an heuristic possibility for the real world? Or is there a problem with the mimicry? More work is required to find it out.

Mimicking Yet Other Evolutionary Phenomena

Many other problems might be illuminated using these methods. One example is the matter of multiple dispersal, migration and hybridisation events. Thus, Evans *et al.* (1999) present molecular evidence for multiple dispersal, migration and hybridisation in the macaques of Sulawesi and the Sunda Shelf. How would their results compare with the possibilities that can be presented through mimicry of dispersal, migration and hybridisation? All of these phenomena can be mimicked with Wessen's programme.

A second example is the question of survival or extinction. Thus, Kaplan (1999) is concerned about the extinction of orang-utans. Up to now, we have taken little notice of studies in which entire groups became extinct; we have concentrated on the ones that survived. Perhaps we should look at the statistics of mimicked extinctions instead of mimicked survivals. This would require modification of Wessen's programme because, other than noting them, data are not fully reported about extinct branches. Perhaps paying attention to this might give further insight into problems like those of Kaplan.

A third example is the question of multiple sequential tectonic events as discussed by Metcalfe (2000). This could be studied by appropriate modifications of the 'continents'. For example, the effects, on the evolution of a series of related species, of a sequential series of tectonic events, might provide insights into the real series of Tethys seas (early, middle and late) that waxed and waned during geological time. This could be examined by allowing migrations, then preventing them, then allowing them again, and so on, in sequence.

Criticism and Challenge

Some of the findings above seem obvious; others are surprising. It is therefore necessary to criticise, to challenge the processes of mimicry. Are there features of these processes that may be creating artefacts?

A first point is to do with biology, the assessment of phylogeny from fossil and living species. We have already determined that phylogenetic reconstruction from the characters of living forms and fossils assesses many more of the fossils as being on lineages leading to living forms than is actually the case. We have also determined that they recognise far fewer fossils as lying on lineages that go extinct. They especially place common ancestors of all living forms much more recently than is actually the case. They usually identify fossils themselves as common ancestors. They consistently get the numbers and times of migrations wrong. In every one of these cases, the evolutionary mimicry provides alternative hypotheses for testing.

Such matters have important implications for the study of human origins. For example, these results imply that the common ancestor of humans and chimpanzees is likely to have existed considerably earlier than the 5 million years generally used for calibrating molecular studies. They suggest also that there may have been much more genetic interchange over the millennia from the original diaspora of the genus *Homo* than is currently accepted. The recent out-of-Africa scenario may have been much more complex than currently believed. There may have been many more migrations of modern humans between and among the various continents than is generally accepted. These are, however, matters to which will return in the next chapter.

A second point relates to the mimicking methodology. It is often assumed that computer studies are overly simplistic and cannot possibly relate to the real world. That can indeed be the case where simulated evolution is based upon some mathematical function. This is not the case here. It is true, of course, that the mimicry is inevitably simpler than the real biology. Yet the mimicry here is actually much more complex than a theoretical mathematical model. More importantly, however, it even allows for much greater complexity than is possible in most practical fossil and molecular studies. Permitting study of the effects of different extinction and fossilisation rates, of different group splitting and interbreeding rates, of variations in migration and back-migration rates, of different blends of the hereditary and non-hereditary characters used in producing phylogenetic relationships and, perhaps most exciting of all, the possibility of looking at interactions among these various effects, is totally new. These things are not included in the usual theoretical modelling investigations. They are almost totally excluded in the usual practical studies of fossils and molecules. This, too, is also examined in the next chapter.

A third point relates to how phylogenies are affected when new fossils are found. In other words, how does the partial and changing nature of the fossil record affect assessments of phylogenies? In the general run of things, if assessments derived from known fossils at a particular point in time are good, then the later interpolation of new fossils should usually not drastically change the assessments. In fact, they should usually strengthen them. When the later interpolation of new fossils does cause drastic changes in our assessments of phylogenies we need to be able to understand why. One very good reason would be that this was because the new fossils contained totally new information. Some workers believe that this is the case for the tiny fossils from Flores (see later chapter). That possibility cannot (must not) be denied.

However, drastic changes could also occur because prior information from fossils already known had not been recognised. This is what happened in relation to the first australopithecines where clear evidence of arboreality in the original fossils was not recognised in the process of identifying human type bipedality and relationships with humans.

It is also true, however, that a new fossil may truly open up to us knowledge about evolutionary pathways that had genuinely been hidden because it was missing. Such cases, where discoveries of members of the genus *Homo* much further back in time than those previously known, requires us to remove younger but clearly less similar fossils from the direct human lineage. Yet we find great difficulty in doing this. For example, despite clear evidence from a number of laboratories that the australopithecines were arboreal as well as terrestrial, and that contemporaneous members of the genus *Homo* were not, we have been unable to remove the australopithecines from human branches. When we see arboreal adaptations in their structure we lay them at the door of prior holdovers rather than present functions.

Should we include such possibilities in our thinking? Evolutionary mimicry suggests that we should.

A fourth point relates to the employment of the cladistic method. Cladistics assesses characters for the determination of phylogenies. The evaluations depend upon whether the individual characters are primitive, shared-derived, uniquely derived, etc., (for the groups being examined in relation to some other closely related out-group). However, in reality, characters are much more complex than this. A character is really only an observable feature. Observable features will themselves most often be the expression of combinations of underlying complexes of characters, every one of which may be primitive, shared-derived, uniquely derived or something else. This means that the characters (really observable features) of cladistics are only very rarely fully primitive or fully derived, etc. Much more commonly, each observable feature (the character of cladistics) will be a complex mix of the primitivenesses, shared-derivednesses and uniquely-derivednesses of the various underlying true characters of which it is a conglomerate. In fact, the one designation for a character (in the cladistic sense) that we can almost certainly assume will not be true is that it be wholly primitive or wholly derived. Yet that is all that is allowed in cladistics as it is currently practised.

The studies employed here allow us to include these levels of complexity. Thus, we have 'hereditary characters' that determine phylogenetic relationships. We have 'non-hereditary characters' that add noise to the character complexes. The mimicry even includes other possibilities. Thus, characters can be directed to display a degree of unidirectional change, that is, change that can continue for a period of time in the same direction. This mimics adaptations to (say) continuous unidirectional change in environmental factors over time (such as a developing ice age, or gradually increasing dryness) and these contribute directly to lineages. A set of characters could be set up to mimic those kinds of adaptations that are actually ontogenetic responses to unidirectional environmental factors. Such features do not immediately contribute to lineages though they may eventually come to do so through developmental/environmental interactions. The complexities of characters permitted in the mimicry can thus display levels of complexity tending much more closely than traditional cladistics towards what really occurs naturally. Such information allows us to see the difference between the 'clear' true phylogenies and the 'clouded' reconstructed phylogenies.

Thus, these studies may allow us to assess the implications of 'getting it wrong' when characters are assessed for cladistic analysis. For instance, the implications of treating a character as primitive or derived, when that 'character' is actually an observation that is a compound of several different underlying characters, each in a different state, could be assessed. The effects of the use of different out-groups in cladistic analyses could be determined. It is possible that factors like these could be the reason for some of the controversies that exist in cladistic studies.

Eventually, it is my hope that we may be able to study the degree to which cladistic methods of handling the characters of living and fossil data can be modified to disentangle these various components of characters. It is possible that this might be done through recognition of the interdependences between characters (observable features) and the definition, therefore of new underlying 'characters' defined from the independent axes of the multivariate handling of character information. Recognition of the greater complexity of 'characters', especially recognition of their interactions using perhaps the methods of statistics, may help provide new ways of disentangling or partitioning character data pertaining to evolution.

This 'double information content' of morphological variables has already been achieved to some degree. Thus in studies of primate size and proportions, one view of variables has shown adaptive content, another view (of the same variables) evolutionary content, and a third underlying developmental content (Oxnard, 2000, see also Chapter 4).

References

Arnason U, Gullberg A, Schweizer Burgette A, Janke A, Molecular estimates of primate divergences and new hypotheses for primate dispersal and the origin of modern humans, *Hereditas* **133**: 217–228, 2000.

Avise JC, *Phylogeography: The History and Formation of Species*, Harvard University Press, Cambridge, 2000.

Ayala FJ, The myth of eve: Molecular biology and human origins, *Science* **270**: 1930–1936, 1995.

Broom R, The dentition of the Transvaal Pleistocene anthropoids, *Plesianthropus* and *Paranthropus*, *Ann Transvaal Mus* **19**: 303–314, 1938.

Broom R, Robinson JT, Swartkrans Ape-Man: *Paranthropus crassidens*, *Transvaal Mus Memoir* **6**: 1–123, 1952.

Cartmill M, Charles Oxnard: An appreciation, in Anapol F, German RZ, Jablonski NG (eds.), *Shaping Primate Evolution*, Cambridge University Press, Cambridge, pp. 1–10, 2004.

Carroll SB, *Endless Forms Most Beautiful*, Weidenfeld and Nicolson, London, 2005.

Dene HM, Goodman M, Prychodko W, Immunodiffusion systematics of the primates, *Folia Primatol* **25**: 35–61, 1976.

Evans BJ, Morales JC, Melnick DJ, Macaques in Sulawesi and the Sunda Shelf: A molecular analysis of dispersal, speciation and gene flow, in Metcalfe I (ed.), *Where Worlds Collide: Faunal and Floral Migrations and Evolution in SE Asia–Australasia*, pp. 25–26, 1995.

Fu Y-X, Li WH, Estimating the age of the common ancestor of men from ZFY intron, *Science* **272**: 1356–1357, 1995.

Gebo DL, MacLatchy L, Kityo R, Deino A, Kingston J, Pilbeam D, A hominoid genus from the early Miocene of Uganda, *Science* **276**: 401–404, 1997.

Goodman M, *Macromolecular Sequences in Systematic and Evolutionary Biology*, New York, Plenum, 1983.

Groves C, Mammals in Sulawesi: Where did they come from and when, and what happened to them when they got there, in Metcalf I, Smith JMP, Morwood M, Davidson I (eds.), *Faunal and Floral Migrations and Evolution in SE Asia–Australasia*, Balkema, Rotterdam, 2000.

Hasegawa M, Horai S, Time of the deepest root for polymorphism in human mitochondrial DNA, *J Mol Evol* **32**: 37–42, 1991.

Hasegawa M, Yano T, Kishino H, A new molecular clock of mitochondrial DNA and the evolution of the hominoids, *Proc Jpn Acad* **60**: 95–105, 1984.

Kaplan G, Survival, adaptability and evolution: The case of the orang-utan, in Metcalfe I (ed.), *Where Worlds Collide: Faunal and Floral Migrations and Evolution in SE Asia–Australasia*, pp. 41–42, 1999.

Long JA, Buffetaut E, A biogeographic comparison of the dinosaurs and associated vertebrate faunas from the Mesozoic of Australia and Southeast Asia, 2000.

McBreatty S, Jablonski NG, First fossil chimpanzee, *Nature* **437/1**: 105–108, 2005.

Metcalfe I, Palaeozoic and Mesozoic tectonic evolution and biogeography of SE Asia–Australasia, in Metcalf I, Smith JMB, Morwood M, Davidson I (eds.), *Faunal and Floral Migrations and Evolution in SE Asia–Australasia*, Balkema, Rotterdam, pp. 15–34, 2000.

Ovchinnikov IV *et al.*, Molecular analysis of Neanderthal DNA from the northern Caucasus, *Nature* **404**: 490–493, 2000.

Oxnard CE, The place of the australopithecines in human evolution: Grounds for doubt? *Nature* (*Lond.*) **258**: 389–395, 1977.

Oxnard CE, The challenge of human origins: Molecules, morphology, morphometrics and modeling, in Brenner S, Hanihara K (eds.), *The Origin and Past of Modern Humans as Viewed from DNA*, in *Recent Advances in Human Biology,* Oxnard CE (series editor), Vol. 1, World Scientific, Singapore, New Jersey, London, Hong Kong, pp. 11–30, 1995.

Oxnard CE, The time and place of human origins: Implications from modeling, in Clark GA, Willermet CM (ed.), *Conceptual Issues in Modern Human Origins Research*, Chap. 26, Aldine de Gruyter, New York, pp. 369–391, 1997.

Oxnard CE, Morphometrics of the primate skeleton and the functional and developmental underpinnings of species diversity, in O'Higgins P, Cohn M (eds.), Development, Growth and Evolution: Implications for the Study of the Hominoid Skeleton, Chap. 10, *Linnean Soc Symp Ser* **20**: pp. 235–264, 2000.

Oxnard CE, Wessen K, Modelling divergence, interbreeding and migration: Species evolution in a changing world, in Metcalfe I, Smith JMB, Morwood MJ, Davidson I (eds.), *Faunal and Floral Migrations and Evolution in SE Asia–Australasia*, Swets and Zeitlinger, Balkema, Lisse, pp. 373–385, 2001.

Raup DM, Gould SJ, Schopf TJM, Simberloff DS, Stochastic models of phylogeny and the evolution of diversity, *J Geol* **81**: 525–542, 1973.

Sarich JM, Wilson AC, Rates of albumin evolution in primates, *Proc Nat Acad Sci USA* **58**: 142–162, 1976.

Sepkoski JJ Jr, Kendrick DC, Numerical experiments with model monophyletic and paraphyletic taxa, *Paleobiology* **19**(2): 168–184, 1993.

Sibley AG, Ahlquist JE, The phylogenetics of the hominoids as indicated by DNA-DNA hybridization, *J Mol Evol* **20**: 2–22, 1984.

Simpson GG, *The Major Features of Evolution*, Columbia University Press, New York, 1953.

Smith AB, Gale AS, Monks NEA, Sea-level change and rock-record bias in the Cretaceous: A problem for extinction and biodiversity studies, *Paleobiology* **27**: 241–253, 2001.

Tavare S, Marshall CR, Will O, Soligo C, Martin RD, Using the fossil record to estimate the age of the last common ancestor of extant primates, *Nature* **416**: 726–729, 2002.

Wessen K, Simulating the origin and evolution of ancient and modern humans, Doctoral Thesis, University of Western Australia, 2002.

Wessen K, *Simulating Human Origins and Evolution*, Cambridge University Press, Cambridge, 2005.

White TD, Suwa G, Asfaw B, Australopithecus ramidus, a new species of early hominid from Aramis, Ethiopia, *Nature* **371**: 306–312, 1994.

White TD, Suwa G, Asfaw B, Australopithecus ramidus, a new species of early hominid from Aramis, Ethiopia, *Nature* **375**: 88, 1995 (letter of correction).

Wood B, Constantino P, Human origins: Life at the top of the tree, in Cracraft J, Donoghue MJ, (eds.), Assembling the Tree of Life, Oxford University Press, New York, pp. 517–535, 2004.

Wood B, Collard M, The human genus, *Science* **284**: 65–71, 1999.

Chapter 7

Modern Humans and Heresies

African Eve and Adam

This discussion of modern human evolution logically follows the ancient human origins of the previous chapter. While it seems highly likely that ancient humans arose in Africa, it is also clear that they must have spread to the various continents of the world in very early times, in something like the last two or possibly even more millions of years. Modern humans, likewise, are believed to have arisen in Africa but in this case perhaps a hundred and fifty thousand or so years ago, and spread, again, to the rest of the world. At about the same time the more ancient groups in the world seem to have disappeared. Did they die out? Were they competed out? Or were they assimilated? Surely there were elements of all three! There remains considerable controversy about modern human origins.

Some workers aver that what happened was replacement, without any genetic admixture, of older humans in every continent by the more recent exodus from Africa. This has been called the African Replacement Hypothesis of Modern Human Origins (or the Noah's Ark Hypothesis). Others hold an opposite view, that the original ancient spread of the genus *Homo* from Africa to the other continents has given rise in all those continents to the humans of today. This was called the Regional Hypothesis of Modern Human Origins (or the Candelabra Hypothesis). This latter view is now believed rather unlikely in its extreme form. It has therefore been modified to allow that, while the origins of modern humans were from ancient humans on each continent, there had also been intermittent and possibly even relatively continuous intermixture across the continents, including especially the recent migrations from Africa. This modification is termed the Multiregional Hypothesis of Modern Human Origins. All this has been examined in the previous chapter. But it also forms the basis for the material in this chapter through a different set of studies.

The evidences for these ideas, contradictory though they seem to be, were based originally upon the morphology of fossils of the genus *Homo* and their geological, geographic and ecological contexts. For example, Fig. 1 shows a few specific fossils as exemplars of human relationships in the African Replacement Hypothesis. More recently, this view is supported by genetic and molecular evidences. For example, studies of mitochondrial DNA (inherited from mothers to all progeny, sons and daughters alike — leading to the idea of the so-called 'African Eve') and Y chromosomes (coming from fathers to all sons — apparently implying an African 'Adam') support the African Replacement Hypothesis (Figs. 2 and 3).

However, data from the very same fossils have been interpreted by others in a different way. Thus, Fig. 4 shows another view of the relationships in Fig. 1. These relationships that support the Regional and Multiregional Hypotheses of Human Origins yet are

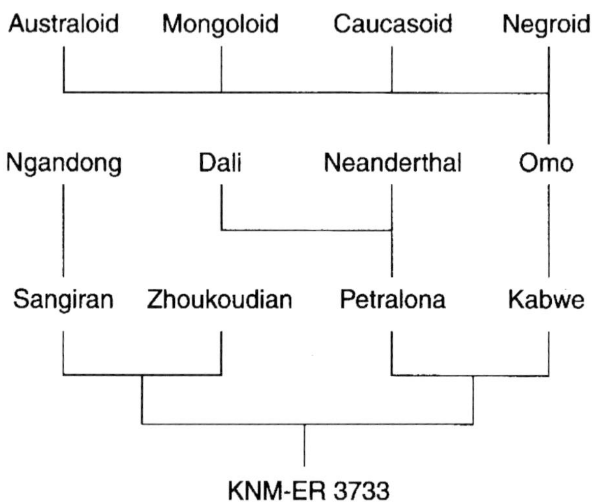

Fig. 1. The Out-of-Africa view of modern human origins with examples of specific fossils (a simplified view).

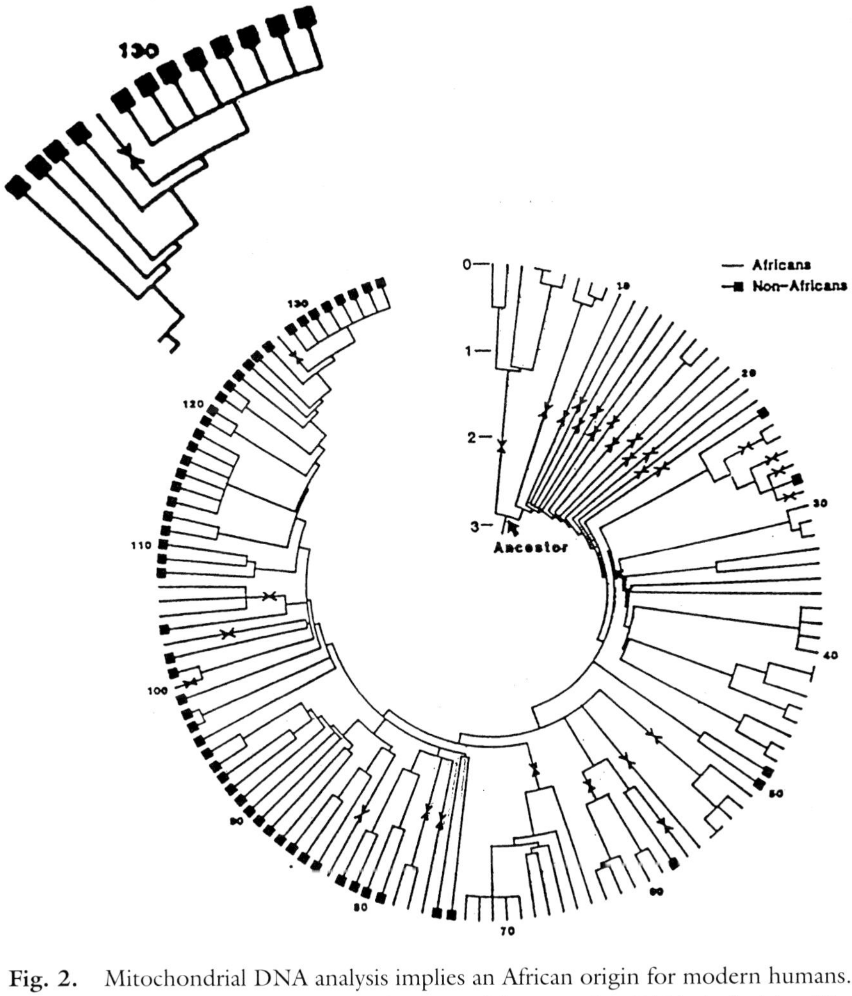

Fig. 2. Mitochondrial DNA analysis implies an African origin for modern humans. Only African individuals are close to the root of the tree of mtDNA similarities. The enlarged inset of a portion of the plot, above left, allows one to more easily see the nature of the relationships.

based upon the same fossils (and many others, of course, not shown in the figure)!

Likewise many of the molecular investigations that seem to so strongly support the African Replacement Hypothesis can also be reinterpreted. For example, many of the original mtDNA studies

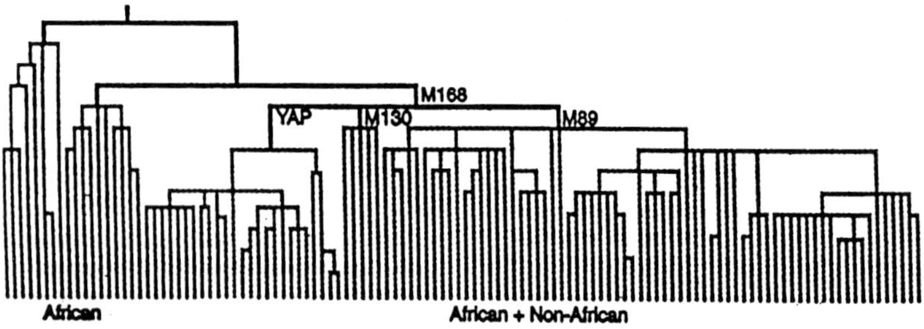

Fig. 3. Y chromosome analysis implies an African origin for modern humans. Only African individuals are close to the root of the tree of Y chromosome haplotype similarities.

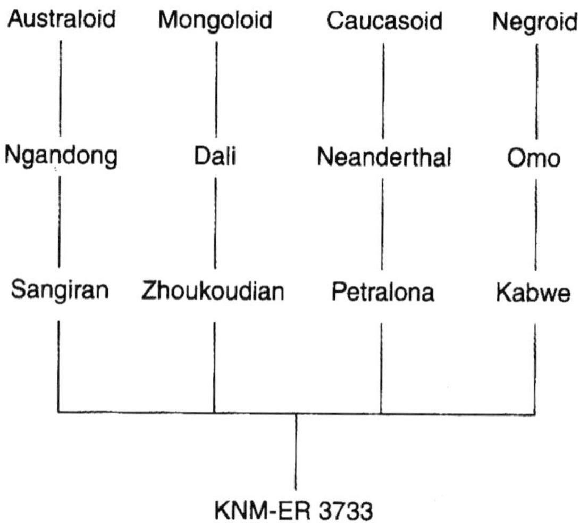

Fig. 4. The regional and multiregional views of modern human origins as shown by the examples of the same fossils in Fig. 1 (a simplified view).

were calibrated according to a human/chimpanzee common ancestor at 5 million years ago and it was this that supplied the 150,000 year timing for modern human origins through the African Replacement Hypothesis. Ayala (1995) has shown, however, that if the human/chimpanzee common ancestor were only slightly older, at say 6.3 million years, then these mtDNA data would extend modern human origins to 300,000 years. He further shows that at 7.5 million years for the common ancestor, modern human origins would be pushed out to 600,000 years, and at 10 million years, to close to a million years. The previous chapter discusses the effects of these changing possibilities for that human/chimpanzee common ancestor time (Oxnard, 1977; Arnason *et al.*, 2000). This is a degree of uncertainty about the time of modern human origins that is far greater than that generally implied by the mtDNA studies.

Greater understanding of the Y chromosome investigations also implies that the congruence with the mitochondrial DNA studies at around 150,000 years for the modern African exodus may be open to reinterpretation. Thus, Fu and Li (1995) have suggested that the overall number of breeding males in the ancient populations could have influenced the calculations of time. If there were only an average of 2500 breeding males, the mean time of modern human origins could be as little as 60,000 years (which seems unlikely). They show, however, that this is a mean that ranges from 31,000 to 219,000 years (and could, thus, cover 150,000 years of the original mtDNA studies). Increasing the number of breeding males (on average) to 10,000 (one would think still a very conservative estimate) pushes the mean origin time out to 313,000 years but with every possibility from 114,000 to 721,000 years! Tripling the number of breeding males to only 30,000 (still a fairly small number) sends the mean to 703,000 years and gives possible limits (not statistical limits) from 284,000 to 1.5 million years!

All of this suggests to me that we do not really have any good figures from either fossil or molecular data. These data do not necessarily falsify either of the major extant hypotheses.

Finally, it has started to be recognised that, in any case, relationships of individual genes, groups of genes and post-gene molecular factors, are very different from relationships of individual people and populations of people (Fig. 5 and Avise, 2000). One way of sorting out these different possibilities is better investigation of the fossils, especially investigating the increased numbers of fossils that modern studies are supplying. But this has difficulties. The last chapter (and Oxnard and Wessen, 2001) implies that determining common ancestors from the characters of the fossils may result in estimates that are wrong by factors as great as ×2 to ×4. This is independently confirmed by computer modelling studies (Tavare *et al.*, 2002). A second way of studying these matters is more sophisticated understanding of the

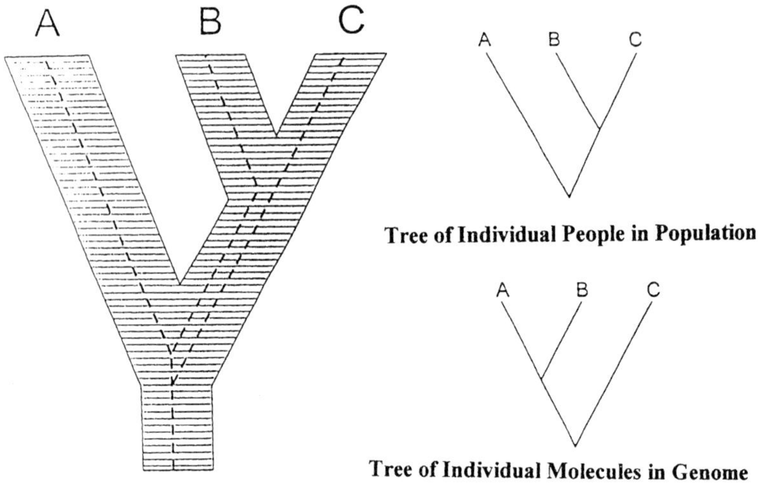

Fig. 5. An original population of individuals dividing into populations A, B and C as shown. This diagram illustrates a difference between the tree of relationships of changing individuals (the shaded tree) and the tree of relationships of changing molecules (the dotted tree).

complex differences between genetic and morphological data and of how they may be viewed. For example, it is now well recognised that population bottlenecks may greatly influence data from both molecules and morphology, and indeed, may possibly limit how far back these methods can 'see'. Third, it is beginning to be recognised that reproductive phenomena (e.g. numbers of breeding males, different mating patterns), may also markedly modify molecular results. As a result, it is now starting to be understood that molecular studies do not provide information about the evolution of individuals and populations, but about the evolution of molecules in individuals and populations. As Ayala (1995) has so succinctly pointed out, these are not the same.

Lineages of Individuals

It was for these kinds of reasons that I started to be interested in whether or not one could understand some of these complexities by mimicking genealogical lineages. This logically followed from my initial attempts to mimic species and subspecies phylogenies (Chap. 6). This was what the organisers of that Kyoto workshop in 1995 (Chap. 6) had originally wanted and which I felt, at the time, I could not do. Since then however I have tried, and, like my first attempts at mimicking species phylogenies, my first tries at individual lineage mimicry were simplistic in the extreme.

I again initially used pencil-and-paper, and a dart (!) and, later, a random numbers table. This allowed me to draw sample genealogies and identify the 'mothers-of-mothers' (compare with mtDNA studies) over several generations. With a little more thought, I realised that I could also identify the 'fathers-of-fathers'. Although I did this (Oxnard, 1997) before I had actually read about Y chromosome investigations, it neatly foreshadowed them. My first lineages were so small and simple that the results have no real value. But they do have heuristic value in that they show that it can be done. Further, because the initial implications were surprising (at least to me) they lead me into trying something better.

Thus, this first attempt (Oxnard, 1997) included only a very small number of generations. In addition, I had to do several special things in order to prevent the whole show going extinct almost immediately. These included: allowing sibling mating when necessary, migrating only very few individuals to another continent (in this context perhaps better called community), and allowing a larger number of progeny than one might be deemed usual. Figure 6 shows one attempt. Two individuals in the first community (A) migrated to a second community. The remaining people in community A were not followed in order to reduce the work. The people who migrated to the second community (B) were joining individuals from earlier generations already in that community. For similar reasons, these earlier generations, too, were not followed. Interbreeding of the migrants was then allowed with individuals in the second community. At a slightly later stage, a migration was made into a third community (C). This community was empty; so this migration did not involve new interbreeding though it separated the progeny from those in the second community. Finally, a second migration from the second to third communities was permitted very near to the present day. The entire

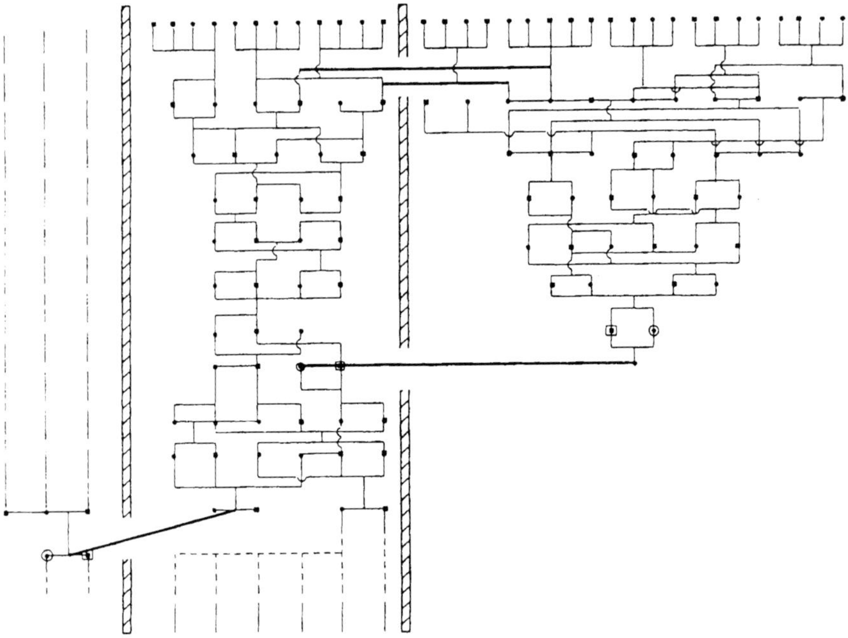

Fig. 6. The original mimicry of male and female individuals, their lineages and migrations (Oxnard, 1997). As by usual convention, males are squares and females circles. The lineage starts at the bottom of the diagram in the population on the left. The rest of the individuals in the population at the left are not shown (dashed lines). One pair from the left population migrates to the middle population which already has individuals in it. Again, however, the earlier individuals are not shown. In the fourth generation after the migration, a second migration occurs into the empty community on the right. Breeding continues to the present day with a third migration, also from the middle to the right population occurring one generation before the present.

simulation was only 13 generations long. Of course there were no characters for these individuals.

The sexes of all individuals were known. This meant that I had documentation of who were mothers and fathers for every individual at the present day, and of grandmothers and grandfathers, great-grandmothers and great-grandfathers, and so on, back through to the starting generation. The whole thing is so simplistic that, were it not for its exemplar value, I would not present it.

What, then, did it show?

First, I took a sample of the present-day individuals (not every individual) and drew the links backwards from their mothers, through their grandmothers, etc, to the base of the genealogy. This allowed me to identify the ultimate mother (Fig. 7). In those days I assumed that this would relate to a mitochondrial DNA lineage but, of course, I know today that this is not the case. It was immediately apparent that there were several common maternal ancestors, one related to every living individual in each (respective) community, and one overall. These 'common mothers' were located at generation 4 in the third community, and generation 6 in the second community. There was a third 'common mother' in the second community that was further back in time. She was the 'mother' common to everyone from the second and third communities. There was a final common ancestral 'mother' at generation 13 in the first community.

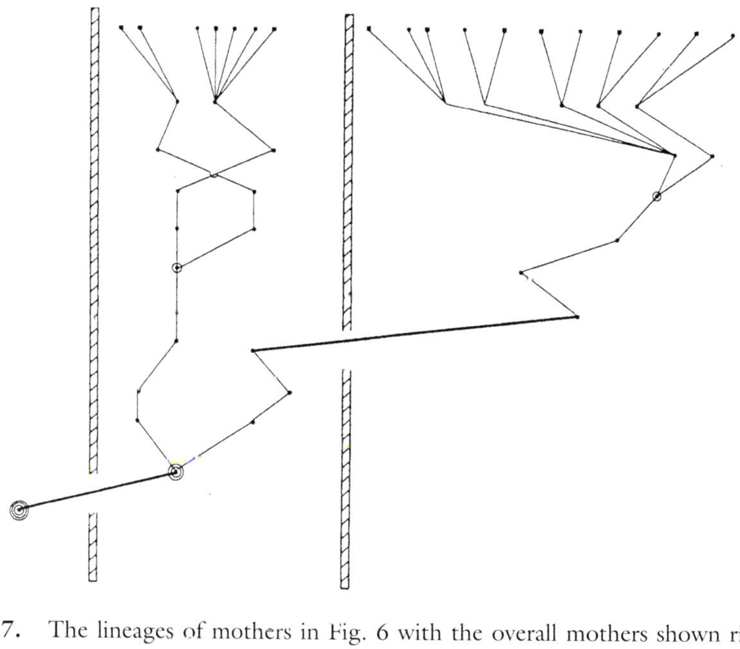

Fig. 7. The lineages of mothers in Fig. 6 with the overall mothers shown ringed (one ring, the ancestral mother in each community; two rings, the ancestral mother of individuals in both communities; three rings, the mother of us all). This can be checked by superimposing this picture on the total lineage of Fig. 6.

Next, I took the sons only from the same sample of individuals at the present time. I did this because I wanted to trace the corresponding fathers. Again, I was able to find the fathers and grandfathers, and so on, tracing back through the lineage in the same way (Fig. 8). The 'fathers-of-us-all' in the second and third communities were just as readily identified as the 'mothers-of-us-all', but, interestingly, they lay two generations further back in each (7 and 8 respectively) than the corresponding mothers. The ultimate 'father-of-the-father-of-us-all-of-us-all' was even more different from the 'mother of us all'. He not only lay in a different generation (11 rather than 13) but also in a different community (the second rather than the first).

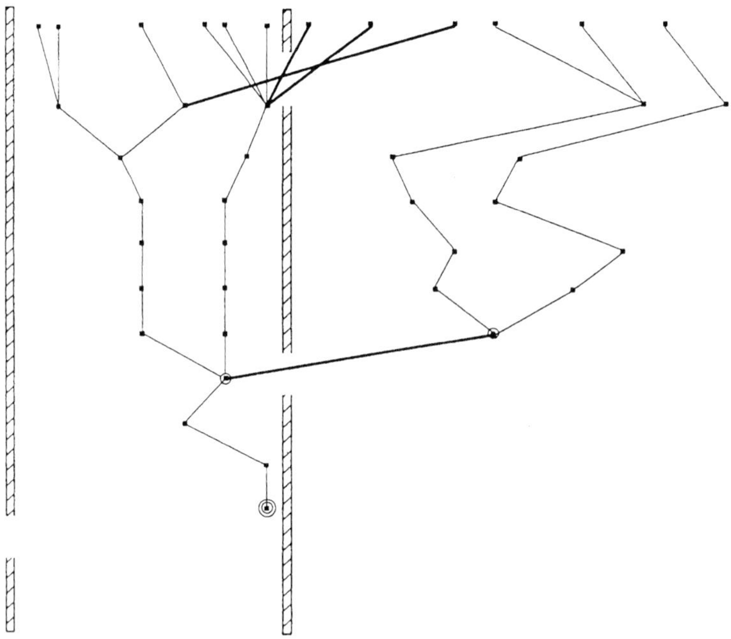

Fig. 8. The lineages of the fathers of sons in Fig. 6 with the overall fathers shown ringed (one ring — the ancestral father of the present-day sons in each community; two rings — the ancestral father of sons in both communities, three rings — the father of us all). This can be checked by superimposing this picture on the total lineage of Fig. 6.

One's immediate irreverent thought about these various 'ancestral grandparents', was: 'it's a long way and a long time to go for sex'.

Ken Wessen's Lineage Programme

Once the collaboration with Ken Wessen (described in Chap. 6 and in Wessen, 2002; Wessen, 2005) was initiated, it was clear that far better things could be done. Wessen was easily able to elaborate a computational programme to mimic genealogies. How his programme works can be initially explained through the small explanatory run shown in Fig. 9.

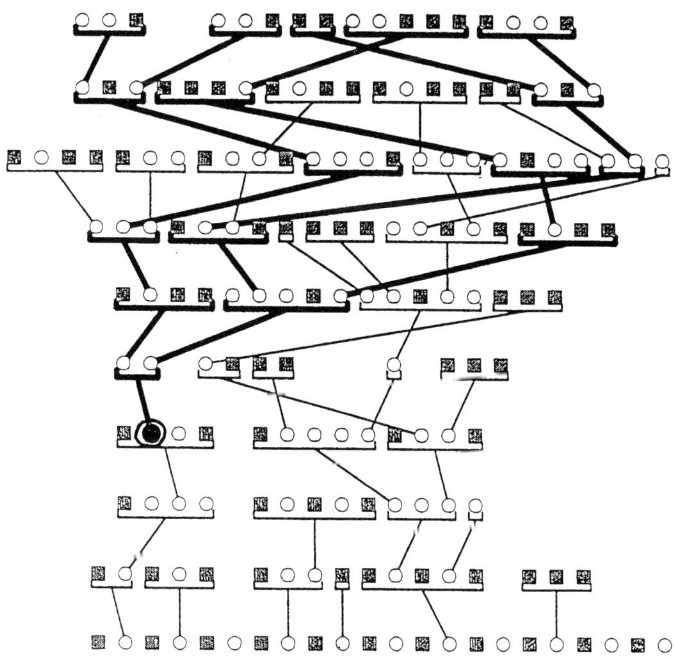

Fig. 9. A very small portion of a Wessen produced lineage to show how it works. There are 10 generations starting with only 11 sex pairs at the foot of the diagram: closed squares — males; open circles — females. The mother of everyone in the present generation is the closed encircled female on the extreme left seven generations from the present.

This figure shows a matriline leading back from every member of the present generation (top of the diagram) to a single ancestral mother who is only seven generations back. If, with the same genealogy, we choose as the starting generation the second generation back from the present rather than the present-day generation we obtain a different mother. She is on a matriline that is rooted further back in the past even before the start of the process (Fig. 10). Yet, there is no

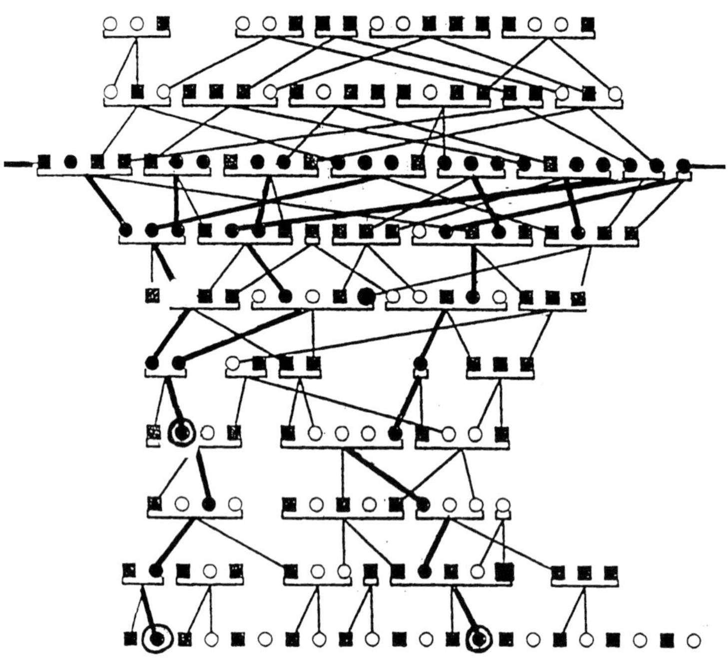

Fig. 10. The same lineage as Fig. 9 but with the maternal ancestors plotted for the generation two back from the present day (i.e. it shows what would have happened if we had done this for our grandparents two generations ago rather than for ourselves today). Now we can see two ancestral mothers. One is the same as in Fig. 9 (with extra bold marking showing the position in the bottom generation where that lineage started). The second is due to the extra individuals in the third generation that did not leave the offspring proceeding through to the top generation. It leads back to a different mother in the 10th generation widely different from the mother in the 7th generation. In order, now, to obtain the single mother of us all we would have to go back well beyond the bottom generation in the diagram. Yet this is the identical genealogy to Fig. 9.

other difference in these two diagrams. What, then, is the implication of finding the female common ancestor — I suspect very little. Certainly it does not herald the start of anything. The community still exists around it. It is clearly necessary to study these matters in much more detail.

In carrying out the genealogical mimicry Wessen's aim was to allow questions to be asked about the genealogies. How are lineage relationships affected by mating factors such as monogamy, polygyny, polyandry (and even polyandrogyny), together with the allowance for male and/or female infidelity? How are relationships affected by altering breeding factors such as reproductive chance, average number of offspring, percent of male offspring, and adult (i.e. breeding) sex ratio? How do different kinds of migrations and back-migrations between communities (analogous to the continents of the species studies but better now called communities) affect lineages? What are the effects of migrations involving primarily males (e.g. a war party), primarily females (e.g. a war party returning with female 'slaves'), both sexes (e.g. a genuine migration of families), or only small numbers (mimicking, in the human milieu, the effects of itinerants or travelling vendors)? Finally, how do a variety of miscellaneous phenomena, such as changing selective advantage and occasional population disasters, affect genealogies? All of these can be independently varied at different times and in various communities.

Putting this into operation was a good deal more difficult than for the species programme. It works in the following way. Wessen defines three separate communities. Start with an existing population (not a single species as in the species and sub-species simulations). Define such migrations and back-migrations to and from other communities at the rates specified for a particular run. Apply the 'natural disasters' component to mimic chance events. Form 'mating groups' from the list of remaining individuals in that generation. 'Cull' these individuals to obtain the adult sex ratio that it is desired to simulate (e.g. in simulating polygyny, some males are culled, for polyandry, some females, and so on). Then examine the new generation and remove the individuals that have died (i.e. have no progeny for whatever reason). This process is repeated with the next generation and the next until the present generation is achieved.

Once all the generations have been simulated (number as specified at the beginning), we can examine the results in two ways. One is to take a sample (randomly determined) of male individuals from each community and find the generation and individual of the male ancestor of all the 'living' individuals. The other is to determine the common female ancestor in the same way. This is then repeated for each community. The community data are then combined, and the male and female ancestral generations and individuals determined for all communities.

The whole thing can then be repeated with hundreds or even thousands of generations, and hundreds and even thousands of times, in order to obtain statistical estimates of many different possibilities. It means that the individual relationships in Figs. 9 and 10 cannot be diagrammed. Instead, the figures only show whole generations as horizontal lines. But, all the data for every generation and every individual are available in the detailed written report. This is possible in our mimicked genealogy, but could never be available in real life.

Offspring and Chance of Survival

One of the first matters that Ken Wessen looked at (Wessen, 2005) was the combined effect of the number of offspring and their chance of survival. Many runs were carried out for each condition in order to obtain statistical information. Thus, runs with an average of 4 children with 100% survival provided a paternal ancestor 58 generations back, but a maternal ancestor 110 generations back, almost twice as far back. Eight offspring with a survival rate of 50% left the paternal generation at approximately the same level, 58 generations back, but found the maternal common ancestor now at only 26 generations back. Twelve offspring with only 33% survival placed the paternal ancestor at 14 generations back and the maternal at 35 generations back. Thus, different totals in the number of offspring and differential survivability changed the times of the maternal and paternal ancestors enormously. These are the factors that are rarely taken into account in mathematical models of common ancestry.

These findings can be seen visually in single runs that I have carried out. Thus, Fig. 11 shows a single run of the difference between 4 children with 100% chance of survival, 8 children with 50% chance of survival and 12 children with 33% chance of survival. The generations of the ancestral mothers and fathers differ completely, and the proportion of individuals who do not reach the present generation are, naturally, also different.

The obverse, the chance of not surviving, may also be important (see later section).

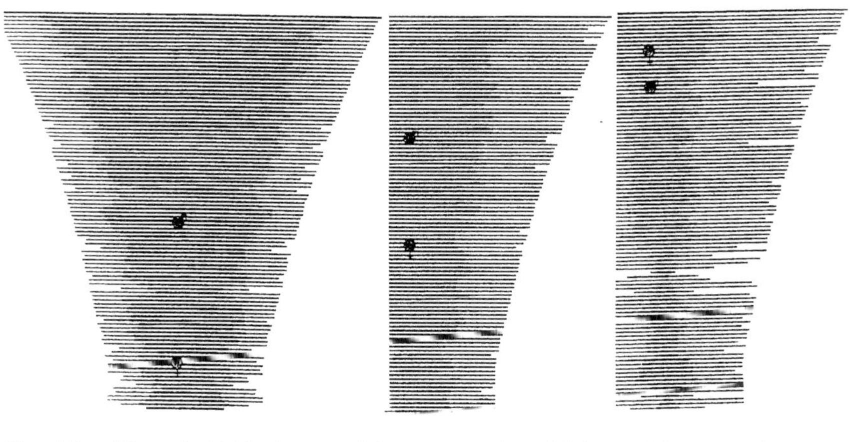

Fig. 11. Three individual runs with averages of 4 children with 100% of survival (first frame), 8 children with 50% chance of survival (second frame) and 12 children with 33% chance of survival (third frame). The runs start at the bottom and the present-day generation is at the top. Of course, there are now so many individuals and generations that they can only be shown visually as a line for each generation. But the full data about all the characters of every individual in every generation are available in the printed output. The first run is shown in its entirety. The second and third are shown as half-plots (because the figures are bilaterally symmetrical). Note the differences in the position of the ancestors and also, of course, the difference in the number of individuals that are discarded (light outer parts of the output) compared with those retained (dark central parts of the output).

Breeding Patterns and Sex Ratios

Similarly, Wessen found that different mating patterns and different adult sex ratios have enormous statistical effects on ancestors in relation to monogamy, polygyny and polyandry. In an average monogamous situation with an approximately 50–50 adult sex ratio (as in many human groups) the generations for male and female ancestors were approximately equal. In obligate polgyny with unbalanced sex ratio (more females than males as in the great apes), the paternal ancestors were on average only 6 generations from the present but the maternal 128 from the present. In polyandry and unbalanced sex ratio the other way (as in some marmoset breeding groups for example), on average the paternal generation was 59 generations before the present and the maternal 82 before the present. It is thus evident that breeding patterns and sex ratios are enormously important in understanding the difference between maternal and paternal lineages (Wessen, 2005).

Again, these findings are evident in single runs (Fig. 12) that I carried out. Thus, Fig. 12 shows, by the usual symbols, the differences between maternal and paternal ancestors in a monogamous run with 50/50 sex ratio, polygyny with 25/75 male to female ratio and polyandry with 75/25 male to female sex ratio. For monogamy, the paternal and maternal common ancestor generations are more or less in the middle of the run at the 55th and 40th generation levels back from the present day. For polygyny, the paternal and maternal common ancestor generations were much later in time and reversed in position at the 3rd and 13th levels back from the present day. Furthermore, the numbers of individuals that had descendants at the present day were much greater for monogamy than polygyny. For polyandry, the paternal common ancestor was almost at the bottom of genealogy (97 generations back), the female about the middle (54th generations back).

How does this look from the viewpoint of known differences among primates.

All the great apes have 2 or 3 to 1 sex ratios in breeding adults (although the way this is achieved is different in the different apes,

Fig. 12. Three individual runs with monogamy and a 50/50 sex ratio (first frame), polygyny and a 1:3 male to female sex ratio (second frame), and polyandry with a 3:1 male to female sex ratio (third frame). The conventions are as in Fig. 11. Note the differences in the position of the ancestors and also, of course, the difference in the number of individuals that are discarded (light outer parts of the output) compared with those not lost (dark central parts of the output).

Oxnard, 1987). Both humans and lesser apes have around a 1 to 1 sex ratio as breeding adults. This means that the great apes show obligate polygyny, whereas gibbons and humans, though they can display polygyny, are not obliged to. In other words, in gibbons and humans, polygyny is facultative. Gibbons and humans have the choice; the great apes (chimpanzees, bonobos, gorillas and orang-utans) do not. This choice is especially true of humans who may be polygynous by being serially monogamous as well as simply polygynous *per se*. Even gibbons in special situations can be facultatively polygynous.

The polyandry case is especially interesting in showing the paternal common ancestor very far back from the present. Among primates, polyandry is much less common. However, it exists in some species (e.g. some marmosets).

The two situations (monogamy and polygyny) were run a number of times allowing migrations to three communities. The results are immediate and clear, even though, at this point, we have only

Table 1. Monogamous model: 10 runs, 25 generations, 3 continents.

	Common ancestors
Using random samples	
Female average	12th generation
Male average	16th generation
Using entire population	
Female average	6th generation
Male average	14th generation

Table 2. Polygynous model: 15 runs, 25 generations, 3 continents.[a]

	Common ancestors
Using random samples	
Female average	16th generation
Male average	9th generation
Using entire population	
Female average	12th generation
Male average	7th generation

[a] Note: The polygynous model went extinct 6 times out of 15.

analysed a small number of runs. Thus, in Tables 1 and 2, female common ancestors are found at younger generations than male common ancestors in monogamy, but at older generations than males in polygyny. In addition there was bonus information: in these few runs, the polygynous model went extinct for a third of the time! Perhaps this latter phenomenon might be modified by including additional possibilities (e.g. the presence of all male bands as in chimpanzees). It might also be the case that this result depends upon the fact that the number of males in polygyny is half the number in monogamy (in other words, a 'number of breeding males' effect, as found by Fu and Li (1995). The same number of males and double the number of females might give a different result.

These findings could have marked implications for primates. Differences in a variety of reproductive factors could be examined. These include differences in sex ratio; most great apes have adult sex ratios of 2 or 3 females or more to each male; sex ratios in humans may have differed from the current ones at different times in the past. They include differences in mating patterns. Contrast average mating patterns in orang-utans (polygyny with an isolated male servicing several separate females), in bonobos (with polyandrogyny and free-for-all mating, in fairly large groups), and in humans (with a combination of monogamy, serial monogamy, facultative polygyny, polyandrogyny, etc). They even include differences in the types of 'migration'. In some species and some communities within species males 'migrate' from their natal groups, in others, females 'migrate' (or are taken). Many other such factors could be investigated.

Of course, let me re-emphasise: these results stem from very few runs; they are little more than exemplars. Would a thousand replications be made, the figures might be completely different. Nevertheless, there are enough surprises here to keep us working hard.

The Effects of One-Child Policies

Finally, I decided to mimic one-child policies. One would think that it was not necessary to examine this situation. But it is always worthwhile doing something however apparently obvious; one may get a surprise.

Figure 13 shows that, with a one-child policy (on average) and with that child being male (on average), extinction of the community occurs about 80 times out of a 100 runs after only 10 to 35 generations. Surprisingly, however, 15 times out of a 100 runs, the population makes it through all the way to 100 generations, but, as the example shows, only just. As a result both paternal and maternal ancestors of the living individuals are very close to the present. In only 5 runs out of a 100, did this policy make it totally successful to 100 generations! In these few cases, the paternal ancestors were

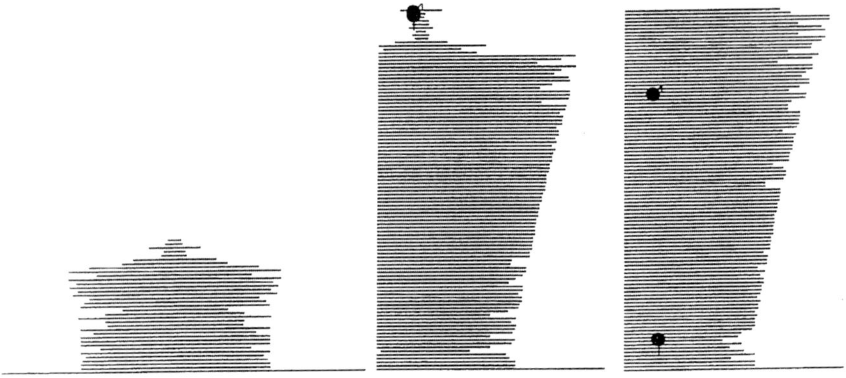

Fig. 13. Three individual runs that mimic the one-child policy. On average, there is a single child and on average that child is a male. However, male and female infidelity is allowed, males are removed more frequently than females (e.g. accidents, murder, and wars), extra females are added occasionally (e.g. wife stealing, male infidelity, etc). The left-hand run shows an example of what happens 80% of the time: the populations die out after only 5 to about 25 generations; there are, therefore, no common ancestors because these are only shown for the present-day generation. The middle-run shows what happens about 15% of the time: the populations do make it to the present generation but only just; it is clear that they are likely to die out at this level. Both male and female common ancestors of the living form are very close to the present day. The right-hand run shows that, on occasion (about 5%) a normal population is achieved at the present day. When that occurs, the father of us all is not far from the present day, but the mother of us all is close to the base.

generally very close to the present and the maternal very far back. Also, however, in these few runs, the positions of the female ancestors were enormously variable.

In carrying out these runs it must be realised that I included such phenomena as: some female offspring being successfully raised (in other words, mimicking avoiding total female abortion, female infanticide, female 'childicide', or female 'adulticide'). I also changed the breeding factors to mimic such phenomena as some females being introduced from outside sources (simulating wife migration, wife buying, wife stealing, etc.). I included provision for some males being lost (accident, competition, murder, war, etc.). These are extra factors that may have affected the results.

I have not yet moved to the point of being able to see what happens when lineages throw up differences so large that evolution into new subspecies or species can be said to have occurred. That would, of course, be an interface between this and the prior chapter. Wessen looks forward to a most ambitious aim in eventually combining his phylogenetic and genealogical programmes into a unified mechanism. This should provide a very powerful and interesting tool.

It is thus evident that the various factors that we can examine have considerable complexity and cover many features that could be important, and are indeed likely to have been important in human and animal genealogies. Understanding their precise effects depends upon examining runs where the factors are varied both independently and together (Wessen, 2005). Much more can be done; examining the interactions among many combinations of factors will require the efforts of many graduate students.

Lineages of Genetic Materials

Finally, Ken Wessen has modified the simulations to examine the implications for gene and chromosome evolution, especially for autosomes, mtDNA and Y chromosomes. In addition to the effects of the population size and bottlenecks (already known to influence gene and molecular determinations), it is also hoped that there will result a better understanding of many other modulating factors. Thus, differences in reproductive sex ratios, mating patterns, degrees of consanguinity, types of migrations (whether largely of males or females or mixed), may all affect the estimations based upon the different gene studies. That is, we can now examine, not just lineages of individuals, but lineages of genetic materials.

One set of these lineages involves a neutral gene model. It superimposes independent and variable mutations and variable recombinations for six loci in the genealogy. This thus allows specification of genetic parameters for autosomes A1 and A2, mitochondrial DNA mt1 and mt2, and Y chromosomes Y1 and Y2. Their combination rates can be varied for each community. An initial diversity can be specified. All genes are currently implemented as 16-bit strings, each bit of which can vary as mutations. Each locus has an independent mutation rate. This rate may be constant or varying over a range of generations. Recombination of genes can be simulated but, at this stage, there is no intragenic recombination. The output relating to this is very detailed, listing the complete genotype and characters for each member of the final generation. This allows a backward look at the final populations. The lineages of these genetic components can be compared with the lineages for individuals. They show that the evolution of the various genetic factors is not at all the same for each, nor the same as the evolution of individuals.

A second set involves a genotype model. This is distinct from the neutral model and mimics two alleles at a single locus. These may be sex-linked. Selection is mimicked by identifying one of the alleles as favoured. Dominance, mutation and heterozygote fitness factors are all incorporated. These factors can be either constant or varying over a range of generations in each or all communities. In this case, in contrast to the neutral model above, and because of the role of selection, the resulting mimicry is not independent of the genealogy. The survival of individuals is directly affected by their genotype.

I have not yet fully examined these new, molecular, aspects of the mimicry. However, they immediately confirm that individual lineages and genetic factor lineages are completely different from each other. The most recent ancestors of genetic factors and of individuals lie in completely different generations. In particular, none of these indicates a specific beginning of a population of individuals or of a population of genetic factors. Populations go much further back than the common ancestors of either type. Neither the common ancestors of genes nor of individuals are markers of the beginnings of populations. If we were to extend this from populations to species, would it still be true?

Common ancestors of gene changes are not markers of beginnings of populations or species, or of species splitting. Markers of the beginnings of both particular individuals and particular genes are all much more recent than the continuing populations themselves. I wonder, is this also true of subspecies and species? It may not be reasonable to expect genes (from say, Neanderthals) to be still present in today's humans even if interbreeding had been occurring at earlier

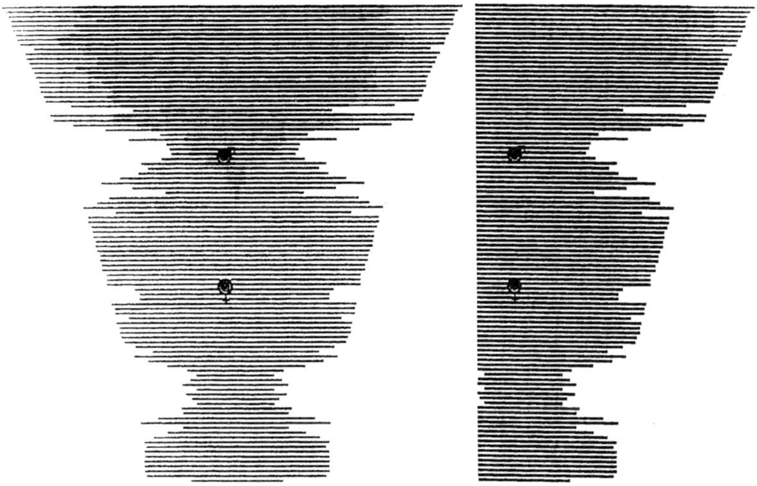

Fig. 14. A run that examines the difference between the two Y chromosome loci. The shape of the distribution and the positions of the common ancestral individuals, are of course, the same. But one Y chromosome (the left-hand plot) is limited to the upper-third of the diagram (the heavily shaded area), the other (right-hand plot) exists almost throughout almost the whole genealogy.

times. Probably not! But possibly so! Neither is it likely that averages of common ancestors of genes will be similar to averages of common ancestors of individuals.

A beginning example of what might be achieved is shown in Fig. 14. This shows a single run for each of the two Y chromosomes. The paternal and maternal common ancestors (individuals) do not change. The shape of the distribution of individuals does not change. But the extension of the Y chromosomes throughout the population (as shown by the darker shading in the plot) is completely different in the two runs, because each was given a very different selective advantage.

Figure 15 is the same run as Fig. 14 but showing the difference between an mtDNA locus and a Y chromosome locus. Again, the shape of the distribution of individuals and the positions of the common ancestors of individuals are the same. However, the gene differences, as shown again by the differences in shading, differ markedly.

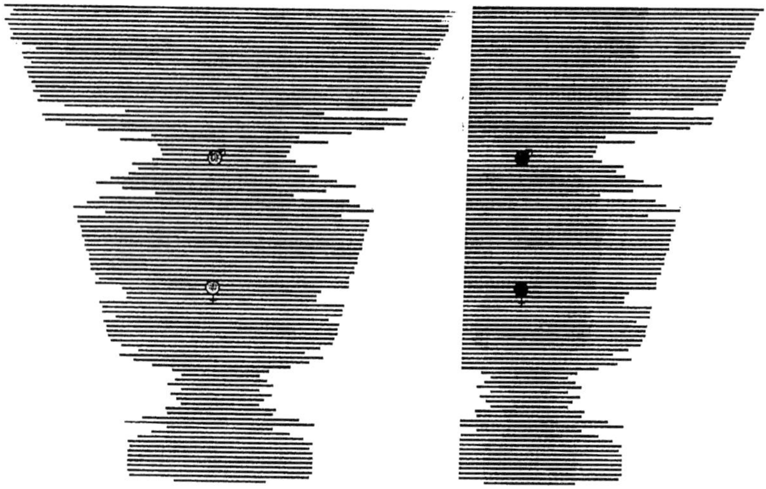

Fig. 15. The same run as Fig. 14, but showing the difference between a mtDNA (left frame) and Y chromosome (right frame). The mtDNA (dark portion of left frame) exists at almost all generation levels. The Y chromosome (dark portion of the right frame) is confined to a much smaller part of the genealogy.

Almost all of the mtDNA elements survive (indeed it is hard to see the ones that do not; they are evident only at the very ends of a few generation levels, the paler parts of the output). In contrast, a very large number of Y chromosome elements are eliminated.

Some Implications for Human Evolution

The heuristics of all this could have implications for human evolution. For example, they could allow the new molecular dates for Neanderthals to be well within the range of modern human origins. They could also imply that major reviews may be needed for assessments of more recent migration dates for the present-day human groups, e.g. African populations migrating into Europe and Asia, Asian populations spreading to Australia and Asians populating the Americas. It is rather likely that such migrations that commenced much earlier than usually postulated, were frequent rather than single or rare events, and involved at least intermittent, and possibly, even, fairly continuous, additions to the gene pools of such people.

To conclude, these results place the 'mother-of-us-all' on one continent, the 'father-of-us-all' on another, and, each in a different generation, i.e. at a different time. This is a result that, with tongue in cheek, means that sex would require both a space machine and a time machine. This is, however, exactly a study of mothers-of-mothers, and fathers-of-fathers, and in this sense it does not replicate the chromosome and gene studies. If they can be confirmed, they would imply a real need for caution in calibrating the molecules in confirming the time and place of origins from fossils.

Such reassessments have serious implications for human evolution. For example, they would allow the new molecular dates for Neanderthals to be well within the range of modern human origins. They also imply that major reviews may be needed for the assessments of migration dates for the present-day human groups, e.g. African populations migrating into Europe and Asia, Asian populations spreading to Australia, and Asian populating of the Americas. It is rather likely that such migrations commenced much earlier than usually postulated, were frequent rather than single or rare events, and involved at least intermittent, and possibly, even, fairly continuous, additions and losses to the gene pools of such peoples. Present-day gene pools may be far more recent than the ages of the first migrating peoples; e.g. DNA from a 60,000-year-old Australian human which seems to differ both from other aboriginal fossils and from modern people (Adcock, 2001) may be merely due to the differences in lineage clocks and gene clocks. Equally, other DNA studies (Ovchinnikov *et al.*, 2000) that imply that Neanderthals show no evidence of having contributed to the modern human gene pool do not exclude the possibility, one might almost say assume a likelihood, that *Homo sapiens* did interbreed with Neanderthals. The evidence may have long since disappeared from modern genetic materials. All of these possibilities are likely to have occurred during the long course of hominoid and human evolution, together with the geological, climactic, ecological, and other changes that went with them.

It is entirely possible that the present-day gene pools may be far more recent than the gene pools and the times of the first migrating people so that their genes do not show. In other words, these results imply that the molecular evidence may have disappeared from modern genetic materials. This is likely to have occurred during the long course of hominoid and human evolution, together with the geological, climactic, ecological and other changes that have occurred.

In other words, mimicry like this may provide a powerful heuristic for examining various factors, not just one by one and as single examples as here, but through the statistics of many runs and in all their various combinations. It may help us to understand that there are many interlocking molecular clocks, not any single clock, and that individual clocks and gene clocks, community clocks and species clocks, may keep different times. The whole problem of timing in evolution is enormously more complex than we generally seem to think. Far more work is required; but it will likely have to be in the hands of my students and my students' students.

References

Adcock GJ *et al.*, Mitochondrial DNA sequences in ancient Australians: Implications for modern human origins, *Proc Natl Acad Sci* **98**: 537–542, 2001.

Arnason U, Gullberg A, Schweizer BAS, Janke A, Molecular estimates of primate divergences and new hypotheses for primate dispersal and the origin of modern humans, *Hereditas* **133**: 217–228, 2000.

Avise JC, *Phylogeography: The History and Formation of Species*, Harvard University Press, Cambridge, 2000.

Ayala FJ, The myth of Eve: Molecular biology and human origins, *Science* **270**: 1930–1936, 1995.

Fu Y-X, Li WH, Estimating the age of the common ancestor of men from ZFY intron, *Science* **272**: 1356–1357, 1995.

Ovchinnikov IV *et al.*, Molecular analysis of Neanderthal DNA from the northern Caucasus, *Nature* **404**: 490–493, 2000.

Oxnard CE, The place of the australopithecines in human evolution: Grounds for doubt? *Nature* (*Lond.*) **258**: 389–395, 1977.

Oxnard CE, *Fossils, Teeth and Sex: New Perspectives on Human Evolution*, Hong Kong University Press, Hong Kong, Washington University Press, Seattle, 1987.

Oxnard CE, The challenge of human origins: Molecules, morphology, morphometrics and modeling, in Brenner S, Hanihara K (eds.), *The Origin and Past of Modern Humans as Viewed from DNA*; in *Recent Advances in Human Biology*, Oxnard CE (series ed.), World Scientific, Singapore, New Jersey, London, Hong Kong, **1**: 11–30, 1995.

Oxnard CE, The time and place of human origins: Implications from modeling, in Clark GA, Willermet CM, (eds.), *Conceptual Issues in Modern Human Origins Research,* Aldine de Gruyter, New York, Chapter 26, pp. 369–391, 1997.

Oxnard CE, Wessen K, Modelling divergence, interbreeding and migration: Species evolution in a changing world, in Metcalfe I, Smith JMB, Morwood MJ, Davidson I (eds.), *Faunal and Floral Migrations and Evolution in SE Asia–Australasia*, Swetz and Zeitlinger, Balkema, Lisse, pp. 373–385, 2001.

Tavare S, Marshall CR, Will O, Soligo C, Martin RD, Using the fossil record to estimate the age of the last common ancestor of extant primates, *Nature* **416**: 726–729, 2002.

Wessen K, Simulating the origin and evolution of ancient and modern humans, Doctoral thesis, University of Western Australia, 2002.

Wessen K, *Simulating Human Origins and Evolution*, Cambridge University Press, Cambridge, 2005.

Chapter 8

Homo floresiensis: A *Very* Cold Case!

So Many Opinions!

Recently, alongside so many young people watching television these days, I have developed an interest in forensic science. This interest has even been recognised by my appointment as Adjunct Professor in the Centre for Forensic Science at the University of Western Australia. My principal role here, of course, is helping fund research through my grants, contributing to the supervision of graduate students and post-docs interested in the assessment of skeletal remains, and, of course, contributing to actual research on bones with Dr Dan Franklin in a forensic science milieu. This new forensic interest has coincided with the discovery (Brown *et al.*, 2004) of the remains of small-bodied individuals at Liang Bua on the Indonesian island of Flores. These remains include much of a single individual dated about 18,000 years ago plus a number of bony fragments of others also dated to within the last 18,000 years or less. They were immediately assigned to a new species *Homo floresiensis* (Brown *et al.*, 2004; Morwood *et al.*, 2005). Yet the matter seems not settled. Are they non-human evolutionary holdouts? Could they be human 'cold cases' of some kind? Their anatomy yields a number of contradictory assessments.

When the find was first published everyone was agog. Of course, very few of us actually had access to the materials. We were mostly dependent upon published accounts, proffered papers at scientific meetings, photographs on the web, and for a very few individuals, access to computer-generated casts, or even the originals. Later, a few more people were able to examine and measure casts, and more post-cranial fragments were found (and some misplaced or damaged!).

Even now the original specimens are not freely available (though it is truly the case that they need protecting, they are very fragile).

I was able to see and handle one of the computer-generated endocasts (through the courtesy of an old colleague in the United States, Ralph Holloway) and I followed this by examining the computer-generated cast that was exhibited that year in the Science Museum, South Kensington, UK. Of course, I could not touch the skull cast, make any measurements, or take my own photographs (though there are excellent pictures in the literature).

Like most people, I suppose, I formed my own immediate opinions based upon what I could see in those two casts, in the first publications, and in the early reported data. Even workers who felt that the fossils were *Homo* of a type, provided pictures and measurements that indicated that the arms were very long in relation to the legs (as in apes and australopithecines) compared with shorter upper limbs relative to that lower in modern humans (Table 1). They also implied that both sets of limb bones were very robust (wide relative to length as in apes and australopithecines) compared with reduced robusticity in humans, particularly in human upper limbs. For these reasons (and others, see below) I was immediately drawn to the position that the fossil creature may have spent time in trees, or was reflecting recent ancestry in trees.

Table 1. Ratio of upper to lower limb lengths in some species.

Living species	Fossil species
Chimpanzee: 100–114	
Bonobo (pygmy chimpanzee) **97–108**	Australopithecus **95–100**
Implication: Locomotor emphasis on upper limb because of climbing, etc.	Implication: Do we assume climbing is still a factor even if generally thought to be bipedal?
Modern humans 67–74	Erectus and Neanderthal humans 67, 72, 73
Implication: Totally non-locomotor upper limb.	Implication: We assume a non-locomotor upper limb!
	Liang Bua estimated **84–90**
	Implication: What can we assume?

Bold numbers = similarities with Liang Bua.

This idea is likewise favoured by a number of other features. Many of these are in the skull. They include the estimated brain size, which, at about 420 cm^3, is totally similar to that of modern apes (300–440 cm^3 for chimpanzees, 380–620 cm^3 for the much larger bodied gorillas) and australopithecines (320–520 cm^3 for *Australopithecus* and 400–550 cm^3 for *Paranthropus*). The Liang Bua skull is much smaller than any human (modern humans have a mean generally well over 1400 cm^3; even most fossil humans are at 900 cm^3 and upwards).

Other skull features include: the position of the temporal line on the vault marking the upper extreme of the temporalis muscle that reaches high towards the midline (as in apes and australopithecines), whereas it is low in humans. Temporalis size can be estimated (to a degree) in the dimensions of the temporal fossa largely containing the temporalis muscle. In this, the fossil resembles apes and australopithecines rather than humans (Fig. 1 and Table 2). The upper canines

Dimensions of temporal fossa containing temporalis muscle

Gorilla	Human today
Australopithecus	Fossil human

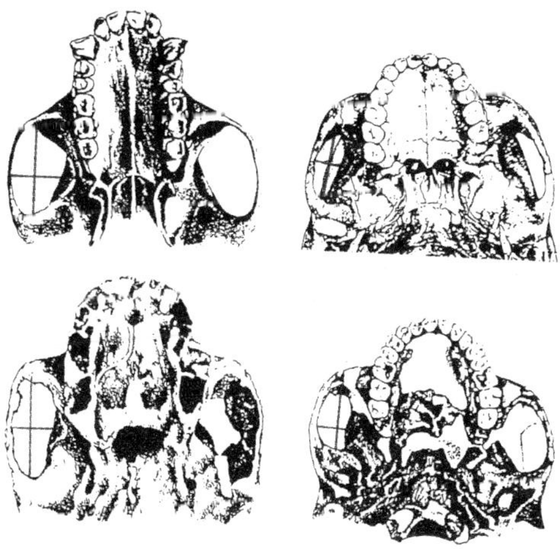

Fig. 1. The shape of the temporal fossa as measured in the ratio of the breadth to the depth of the fossa. This fossa contains, largely, the temporal muscle.

Table 2. Ratio of the depth of temporal fossa to its width (i.e. a measure of the thickness of the temporalis muscle; see another chapter).

Living species	Fossil species
Great apes: **0.74–0.88** Large temporal muscle including large superficial head.	Australopithecines **0.73–0.80** Do we assume large temporal muscle including large superficial head?
Modern humans 0.60 Small temporal muscle without superficial head.	Erectus, Neanderthal humans 0.63–0.66 Presumably we assume small temporal muscle without superficial head?
	Liang Bua estimated **0.73** What do we assume?

Bold numbers = similarities with Liang Bua.

of the fossil are large and their roots are surrounded by strong and long buttresses of bone reaching up towards the margin of the orbit. This is, again, not unlike the situation in apes and australopithecines, but quite different from that in humans. The teeth overall are big relative to jaw size (again as in apes and australopithecines but not humans). The fossil has no chin as in apes and australopithecines (though humans too, on occasions, can have small or even no chins). Many of these features imply ape-like faces and possibly, therefore, ape-like diets.

There were even features of the rest of the skeleton that supported the idea of an ape-like or australopithecine-like relationship. Thus, there is a medial inclination of the neck of the thigh bone (as in apes). There is a lateral flair to the pelvis as found in many creatures that move on four limbs and that climb trees (e.g. apes). These features are also similar to the various australopithecines from which we may infer that these latter could also move in the trees on occasions. A femoral head directed medially is not usually seen in humans in whom the head reaches cranially towards the pelvis rather than medially. A lateral pelvic flair is not usually seen in humans in whom the pelvis spreads markedly anterolaterally rather than simply laterally. These are both features of apes and australopithecines. There is increased robusticity of limb bones (i.e. wider in proportion to the length) that approaches the high robusticity of limb bones in apes and australopithecines.

Taken together, all of these features seem to be primitive, resembling those of creatures like the chimpanzees and bonobos of today, and fossils such as *Australopithecus* and *Paranthropus* of 2, 3 and more millions of years ago (Table 3). In an evolutionary context, this suggests a species with a very early divergence from our own lineage even though the particular specimens are only about 18,000 years old. Though, of course, this fossil cannot have been directly in any of these categories (habilines, australopithecines, paranthropines and African apes are all African forms) it is not impossible that it might have been related to some approximately equivalent Asian parallel. Others have also noted some of these features and posited equivalent possibilities (e.g. Argue *et al.*, 2006). Of course, this would mean a long-continued existence of a very small population isolated on a small island! Luckily, as it turns out, the only people to whom I told about my thoughts along these lines were my graduate students on my return from that research visit.

Table 3. Initial comparisons between the fossil, Liang Bua, apes, australopithecines and humans.

Feature	Liang Bua	Australopithecus	Chimpanzee	Modern Human
Brain size	**Small** **417 ccs**	**Small** **430 ccs**	**Small** **450 ccs**	Large 1400 ccs
Temporal lines	**High**	**High**	**High**	Low
Temporal fossa	**Deep**	**Deep**	**Deep**	Shallow
Canine buttress	**Strong**	**Strong**	**Strong**	Weak
Large teeth	**Yes**	**Yes**	**Yes**	No
Chin	**Absent**	**Absent**	**Absent**	Present[a]
Interlimb ratio	**84–90**	**95–100**	**97–114**	67–74
Pelvis length	**Short**	**Short**	Long	**Short**
Pelvic flair	**Lateral**	**Lateral**	**Lateral**	Anterolateral
Femur neck angle	**Medial**	**Medial**	**Medial**	Cranial
Long bone width/length ratio	**High**	**High**	**High**	Low, especially in upper limbs
Relative foot size	**Big**	**Big**	**Big**	Small

[a] Reduced or even absent chins are found in some normal modern humans.
Bold indicates similarities with the fossil Liang Bua.

Table 4. Could Liang Bua be simply a member of the genus *Homo*?

Feature	Liang Bua	Early *Homo*	Modern *Homo*
Brow ridge, groove	Present	Often present	Present on occasions
Occipital flattening	Present	Often present	Present on occasions
Narrow middle skull	Present	Often present	Present on occasions
Chin	Absent	Mostly absent	Absent on occasions
Bridge of nose	Depressed[a]	Often depressed	Depressed on occasions
Cheek bones	Wide	Intermediate	Narrow
Human-like dentition	Yes	Yes	Yes
Human-like endocast	Yes	Yes	Yes

[a] Damaged — difficult to be certain.

In fact, I quickly decided that my original idea was likely wrong. So many other characteristics of the fossil (e.g. brain form as known from the endocast, cranial, facial and jaw shapes, and especially the form of the teeth, Table 4) seem clearly human. Others have also noticed these facts. This leads to the question: Could Liang Bua be simply a member of the genus *Homo*?

These similarities with various humans are supported by another idea. Thus, some aspects of the reduced size of the Liang Bua fossil (e.g. presumed brain size versus presumed body size) might be interpreted as stemming from island dwarfing, either within some group of modern humans long confined to the island, or in some other island group such as *Homo erectus* lingering much later than *H. erectus* generally. Careful studies of island dwarfing of many mammalian species, however, imply that the reduction in the brain that occurs in island dwarfing does not follow *pari passu* with the reduction in body size (summarised in Martin *et al.*, 2006). Though small, brains are larger in island dwarfs, whatever the species, than expected on the basis of the body reduction. And of course, some creatures show increased size and variability in island situations. Island dwarfing as the total explanation seems unlikely.

Yet, some of the characteristics (e.g. increased robusticity of skull and limb bones, reduced degree of humeral torsion, medial inclination of femoral neck, large feet) provide support for designation as an older form of *Homo*, but one that happened to have survived until very recently on that island. Such individuals might possibly have been descendants of an Asian *Homo erectus* known to have been out of Africa at least 1.8 million years ago or even more (Argue *et al.*, 2006).

These human-like resemblances quickly lead to yet another idea (e.g. Weber *et al.*, 2005; Martin *et al.*, 2006; Jacob *et al.*, 2006; Hershkovitz *et al.*, 2007): the Liang Bua fossils could have been pathological human specimens, to wit genetic microcephalics. In general terms this would seem unlikely because the incidence of such a condition in the general population is said to be between 1:250,000 and 1:500,000 and, therefore, apparently incompatible with the finding together of two small adult mandibles (two small adults at least) at Liang Bua. Some microcephalies are, however, genetic autosomal recessives. In small populations, therefore, especially with high degrees of consanguinity, they might well be present in numbers much larger than indicated by the general population incidence.

The idea of fossil humans displaying pathology has generally a bad history. The earliest Neanderthal skeleton thought originally to be a bear, and then a member of an ancient 'barbarous and savage race', was also designated a consequence of rickets, of horseback riding (e.g. from a Cossack soldier wounded in battle), and so on. The apparently contorted remains of that human fossil also produced a diagnosis of acromegaly (large hand and feet, large faces and jaws). Although, in general, such medical diagnoses have been (properly) rejected, one such diagnosis was shown to be completely correct: one distorted skeleton that gave rise to the image of a bent shambling not quite fully bipedal human ancestor, was shown to be totally due to severe osteoarthritis in an old man (Straus and Cave, 1957). And of course many pathological conditions (e.g. yaws, syphilis, rickets and osteomalacia) have been recognised in human fossils.

To return to the medical diagnosis of microcephaly, of course, there are many different forms. Literature descriptions are available. There is, in particular, a single individual of Lorain's infantilism with pituitary dysfunction for which a complete and highly detailed post-mortem account (carried out many years ago (Hill, 1933) is available. A variety of other genetic microcephalics have also been entered into this controversy (Martin *et al.*, 2006). Most recently of all has been the detailed examination of a large number of Laron syndrome individuals (Hershkovitz *et al.*, 2007). The Laron syndromes are also genetic recessives displaying the effects of primary pituitary growth hormone insensitivity. Being genetic in origin (especially autosomal recessives) these conditions must be taken seriously as being individuals who might have come to represent small long-isolated communities. Table 5 indicates a selection of

Table 5. Features generally common to the Liang Bua fossil and various genetic microcephalies.

Features	Liang Bua	Genetic microcephaly	Lorain infantilism	Laron syndrome
Stature	Small, 106 est.	Small	Small, 121	Small, 95–136
Skull size	Small	Small	Small	Small
Cranial bone thickness	**Normal to thick?**	Thin	Thin	Normal to thin?
Facial height	Reduced	Reduced	Reduced	Reduced
Prognathism	**Considerable?**	Variable	Variable	Slight
Mandible size	Small	Small	Small	Small
Chin	**Absent**	Small	Small	Small
Tooth anomalies	Various	Various	Various	Various
Clavicle	Short	Short	Short	Short
Humerus width	Pronounced	Pronounced	Pronounced	Pronounced
Lateral flaring of ilium	Marked	Marked	Marked	Marked
Tibia long axis	Curved	Curved	Curved	Curved
Limb proportion	Abnormal	Abnormal	Abnormal	Abnormal

Bold = Features of Liang Bua that are different. ? = features that are debatable.

13 features (from a total of 34) in which these various microcephalics seem to resemble the Liang Bua fossil.

Because, however, these various pathologies produce small individuals, and because many of these features are also shared with the children of equivalent height, we cannot immediately jump to the conclusion that this means Liang Bua fossil is one of these genetic pathologies. A comparison of a genetic microcephalic skull, for example, to a normal sized skull from the same ethnic group indicates just how similar they are (Fig. 2). In comparison, Liang Bua juxtaposed to modern human (but, it must be pointed out, not from the same part of the world) seem to be two completely different skulls (Fig. 3).

Many of the characters enumerated in the various hypotheses above also feature in more than one of the hypotheses. Thus, longer arms relative to legs, high robusticity of limb bones, curious torsion of the humerus, varus form of the femoral neck, large hands and feet relative to limb lengths, are found, to greater or lesser degrees, in the entire range of evolutionary possibilities: *Homo neanderthalensis*, *H. erectus*, *H. habilis*, *Australopithecus*, *Paranthropus*, *Pan* and *Gorilla*.

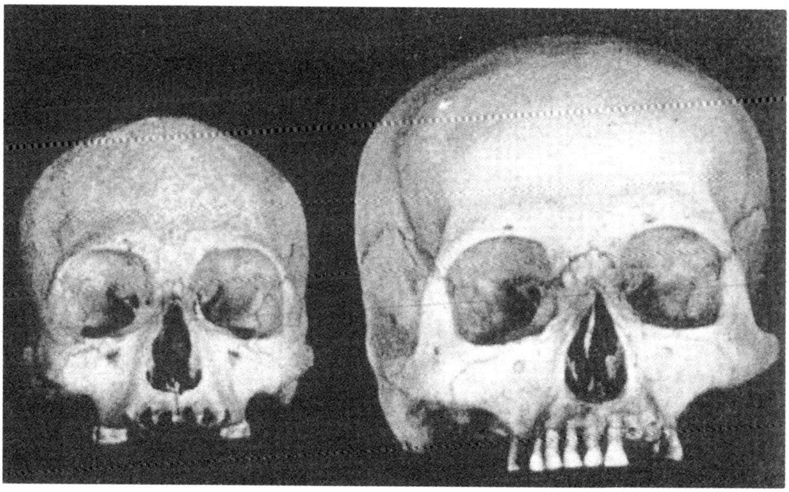

Fig. 2. Comparison of skulls of a microcephalic and a normal human of the same ethnic origin.

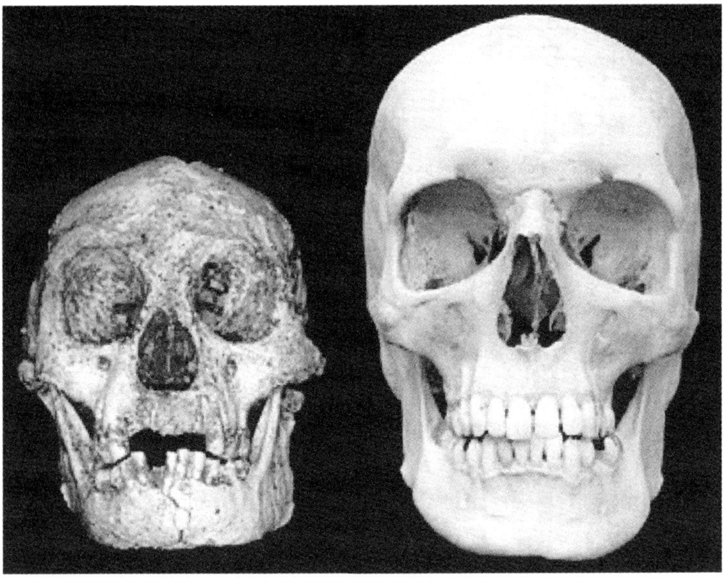

Fig. 3. Comparison of skulls of Liang Bua and a modern human.

Table 6. Comparisons of humerus.

Feature	Liang Bua fossil	*Australopithecus*	*Pan*	*Homo*
Humeral torsion in degrees	**120 (110)**	**120**	**120**	140
Humeral head inclination	**Medial (but damaged)**	**Medial**	**Medial**	Cranial

Bold type indicates similarities.

Most recently, however, several reports have returned us to a very primitive diagnosis for the fossil. Thus, Larson and colleagues (2007) have found primitive features of the shoulder (clavicle, scapula and upper end of the humerus). These include scapular shape (with a curious rounded lower angle), clavicular dimensions (short and thick) and humerus with curious torsion and medially directed head (Table 6).

Equally, Tocheri and colleagues (2007) have found primitive features like those in apes in three wrist bones (trapezoid, capitate and scaphoid, Fig. 4). These imply that the wrist in Liang Bua had a very deep carpal tunnel as is evident in apes but not in ancient or modern humans.

Work is in progress implying similar findings in the lower limb, and a specific item is already available for the femoral head (Fig. 5). I have attempted to figure the humeral head in an equivalent manner

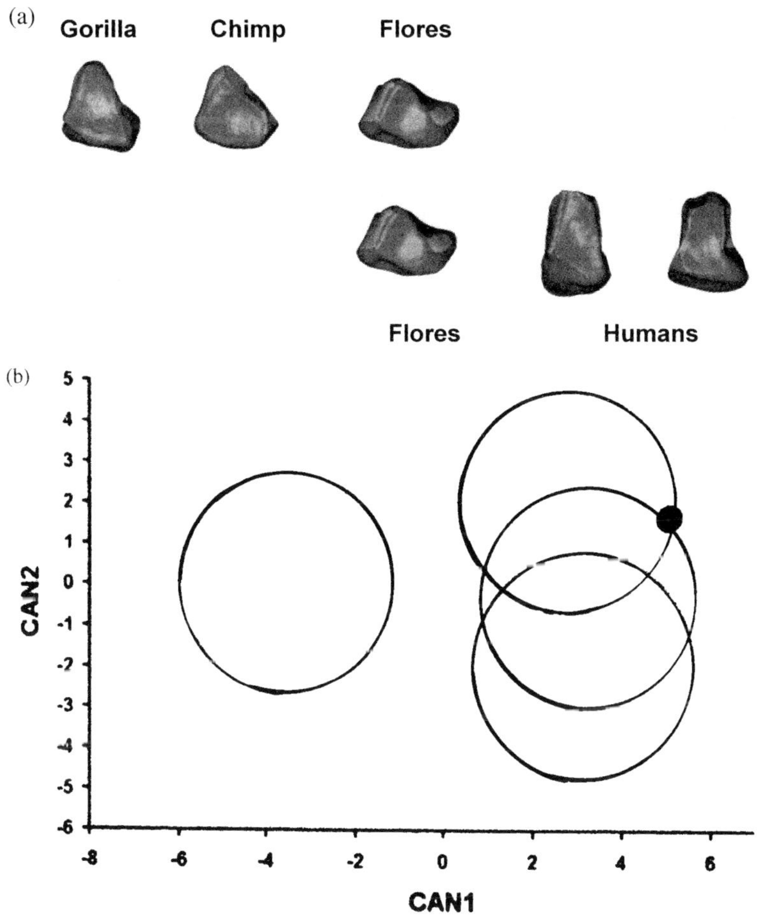

Fig. 4. Tocheri's study of one of the wrist bones, above the trapezoid, and below the data, both visual and statistical that clearly imply that Ling Bua (•) is closer to apes (overlapping circles) than to humans (isolated circle).

(c)

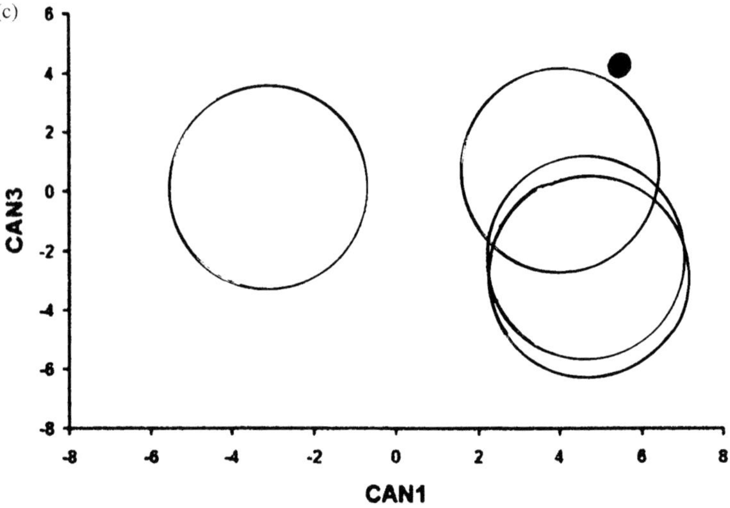

Fig. 4. (*Continued*)

(Fig. 6). Although I do not have a picture of the Liang Bua humeral head inclination, the inclinations of the humeral heads in the other species provide the same result as the femoral head (Fig. 5). Thus, all these features of Liang Bua are described as similar to modern apes and australopithecines, and some of them, to a degree, are said to be reminiscent of very early *Homo*: Nariokotome (Larson *et al.*, 2007). This strongly supports the first hypothesis, an hypothesis that, I explained above, I was tempted to reject.

As a result of all this, there is much argument and controversy. Are we stuck with an evolutionary impasse until more remnants are found? Can the paradox be resolved in any other way?

Yet Another Opinion?

Just at the point at which I changed my mind and went for a human-like diagnosis, mainly on the basis of the human-like teeth, jaws and skull of Liang Bua, another of those accidental, serendipitous, quite unusual, scientific liaisons arose. A modern human ecologist, environmentalist and evolutionary biologist, Peter Obendorf, contacted

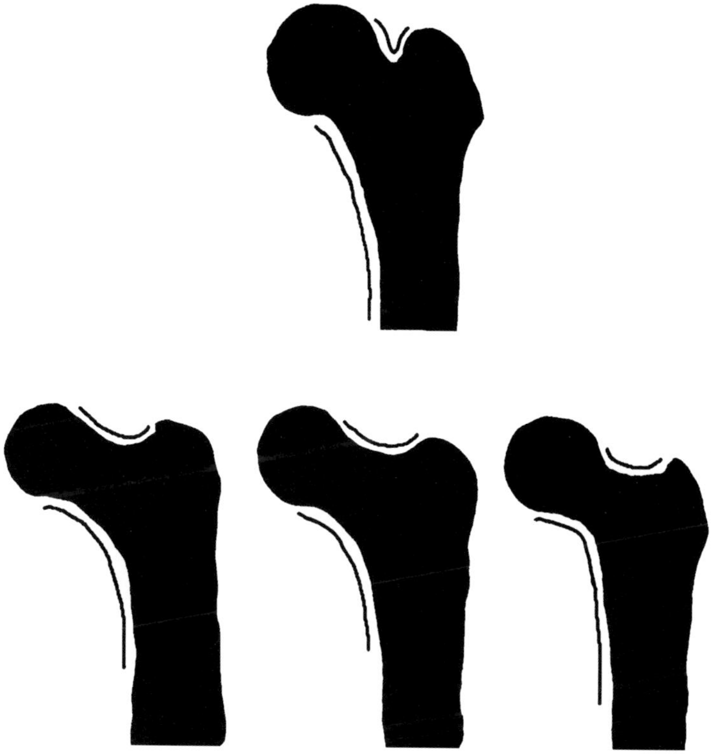

Fig. 5. Comparisons of the outlines of the head and neck of the femur; above: a modern human; below: two australopithecines and Liang Bua.

me, a medical anatomist and human biologist, and, a little later, we met up with an agricultural ecologist, Ben Kefford who was also an ecotoxicologist and biological statistician. A curious combination indeed, and none of us palaeontologists!

First, one of the features of education in ecological and environmental biology is to look for environmental factors in animal and plant anatomy and physiology. In the studies of fossils, in contrast, the disciplines of genetics, development and evolution generally hold sway. Every environmental biologist, though knowing about evolutionary adaptation to environment, is also keenly aware of those many ways in which environment affects, directly, the forms of animals. Careful thought indicated to Obendorf that there was at least one

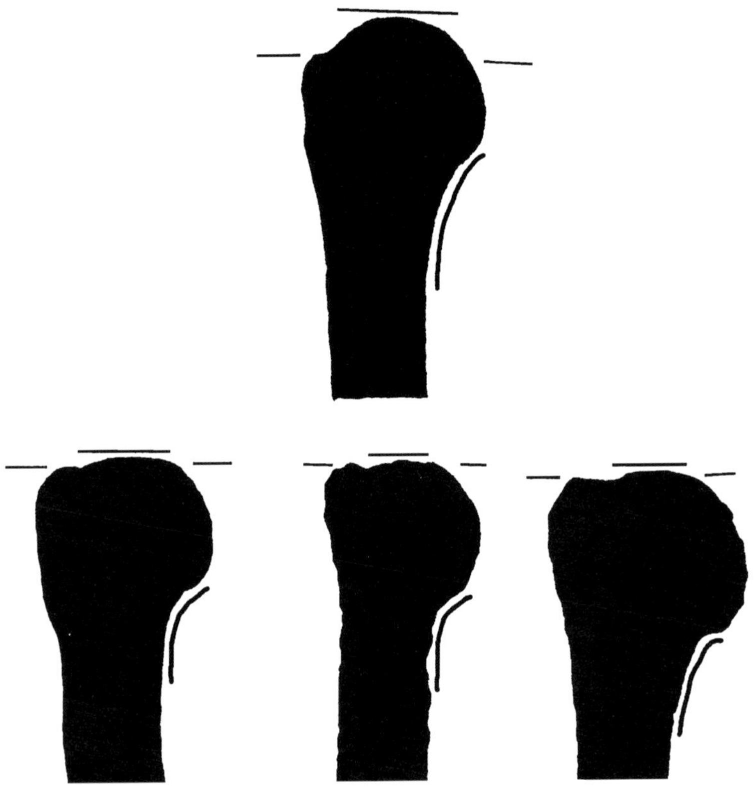

Fig. 6. Comparisons of the outlines of the head and neck of the humerus; above: a modern human; below: two australopithecines and *Homo erectus*. Unfortunately the humeral head of Liang Bua is damaged; the available photographs do not indicate whether or not it had a medially directed head like the australopithecines and *Homo erectus* though Larson says that it had.

environmental condition that might result in many of the characters displayed by the Liang Bua fossils. That is: environmental deficiency of iodine.

Secondly, one of the features of a traditional medical education such as mine was a principle to look for the single diagnosis that covers all symptoms and signs. Of course, more often than not this is honoured in the breech. Many patients have more than one thing wrong with them! Careful thought reminded me, however, that iodine deficiency displays a curious suite of pathological features of bone and bones in cretinism. In addition, because I am old, I actually know about cretinism, and was overtly taught about it nearly 60 years ago, though, as a young medical student, I never actually saw a case. Cretinism does not, should not, exist nowadays because we know what to do about it. As a result it has been long absent in developed countries, and, therefore, materials from such individuals are not easily found. Sixty years ago in England, however, the oldest of such individuals were either still alive, or so recently dead that the condition was still in the English medical curriculum of my time!

Third, one of the features of training in ecotoxicology is to know about plants and their biochemical, especially toxic, features. Thus, Kefford was especially aware of certain plant toxins that potentiate iodine deficiency. In addition, as a keen biological statistician, he was readily able to test the hypotheses through statistical analysis of data, especially using some techniques that were new to me.

The human condition of myxedematous endemic cretinism, of (mainly) environmental origin, produces individuals with small bodies, small brains, reduced intelligence, and lessened mobility, but who often survive to adulthood. They are the offspring of mothers who have severe iodine deficiency. They are found in considerable numbers in communities living in iodine-deficient regions of the world and, though the condition is not inherited, they continue to appear with the generations as long as the environmental deficiency exists and treatment is not provided.

This leads to a plethora of questions. What is the anatomy of cretinism? Are cretins really similar to the fossils? Are they different from genetic microcephalics? How do they compare with modern humans, with fossil humans, with australopithecines and with apes? Is there iodine deficiency in Flores today? Could there have been iodine deficiency in Flores thousands of years ago? Are there plant foods in Flores containing substances that might exacerbate the effects of iodine deficiency? What social and cultural factors might have been involved? Are there any written or oral records (Obendorf, Oxnard and Kefford, 2008)?

The Biology of Cretinism

Reduced dietary iodine produces hypothyroidism in animals and humans. This is a condition whereby decreased levels of thyroid hormone result in subnormal metabolism, and therefore subnormal growth and activity. In the iodine-deficient humans, thyroid enlargement (goitre) may develop as the pituitary attempts to kick the thyroid into producing more thyroid hormone. The condition of myxoedema (a thickening of skin and other tissues, especially obvious in the face) and many other signs and symptoms may ensue. Treatment with iodine (iodine dietary supplementation) is all that is necessary.

In the offspring of the already iodine-deficient mother, the greater human tragedy of cretinism may follow. The early foetus fails to develop a thyroid. The developing pituitary tries (but fails) to kick-start a thyroid gland with the result that the struggling pituitary itself becomes greatly enlarged. This is to no avail. The child remains without a thyroid and the total lack of thyroid hormone determines that it becomes a cretin. There is retardation in time and degree of attainment of all growth milestones, and, if really gross, of overall developmental arrest. Treatment with thyroid hormone early enough reverses many of the features of the condition.

In terms of the skeleton (after all, it is the knowledge of the skeleton that is paramount in fossils) the interference relates to the delay in the formation and development of cartilage and bone. All cartilage and bone growth is markedly slowed but, perhaps paradoxically, continues longer into adult life. Those bones that are most dependent upon prior cartilage development and later replacement of cartilage by bone (in other words, those bones that form 'in cartilage' i.e. undergoing endochondral ossification) are most affected (e.g. the skull base, long bone length). But bones that grow sub-periosteally (e.g. long bone widths) or that ossify intramembranously (e.g. bones of the skull face and vault) are also strongly affected, though to somewhat lesser degree. As a result of different levels of interference in these different processes, growth is not only reduced but also changed in proportions!

The Cretin Skull

First let us examine the skull. Faces consist largely of 'membrane bone' (bone developing in a connective tissue membrane that can only grow by apposition, i.e. on the surface). Facial growth, though reduced in cretins compared with normal, is not reduced as much as the cranial base. The cranial base is largely 'cartilage bone' (bone replacing a prior cartilage model during development). Cartilage grows interstitially, i.e. in three dimensions, and therefore normally much more quickly than the two-dimensional appositional growth. As a result, in cretins, the face, though absolutely small, appears relatively large compared to the base, the growth of which is even more reduced; thus the face is robust and protruding. In this it contrasts with both normal children and microcephalics in whom the face is usually relatively small and delicate compared to the head as a whole. Many of the teeth of cretins are of normal size, and this makes them large relative to the small jaws. They often project forward like the jaws, and are usually reduced in number and with anomalies (there is less space for near normal sized teeth in small jaws). The growth delays mean that some milk teeth roots do not resorb; all the milk teeth are not shed and so some persist with longer roots than normal. Accordingly, the adult jaw may bear some milk teeth (a mixed dentition, both milk and permanent teeth).

These various features are not generally characteristic of adult modern humans, modern apes, or the various fossils (e.g. *Australopithecus*, *Paranthropus*) said to be related to human evolutionary lineages, though some individual features may be found in some species.

The delays in growth, together with very slow yet continued growth into adult life, further affect the individual parts of the skull. This is especially seen in the joints between the various skull bones. The fibrous joints (sutures) of the vault of the skull remain open very much longer than normal. Some sutures between these membrane bones may never completely close even in adults. Thus, even in adults, there is often a midline suture dividing the frontal bone; the various fontanelles may remain slightly open; and one or more sutures may separate the different developmental elements of the vault portion of the occipital bone.

Such problems also affect the cartilaginous growth plates between the bones of the cranial base, and, because this process involves cartilage, to an even greater extent. Thus, even in adults, cartilage may still exist between the sphenoid and occipital bones (a persistent sphenoccipital synchondrosis), and between the various developmental components of bones of the skull base (e.g. within the various portions of the sphenoid and occipital bones) that normally fuse.

Open sutures of the skull vault and open synchondroses of the skull base do not characterise adult members of any of the large primates whether ape or human.

Various bony parts (processes) projecting from the skull are also involved. These processes are derived from separate centres of ossification that normally fuse later with the skull. In cretins, delayed ossification means that these processes may remain separated from the skull, and may sometimes even remain cartilage, never being converted into bone. As a result, in the preparation of the dried skull, these processes are lost, or they may be very small. Examples are the styloid and vaginal processes that project downwards underneath the skull, and the anterior and posterior clinoid processes around the pituitary gland that project upwards inside the skull.

Let us particularly look at the styloid process. Normally this structure projects downwards from the skull base towards the hyoid bone. We can feel it in our own necks, as we gently move the two styloids from side to side with our fingers. An uncomfortable feeling it is too! In individual dried skulls the styloid process is often broken off, but one knows when it was present; its broken cross section can still be seen on the skull base. The styloid process develops in an interesting way. Normally four separate centres of ossification fuse and form the long and complete styloid process attaching to the skull. Even in normal humans, however, this process is quite variable. In some people only a smaller number of the centres may fuse giving a shorter styloid. In others, ossification including all centres may even extend all the way down to the hyoid so that the hyoid may be attached to the skull. If this occurs on both sides there may actually be a bony ring encircling the respiratory (larynx) and gut (pharynx) tubes in the upper neck.

In cretinism, cartilage growth and subsequent replacement with bone is so delayed that a bony styloid attached to the skull may never form. There may be left only a tiny point of bone, or in some cases, nothing at all, projecting from the skull base. Of course, all parts of the styloid are present; the structure itself is not missing. After all, it is a fundamental part of the vertebrate skull: a component of the second gill arch. It just appears to be lost in cretins because it only consists of pips of cartilage or bone that never fuse. They separate and drop off after death so that the dried skull has, apparently, no styloid process.

The styloid process, if only because it is a fundamental component of the vertebrate skull exists in all larger primates (whether ape or human).

This problem also occurs in the other skull processes mentioned above. The vaginal crest (nothing to do with the female genitalia) is a short sharp fold of bone around the base of the styloid process. In cretins it may remain cartilaginous, thus becoming absent in the dried cretin skull. The four clinoid processes, like sentinels around the pit containing the pituitary gland (sella turcica) inside the skull, may, likewise, fail to pass beyond the cartilaginous state and so appear to or absent in cretins. Indeed, in two cretin skulls I examined, this was quite explicit, the state of preservation being such that the clinoid processes were actually present as small, dried, wizened twists of cartilage which happened not to have disintegrated or fallen off.

Interestingly a similar mechanism occurs in the crista galli, a midline bony process at the front of the inside of the skull giving attachment to the powerful dura mater sheet supporting and protecting the brain, and in the coronoid process of the mandible, to which the temporalis muscle is attached, a powerful muscle of mastication. There is a separate centre of ossification in the tip of each of these processes. Many anthropologists use the tips of such processes as 'defined points' for their measurements not realising that, in some otherwise normal skulls, the tip may not have fused and may have fallen off. Careful examination with a hand lens readily demonstrates, in a reasonably well prepared skull, when the tips are missing, and when, therefore, the points will not be equivalent to those in other skulls. If this is not realised, such measurements will be short. For our purposes, however, it is enough to record that 'lost' tips or 'twisted cartilaginous rudiments' are common in adult cretins.

All these processes are fully present in normal primate, human and nonhuman, skulls.

Differences in various spaces in skull bones also characterise cretins. The frontal air sinus and even other air sinuses may be small, very small or absent. The bone around them is thicker relative to normal. The mastoid air cells may be reduced or totally absent but are, again, surrounded by a relatively thick bone. The marrow spaces (the diploe) may be very much reduced. But they are surrounded by inside and outside bony 'tables', and these may be almost contiguous and very thick (relative to small skull size). All of these bony thicknesses may display osteosclerosis, that is, they are not only thick but also contain denser (more calcified) bone tissue than usual. This is readily evident on radiographs.

Again, almost none of these cretin features are found in normal adult humans, in human children or in human microcephalics. When spaces are small in children and human microcephalics, they may be reduced but are surrounded by very thin cortical bone. Small air sinuses, less diploe, reduced or absent mastoid air cells, together with thicker surrounding bones are evident in modern apes and australopithecines. But of course, these are both smaller and much more robust skulls than humans.

Cretin Limb Bones

Cretins also display a number of curious features of their limb bones. Thus, the ratio of the length of the arm to that of the leg is higher than in non-affected humans. This is so to a degree not very dissimilar to that in apes and quite similar to that in australopithecines. This is not because of adaptations to arm-hanging and climbing as it is in great apes (and possibly also to arboreal activities of australopithecines) but because of the reduction in cartilaginous growth in length of the weight-bearing legs which is greater than that of the non-weight-bearing arms. Though the arms in cretins are long relative to the legs, both limbs in cretins are short relative to the trunk.

A similar effect produces cretin limb bones that appear more robust (much greater in breadth than in length) than normal. Again, this is because breadths of cretin limb bones are reduced (due to a lesser subperiosteal growth deficit) less than their lengths (due to greater cartilaginous, epiphyseal plate, growth reduction), and not because they mirror the true robusticity of ape (and australopithecine) limbs involved in powerful locomotion.

The spurious nature of the robusticity of limb bones in cretins is made clear by the fact that the muscular markings on them are ill defined, even absent. Thus, the bones look robust (wide relative to length) and measure as robust (in the width/length ratio), but their various stiffening pillars are reduced and rounded (e.g. the posterior ridge on the femur, the supracondylar ridges on the humerus). Further, in cretins the actual bony surfaces are smooth, and the muscle markings are faint or even absent. The implication is that, robust though they appear, the musculature is reduced and of lesser power. This is indeed evidenced by the reduced physical abilities of cretins.

There are yet other curious features of limb bones in cretins. The humerus and femur display shapes medically described as varus deformations. This means that, at the shoulder and hip, whereas human humeral and femoral heads normally reach up towards the scapula and pelvis, in cretins they are much more medially inclined. In these features cretins resemble apes and australopithecines (Figs. 7 and 8). In apes (and presumably australopithecines) such features relate to the locomotor functions of shoulders and hips (in apes certainly through quadrupedal weight-bearing, in australopithecines likely also so, even though these latter are also bipedal after a fashion). Such characters may be defined as primitive compared with humans. In cretins, however, they only appear to be 'primitive'. Cretin varus deformities are due to disturbances of the growth of the humeral and femoral epiphyses (Ortner and Holtz, 2005).

Likewise, long bones in cretins show a variety of torsions and bends. Some of these (torsion in the humerus) can be, again, apparently similar

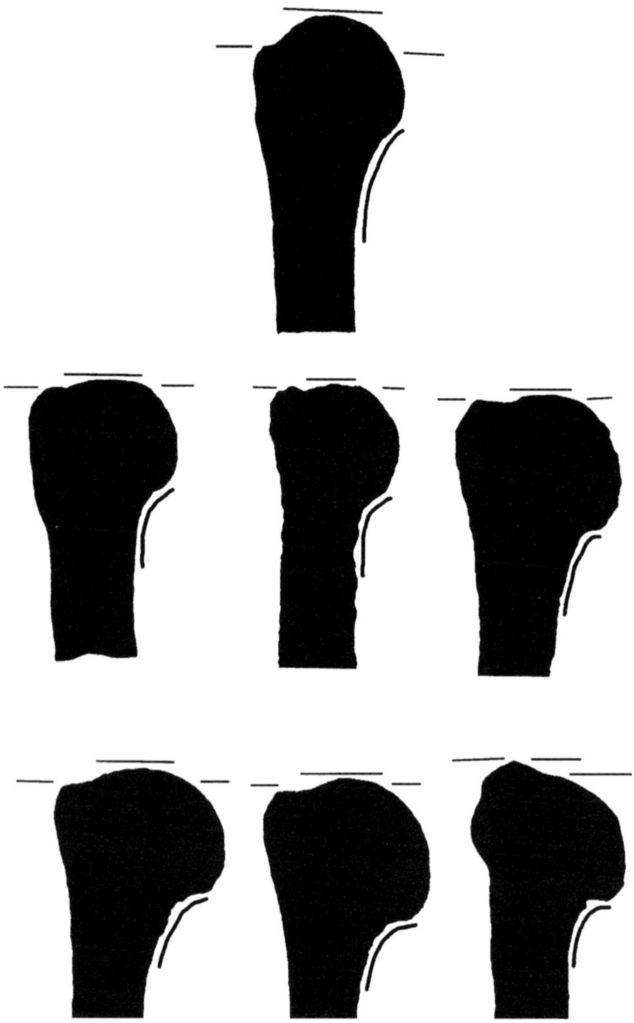

Fig. 7. The same comparisons of the outlines of the humeral head and neck as Fig. 6 with three cretins added at the bottom row. The upper end of the humerus of Liang Bua is too damaged to permit its inclusion in this comparison.

to what we find in apes and australopithecines. Other torsions (e.g. in the femur) are not generally found in any ape or human. In apes and australopithecines the humeral torsions relate to the abilities to use the upper limbs above the head in climbing and might be thought to

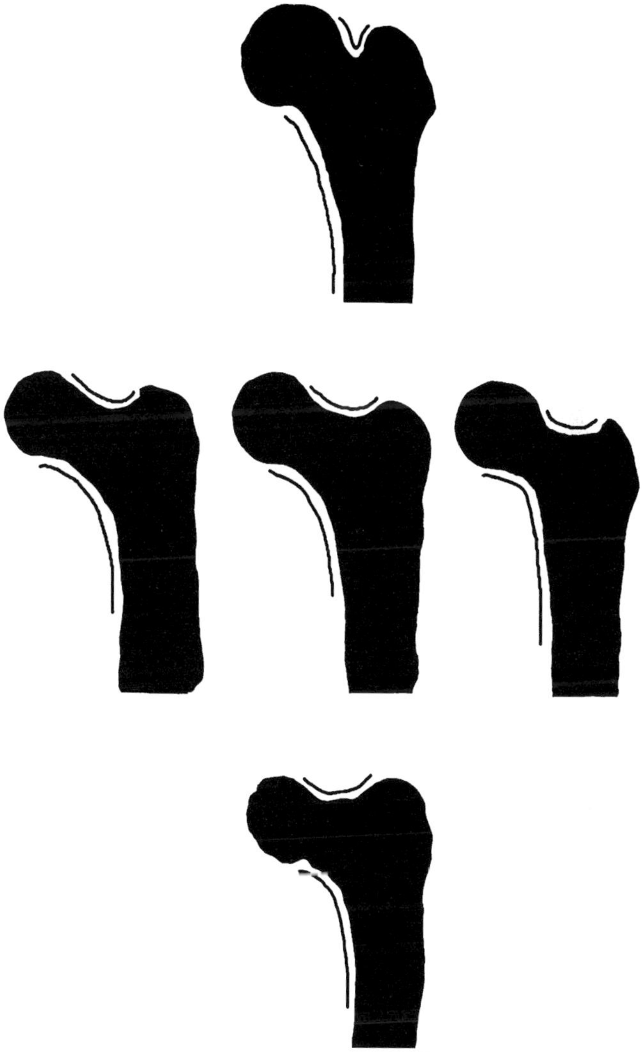

Fig. 8. The same comparisons of the outlines of the femoral head and neck as Fig. 5 with cretins added.

be primitive characters. In cretins they are again merely evidence of the hypothyroid problem in bone growth.

Finally, there are special features of the bones of the hands and feet that characterise cretins. Cretin extremities are, of course, absolutely small compared to normal humans. In this respect they resemble all the genetic microcephalics. This is because cretins, like the microcephalics, are small. Relative to the other limb bones, however, adult cretin hands and feet are large and wide, thus contrasting with the delicate hand and foot bones of genetic microcephalics. Cretin feet are so relatively large that their length may be as much as 64% of tibia length, though only 50% in normal humans. The breadths of all the bones of the fingers and toes are large relative to their length, and this is especially evident in radiographic comparisons. Such differences may relate in part to the special growth of hands and feet. These elements, to a degree more than other skeletal elements, continue to grow throughout life.

The bones of the wrist (carpals) are a special case. Some of the carpal bones actually start as two separate centres of ossification, one dorsal, one ventral, in a single cartilage precursor. They then fuse to give a single carpal bone. This may be delayed in cretins so that either one of the centres (usually the ventral) never becomes a bone and so that carpal appears to have only a dorsal part, or both appear but never fuse so that a bipartite carpal is found. Although Tocheri *et al.*'s (2007) paper implies that the trapezoid is really like that of chimpanzees and gorillas, careful examination at the correct magnification indicates that it is actually a carpal with the ventral portion missing. This does not occur in apes (Fig. 9).

Finally, the hand and feet phenomena in cretins may relate to the following. Thus, though growth is slowed everywhere in cretins, it is slowed relatively less in those parts that, even in normal humans, continue to grow throughout life, hence relatively larger hand and foot skeletons (and also, as it happens, relatively larger lower face and especially lower jaw). This also applies to the cartilaginous parts of ears and nose. All these parts are relatively large in cretins but are small and delicate in children and in adult microcephalics (though, of course, some of this is not apparent skeletally). Some of these features are shown in Figs. 10, 11 and 12.

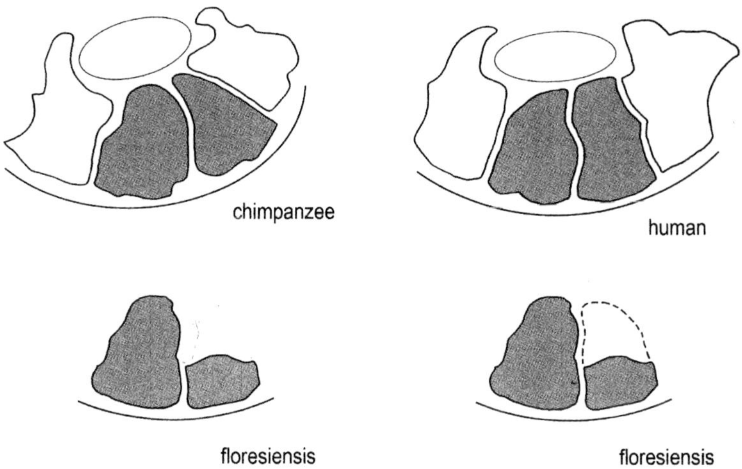

Fig. 9. Reconstruction of the trapezoid in relation to the capitate (from Tocheri *et al.*, 2007) with an additional multiplication factor kindly provided by Tocheri, and articulated alongside diagrams of Lewis' (1965) cross-sectional views of dissections of ape and human articulated wrist bones.

Fig. 10. Two cretins. Note the huge hands and feet, and the curiously twisted leg.

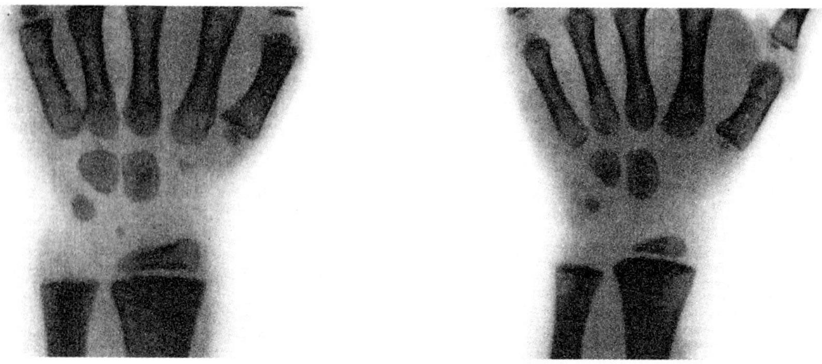

Fig. 11. Radiographs of normal and cretin hands at the same radiological stage. Normal: 5 year's old; cretin: 11 year's old.

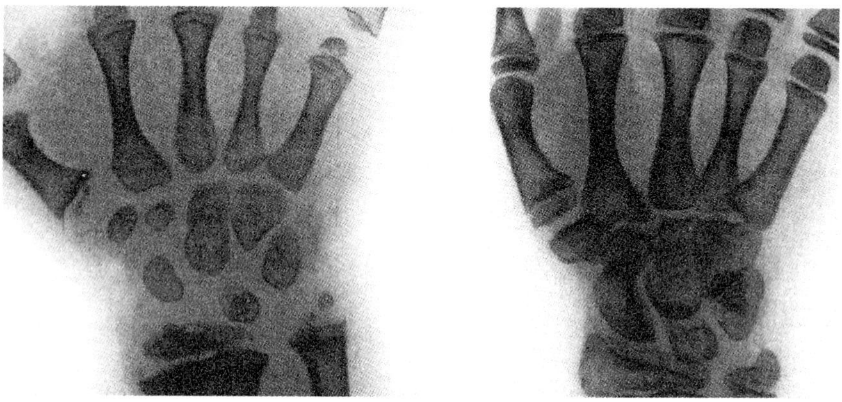

Fig. 12. Radiograph of (left) an adult cretin before and (right) a few months after treatment with thyroid hormone.

The Trunk Skeleton in Cretins

Though no cretin skeletons have been sighted, literature descriptions and figures show that the changes in the trunk parallel those in the head and limbs. Thus, in the spine the vertebral epiphyses often do not fuse with the vertebral centra, indeed, and may remain entirely cartilaginous, and, again, tend to be lost when the cadaver is skeletonised. As a result, the height of each vertebra is shortened relative to the other two dimensions. Any calculation of trunk length derived from the summations of the heights of individual vertebrae gives, for this reason, a considerably shorter trunk length than actuality. The few radiographs of cretin spines make this clear (Fig. 13). The intervertebral foramina appear smaller in cretins (and in this sense mirror what is found in some of the foramina of the skull, especially the small external auditory meatus of cretins). These are not features of normal humans, apes or australopithecines.

The adult cretin sternum is the same. The human sternum is normally a single bone that develops from the coalescence of its several separate developmental segments (sternebrae). In adult cretins, these segments often remain separate. This is especially obvious in the sternum that I was able to examine at the Royal College of Surgeons of England. Again, separate sternebrae occur in apes and monkeys.

Normal and Cretin at 8 years

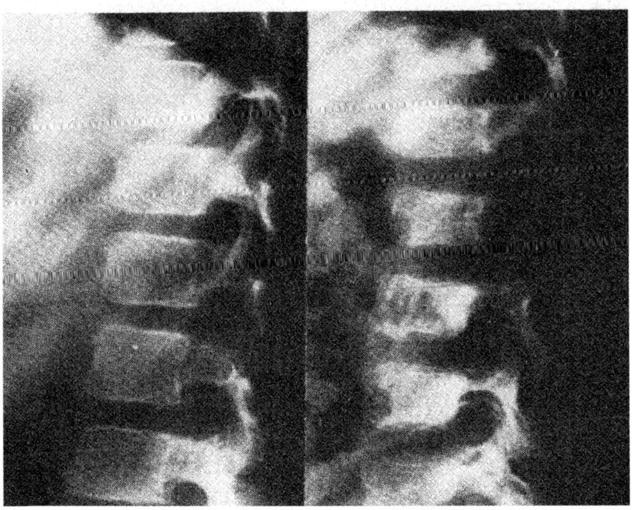

Fig. 13. Radiograph of normal and cretin spines (of same age).

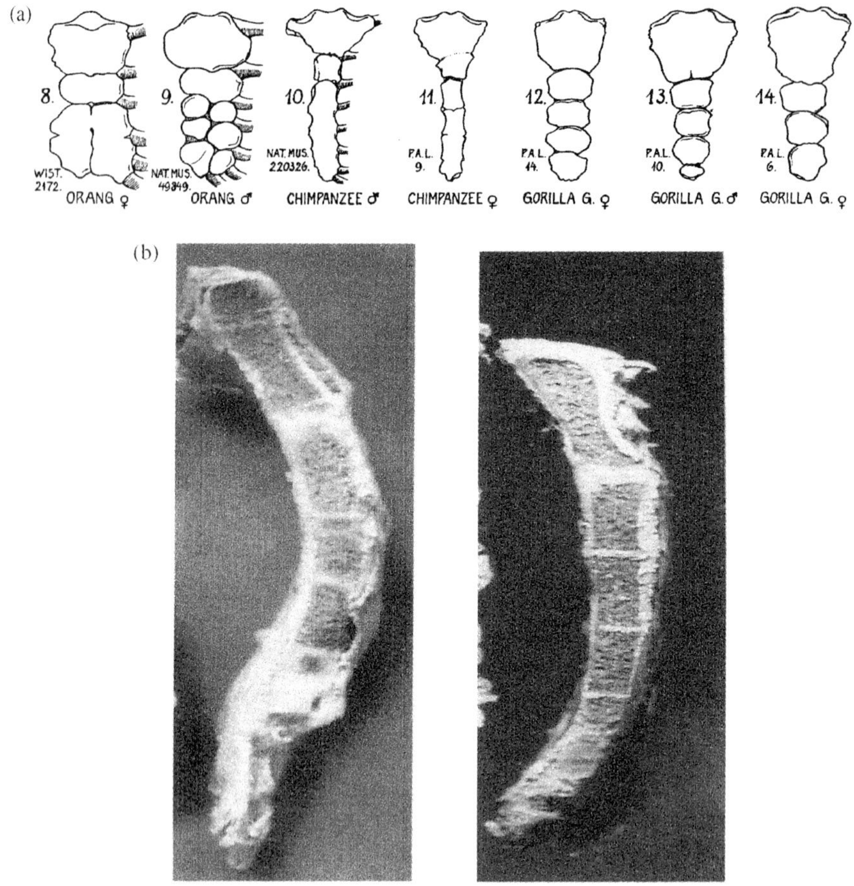

Fig. 14. Radiographs of cretin sternums. Diagrams of sternums of apes and humans (from Schultz, 1930).

Though this might be thought to be a 'primitive' condition, its aetiology in cretins is completely different (Fig. 14).

Even the rib cage in cretins differs from that in normal humans. There are sometimes fewer bony ribs. But of course, it is not that ribs are missing; it is that the last rib or even the last two ribs may be wholly cartilaginous and thus be lost in the preparation of the skeleton. Again, though there may be different numbers of ribs in both humans and apes (for example, ribs on cervical and lumbar vertebrae due to alterations in the developmental series in individuals), it is always evident when ribs are present, both because they exist, and because of their joint surfaces with appropriate vertebrae.*

Summary of the Cretin Skeleton

In Table 5 the Liang Bua fossil seems similar to genetic microcephaly, genetic Loraine infantilism, genetic Laron syndromes and cretinism, in anatomical features due to the specimens being small. For example, all have small faces, small crania, small hands and feet, and so on. The common factor in all of these shared characteristics is that they are definitions of smallness, of dwarfism and are shared among all the small specimens of whatever diagnosis. In contrast, in Table 7, the Liang Bua fossil is compared with normal adult humans, genetic microcephalics and cretins in anatomical features that are due to the special growth deficiencies that characterise cretinism. Most of these features are shared by Liang Bua and cretins alone. They include, for instance, open sutures, nonfused synchondroses and cartilaginous epiphyseal plates. They include small or absent styloid, vaginal and clinoid processes. They include, further, disproportions of long bone lengths, orientations and torsions, and so on. It cannot fail to be of interest that these features are not found in normal adult humans, children or microcephalics.

In summary then, there are at least 12 features in which the fossil from Liang Bua resembles chimpanzees, bonobos, australopithecines and paranthropines (plus the smallest of the habilines) but not modern humans (Table 3). It is possible to read this as supporting the idea that the Liang Bua fossil is a new species.

There are at least eight features, possibly more, in which the fossil from Liang Bua resembles *Homo erectus*, archaic *Homo* and modern *Homo* (Table 4, earlier). These are largely features that are present to increasing degree from the palaeontologically older to younger species. They seem to imply that the Liang Bua fossil belongs to the genus *Homo*, if not *Homo sapiens* itself.

There are 13 characters in Table 5 in which Liang Bua resembles the various genetic microcephalies (including the one Loraine dwarf

*Since these words were written, and thanks to the help of Dr Gerhard Hotz and the Naturhistorische Museum Basel, Switzerland, I have fully examined six cretin skeletons. This entirely corroborates and extends the above results (Oxnard, 2008).

Table 7. Anatomical features in the Liang Bua fossil, genetic microcephaly, cretinism and normal humans.

Feature	Liang Bua fossil	Genetic microcephaly	Cretinism	Normal adults
Origin		Genetic	Environmental	
Brain size (cm^3)	417	380–900, $n = 33$	570–1070, $n = 7$	Mean 1400
Endocast	Normal?	Normal?	Normal?	Normal
Skull form	**Marked asymmetry**	Symmetric 5 but some asymmetric	**Marked asymmetry**	Slight asymmetry usual
Temporal lines	**High**	Low 2, medium 2	**High**	Very low
Temporal fossa	**Deep, 0.73**	Shallow, 0.66	**Deep, 0.70–0.76**	Shallow, 0.60
Fontanelles	**Open?**	Closed in adult	**Open in adult**	Closed
Cranial base	**Sutures open?**	Sutures closed	**Sutures open**	Sutures closed
Ear hole	**Small**	Normal	**Small**	Normal
Styloid process	**Absent**	Present	**Absent**	Present
Vaginal crest	**Absent**	Present	**Absent**	Present
Ant. clinoids	**Absent**	Present	**Absent/small**	Present
Post. clinoids	**Absent**	Present	**Absent/small**	Present
Sella turcica	**Very large**	Small	**Very large**	Normal
Crista galli	?	Normal	Small	Normal
Diploe	**Thin**	Normal	**Thin**	Normal
Vault	**Thick**	Thin	**Thick**	normal
Frontal sinuses	**Absent?**	Normal/small	**Absent/small**	Normal/large
Mastoid cells	**Absent?**	Normal/small	Absent/small	Normal/large
Face	**Heavy**	Gracile	**Heavy**	Normal
Teeth/jaws	**Protruding**	Gracile	**Protruding**	Normal
Teeth	**Mixed dentition**	Normal	**Mixed dentition**	Normal
Dental development	**Delayed**	Normal	**Delayed**	Normal
Relative tooth size	**Big/missing**	Normal	**Big/missing**	Normal

(*Continued*)

Table 7. (*Continued*)

Feature	Liang Bua fossil	Genetic microcephaly	Cretinism	Normal adults
Chin	**Absent**	Usually present	**Absent/small**	Present*
Vertebrae	**Height reduced?**	Normal	**Height reduced (epiphyses cartilage)**	Normal
Sternum	?	Normal	**Several sternebrae**	Normal
Lateral flair of ilium of pelvis	**Yes**	Not usual	**Yes**	No — flair anterolateral
Ratio upper limbs to lower limb	**High**	Normal	**High**	Normal
Long bone width/ length ratios	**High (i.e. apparently robust**	Delicate, not robust	**High (i.e. apparently robust)**	Normal
Bone ends	**Varus femur Humerus damaged?**	Usually normal	**Varus femur and humerus**	Normal
Epiphyses	**Open in humerus?**	Usually normal	**Frequently open**	Normal
Bones twisted	**Twist of humerus abnormal**	Usually normal	**Abnormal twists of various long bones**	Normal
Bone surface	**Smooth, muscle markings light**	Surface and muscle marking normal	**Smooth, muscle markings light**	Surface and muscle markings norm
Hands and feet	**Feet large in proportion to thigh**	Hands and feet small and delicate	**Hands and feet large in proportion**	Normal

? = condition not certain because of fossil damage.
Similarities confined to the Liang Bua fossil and cretins are in bold.

(Hill, 1936), and the Laron syndrome sample (Kershkovitz *et al.*, 2007)). When features are categorised as being merely small, they are shared with all the specimens.

However, including cretins in this comparison changes the picture entirely. Thus, if the features are defined in relation to the changed proportions of the growth deficiencies of cretins, then the Liang Bua fossil and the cretins exclusively share 29 features (Table 6). Another seven are likely to be shared but (due to the condition of the fossil or because particular features have not been or cannot be examined) we cannot be certain.

Of course, all cretins are not identical. The effects of the deficiency vary to greater or lesser degree. Their genetic heritages can also be expected to influence the picture. Most of the known cretins for which post-mortem reports are available are from European populations. The cretin (described by Dolega, 1891) for which a sketch is available can be compared with the Liang Bua fossil (Fig. 15). There are some shape similarities, the lack of chin, the large jaw compared with the rest of the skull, the protuberant anterior teeth and possibly open sutures. There are, however, some differences from Liang Bua: the considerably smaller cranium relative to the face, the relatively larger cheek and orbital bones. However, comparison of any two cretins also shows such differences (Dolega, 1891; Knaggs, 1928; Hill, 1936).

There is one structure that is unique in cretins in being statistically significantly larger even than in full-sized humans! That is the sella turcica: the bony fossa inside the skull surrounded by the clinoid processes and that contains the pituitary gland. In cretins the enlarged sella turcica is not directly due to the thyroid deficit affecting growth. It is related to the pituitary gland's attempt to drive thyroid development to produce the missing thyroid hormone. We will examine this in more detail below.

The Brain and Neurological Function in Cretins

The deficits of cretinism as well as being obvious in the reduced and delayed growth and development of cartilage and bone are also evident in a reduced brain and nervous system. One problem in

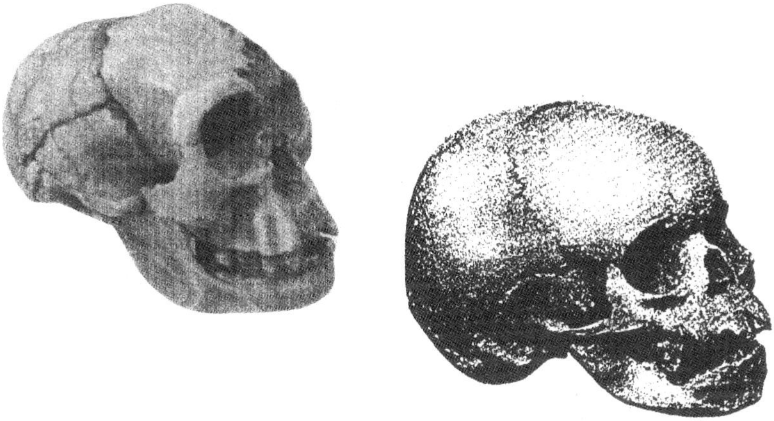

Fig. 15. Liang Bua skull and cretin skull (Dolega) in approximately similar orientation. Relative sizes not known.

humans, neurological cretinism, shows severe deaf mutism, serious intellectual impairment, and major locomotor disabilities, and has much more severe neurological problems than the myxoedematous endemic cretins. Neurological cretins do not fare well and almost all die in childhood. Another cretin problem called sporadic cretinism is anatomically and physiologically identical to myxoedematous endemic cretinism but is not due to iodine deficiency. They are otherwise similar, however, so they provide useful information. Individuals with myxoedematous endemic cretinism (and sporadic cretins) are neurologically only somewhat challenged, and are generally quite capable of living to adulthood, but require some level of support.

As a result, the myxoedematous endemic cretins show only lesser motor, sensory and cognitive problems than neurological cretins. Their motor development is delayed with impaired walking, but they can walk. Their speech is often poor and not easily understood, but they do usually speak. Their intellectual abilities are reduced to greater or lesser degree so that they have difficulty in understanding concepts, but they know the name of the family's dog though not of the country's prime minister (some would say they are lucky in this regard). Thus, with help they can cope, are usually looked after and loved within the family and especially by the mother. In adult life in today's human societies, they usually need family and even community care and shelter for most of their lives.

The Problem of the Small Brain

Despite all of the above, however, there is no doubt that the most striking feature of the Liang Bua fossil is the remarkably small size of the brain endocast, smaller even than all but the very smallest microcephalics (Martin has identified three at about 400 cm^3 out of 33 microcephalics) and smaller than the smallest cretin (at 570 cm^3, this last was the smallest in a very small sample of seven that I have inspected).

In terms of species brains, the Flores brain (endocast, 417 cm^3) rivals only those of nonhuman species: chimpanzees, bonobos and *Australopithecus* and *Paranthropus*. It is not only nowhere close to modern humans at a mean of (say) 1400 cm^3, but also much smaller than all other human fossils that are very much older. For example, at 30,000 to 130,000 years ago, *Homo neanderthalensis* can be as small as 1200 cm^3 (even though as large as 1700+ cm^3, n = 22). Archaic *Homo* (200,000 to 500,000 years ago) has volumes as small as 1100 cm^3 even if they range up to 1450 (n = 17). Even *H. erectus* (about 1,000,000 to 2,000,000 years ago) though having minimum values as small as about 900 cm^3 also ranges up to almost 1300 (n = 28). It is only with *H. habilis* (more than 2,000,000 years ago) that the fossil falls somewhat close (from 500 cm^3 to just under 900, n = only 7). Even for habilines, however, those very small specimens at 500+ cm^3 may actually really be australopithecines! Yet the Flores fossil (at 420 cm^3) is under 20,000 years old.

We can examine brain size in the various pathologies in more detail. There are many different forms of genetic microcephaly, mostly autosomal recessives. Unlike the species brain size distributions (which are normally distributed as would be expected) the skull volumes of microcephalics (n = 33, Martin *et al.*, 2006) are not statistically normal but are strung out like a long sausage (very platykurtotic). This may be partly because they are not one group with a normal distribution, but several different groups of syndromes with data artificially coalesced. It may also be because each of the different conditions may be expressed to greater or lesser degrees, again giving a long-drawn out distribution.

There is also the question of just how small skulls (and brains) can be in normal humans. Skulls and brains are considerably smaller in pygmy or pygmoid people (including some of the highlander people of present-day Flores). Particular well-known small skulls include one human female Melanesian from a small offshore island of New Ireland (790 cm^3 (Schlaginhaufen, 1954)), another female Melanesian at 900 cm^3 (Dart, 1956), two bushmen females (895 and 918 cm^3 (Shellshear, 1934) quoting Marshall) and one female Vedda skull (950 cm^3 (Flower, 1889)). Even two European skulls said to be from normal females are 807 and 830 cm^3. These seven small skulls thus range from 790–950 cm^3, well below 1000 cm^3. On statistical grounds, i.e. being a certain number of standard deviations below the human norm, these might well be counted as microcephalics; certainly they are below several known microcephalics. All save the two Europeans are, however, from ethnic groups with average volumes already reduced to only 1100 cm^3 and containing as many as 5% of skulls below 1000 cm^3. As a result, they can scarcely be accepted as abnormal.

How do cretins compared with all these? Of course, there are not too many records available. As noted before, dimensions of cretin heads are not a good guide to brain size because of the well-documented extra thickness of the skin, the scalp and the bone. As a result of these factors alone, brain size in cretins is considerably less than that would appear on the surface.

Congenital hypothyroidism can reduce brain size by more than 50%. Brains of adult European male endemic cretins scaled with height to 700 cm^3 (De Quervain and Wegelin, 1936) are compared to 1400 cm^3 for their European normals. In a severely iodine-deficient area of China, cretin brains were 52.5% less than normal for age. The brain sizes from cretin skulls and literature that ranged from 570 to 1070 cm^3 (in a sample of only 7) are mainly European cretins whose parent populations are presumably normal for Europeans (i.e. with a mean of around 1400 cm^3). Who knows what the volume of cretin skulls might be if the parent population were pygmoid peoples of Flores (already much smaller than Europeans), perhaps not unlike the pygmoid peoples of other SE Asian Islands (means around 1130 cm^3), or hunter–gatherer populations of 20,000 years ago, perhaps even smaller still? That 50% figure could easily give cretins from these ethnic groups of less than 500 cm^3. It is highly likely that the brains of cretins from SE Asian locations would be considerably smaller than those of European cretins.

The final nail in the coffin regarding possible brain size normality is evident from the recent studies from Palau. There, though brain size of the recent fossils could not be determined directly, it is evident, from extrapolation from skull size, that the average brain of some half a hundred or more individuals must have been of the order of 850 to 950 cc. These individuals could not have been cretins — Palau is a small island with total access to iodine foods. Hence in that area of the world, pygmoid peoples have existed that are so small that, at 52.5% reduction, they could easily have yielded cretins of 400 cc.

In addition, in many populations (and especially, presumably, in prior hunter–gatherers), brain size is further decreased by undernutrition. Careful experimental studies in growing monkeys where reduced brain weights due to reduced growth were produced in only a few months by low protein yet isocaloric diets (Dodge *et al.*, 1975) confirm this. It might especially be expected to reduce brain size in cretins of such groups, individuals already poorly fed and poorly looked after.

An opposite factor affecting brain size is that the very low volume reported for the Flores fossil may be incorrectly small. Already, the volume has been raised by almost 40 cm^3 from 380 cm^3 on first reporting to 417 cm^3 (presumably on the basis of improved reconstruction). Can we add to this the possibility that post-mortem deformation might have reduced the fossil volume because, when discovered, the fossil material was soft and deformable? Can we further add the possibility that, if the fossil were a cretin, the lack of preservation of the non-osseous material, in open synchondroses and sutures, with consequent collapse of bony plates towards one another, could have reduced the estimated volume even further?

As a result of all this, the apparent contradiction of the small size of the fossil and the larger sizes of cretins may not be insurmountable.

The Problem of the 'Turks Saddle'

In contrast to the problem of a small skull in Liang Bua is the problem of a large sella turcica (Turk's saddle) in Liang Bua. This is the bony fossa in the centre of the skull base containing the pituitary gland. In this single feature, cretins display not only relative but also absolute size increase over modern humans and microcephalics (Bellini and Neves, 1956; Anderson, 1961), and even over apes and fossils.

The sella turcica of cretins is even larger than that in normal much larger skulls (Fig. 16) because it contains the pituitary gland that enlarges in the aforementioned struggle to kick-start thyroid development in an iodine-deficient internal environment. The sizes of Chinese cretin pituitary fossae were 14.0 + 3.1 mm ($n = 58$). This completely exceeds the normal human range (8.6 + 1.2 mm, $n = 5$, $P > 0.001$, two-tailed test (He, 1984)) despite the much larger skulls of normals. The Flores pituitary fossa is, likewise, at 12.9 mm, also completely outside the range of normal humans and completely within the range of Chinese cretins.

However, the original investigator has noted in the media that the pituitary fossa of Liang Bua is normal. Another investigator gives its size as about 9 mm. Our measurement of the published computer derived endocast was, as given above, 12.9. How can this be? We have therefore looked more carefully at the anatomy of the cretin pituitary fossa. It turns out that a wide range of fossae are found in cretins. These range, on the one hand, from a normal entrance to the pituitary fossa (of about 8 mm) with a large ballooned pituitary fossa (of about 12 mm) inside the bone, to, on the other hand, a large entrance (of about 12 mm) with a smaller extent of the fossa length (of about 8 mm). We have returned to our original measurement of the Liang Bua fossa. Made from an endocast, the depths of the fossa thus appear as a convexity and this is about

(a)

(b)

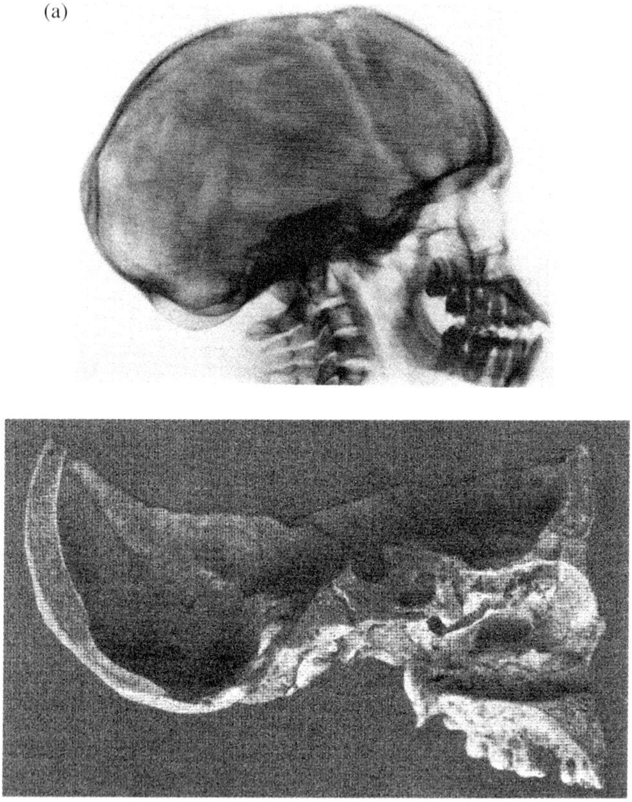

Fig. 16. (a) and (b): Radiograph and drawing of cretin skulls showing wide and ballooned Turk's Saddles (sella turcia) in the centre, (c): Sketches of normal and cretin pituitary fossae. The normal one has approximately the same dimension for the pituitary opening and the depths of the fossa. The cretin examples show: middle, a ballooned internal cavity and a normal size opening; and below, a normal sized internal cavity with a very wide opening. In the middle case the enlarged pituitary must have been within the bone, in the other it must have extruded above the bone.

8 mm. The opening of the fossa (a deep concavity on the endocast) measures about 12–13 mm. These measurements seem to explain both sets of figures and are further evidences of a cretinoid state for the Liang Bua fossil.

It is of specific interest that investigators, thinking that the body's attempt to stimulate the pituitary to produce pituitary hormone in Laron Syndrome, expected that the pituitary gland would be increased

(c)

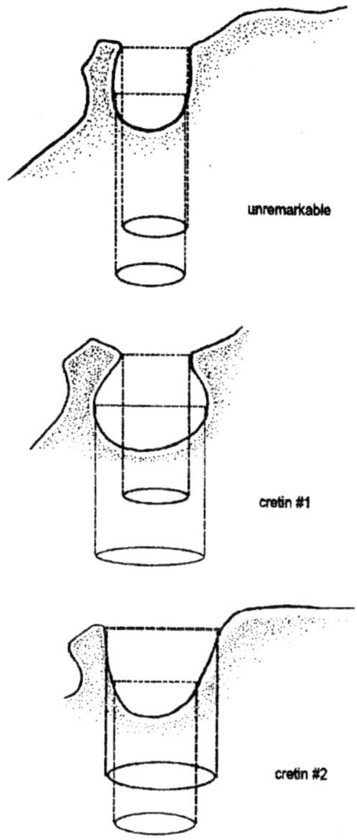

Fig. 16. (*Continued*)

in this condition. To their surprise, the pituitary was smaller than normal in some of these dwarfed individuals (Kornreich *et al.*, 2004).

Of course, there are a variety of conditions in humans that do give a large sella turcica: pituitary neoplasms, blood vessel malformations, local space occupying tumours, other lesions, even some cases of Down's syndrome, and so on. However, a large sella correlated with reduced bone growth is usually found in cretinism. All this, one would think, should settle the matter for Liang Bua!

Skull Measurements

It has also seemed worthwhile testing these ideas with measurements (obtained by me) using multivariate statistical methods (carried out by Ben Kefford). Our base sample followed the measurements of a large sample of humans from the study of Howells (1996). We employed them twice: the full sample as all humans, and a restricted sample of the smaller Andamanese Islanders alone. The measurements we chose were those of Howells that we could replicate (or obtain from the literature) for the small samples of microcephalics, cretins and fossils, a total of 19 variables. In addition, we carried out the analyses on a subset of only nine measurements. These were chosen before the fact because I could identify those measurements that specifically included aspects of the cartilaginous components of the skull (mainly the cranial base). I did this because I knew that this aspect of the growth deficiency is especially affected by cretinism. These particular measurements and analyses are, therefore, a special test of the cretin hypothesis.

Principal component analysis of all 19 cranial variables (Fig. 17) located all the cretins in the first three components (65% of the variation) within the group of normal humans, and particularly close to the Andamanese Islanders. The Flores fossil was located among the cretins. In contrast, the microcephalics we examined were quite separate from normal humans (and therefore also from the Flores fossil). Yet they did not, themselves, fall closely together. This implies that they are not a single group.

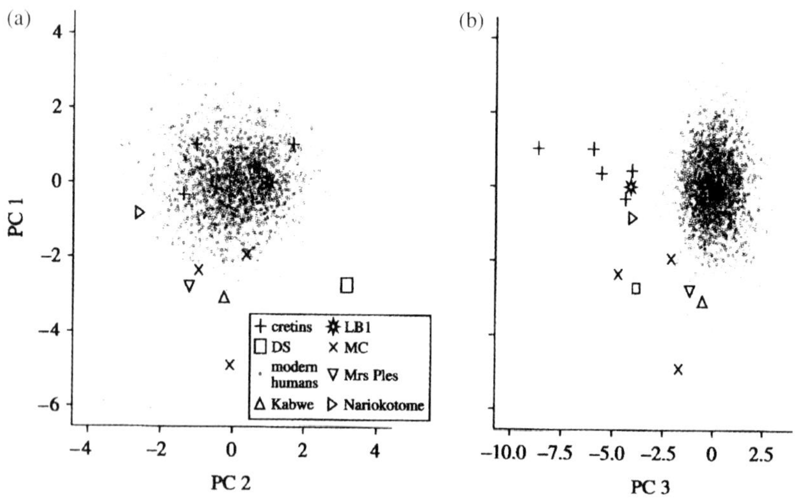

Fig. 17. Principal component analysis of full set of variables measuring all parts of the skull.

Analysis of the nine variables special to the cartilaginous portions of the skull (77.5% of the variation) maintains the integrity of the cretins as a group but separates them completely from modern humans. The Flores fossil is again located among the cretins. This similarity of Liang Bua with cretins and separation from humans is highly significant ($p < 0.001$). The microcephalics are again separated widely both from each other, and from cretins, the Liang Bua fossil, and normal humans. These findings were also true for the sample of smaller Andamanese Islanders (Fig. 18).

We followed these studies by Discriminant Function Analysis of the same sets of data (Fig. 19). This attempts to see how different these individuals are when treated as groups (rather than as individuals in a single data universe). In this case, there are, of course, only three data groups: all modern humans (or all Andamanese Islanders alone, respectively), all cretins and all microcephalics. As a result, there are, therefore, only two discriminant functions. These correctly classify (in each respective analysis) all modern humans ($n = 2523$) and all Andamanese Islanders ($n = 69$) into their respective groups. All cretins are also classified as a single group ($n = 5$). The Flores fossil falls completely with the group of cretins (with a posterior probability

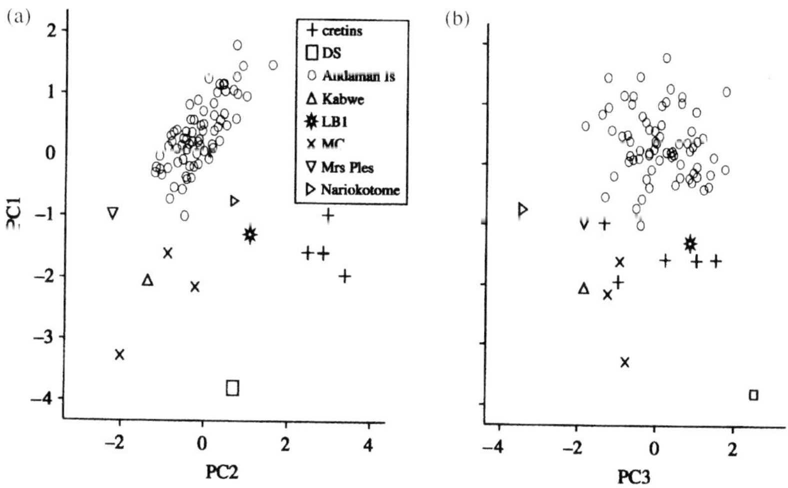

Fig. 18. Principal component analysis of restricted set of variables measuring cartilage bones of the skull base.

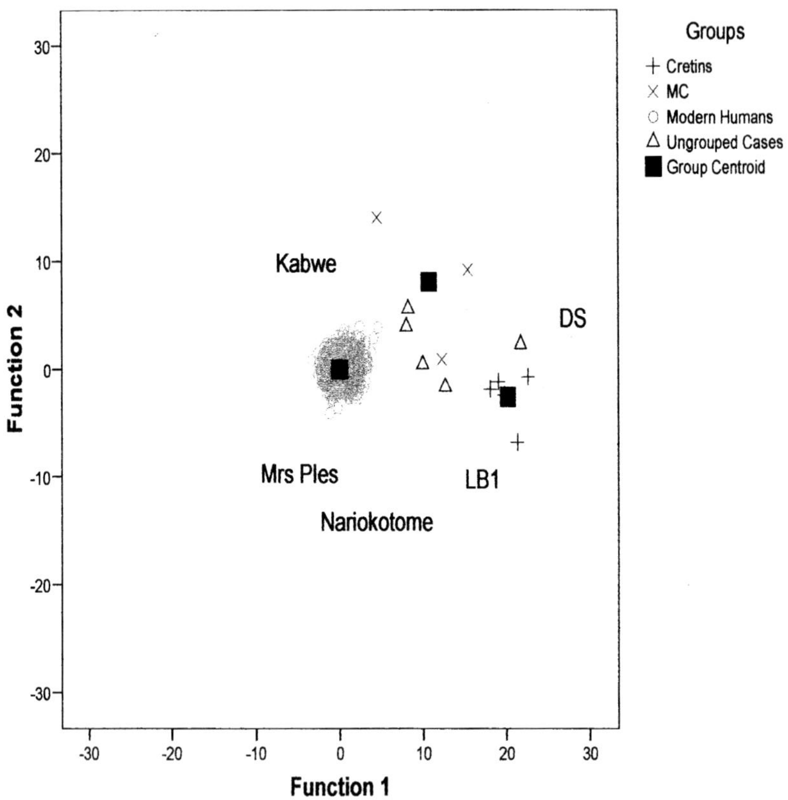

Fig. 19. Discriminant function analysis of skull measurements.

of $P > 0.998$). The four microcephalics, however, do not fall together as a group. They are not only widely separated from all humans and from all cretins, but also from one another. This is not surprising as the various microcephalics are of different aetiology.

In addition, for interest's sake, we included the australopithecine, Mrs Ples, and the hominins, Nariokotome and Kabwe in these analyses. These individual specimens were placed far distant from all the others. This is what we would expect: Mrs Ples, Nariokotome and Kabwe do not belong to any of the aforementioned; they are completely different species. These results thus supply statistical tests supplementing the descriptive and univariate anatomical evidence implying that the Flores fossil is similar to cretins and different from normal humans, from genetic microcephalics, and from the three much older fossils.

General Appearance of Cretins

Though the present problem is concerned with what can be discerned from the bones, there are also features of the soft tissues of cretins that are especially noticeable. Thus, many of the soft tissues show myxoedematous changes (hence the condition of myxoedema in adults, and the term *myxoedematous* endemic cretinism).

This is especially obvious in the head and face. Cretins are small, of course, even as adults but, unlike many genetic microcephalics, tend to be unattractive. They have smaller heads as befits their dwarfed status, but their faces (especially their lower faces) are large in relation to the rest. The head overall appears considerably larger than one would judge from the size of the skull because the skin and the underlying scalp tissues are much thicker (myxoedematous) than normal. The skin of the face is generally thickened and protruding. The tongues are large and lolling and tend to spill out of their large and protuberant lips. These various features are in marked contrast to those of many of the genetic microcephalics who, though they have small faces and small crania, generally have faces that are small and delicate even judging by their crania, i.e. they resemble attractive children.

The myxoedematous thickening of tissues, together with additional deposits of fat, are found elsewhere in the cretin body. Thus, their trunks are corpulent, they are often pot-bellied, and breasts in females are heavy and pendulous. Their backs often show exaggerated spinal curvatures, emphasising greater roundness of the shoulders, forward protrusion of the abdomen, sway back and posterior emphasis of the buttocks.

The soft tissues of the hands and feet of adult cretins, though small compared to normal adults, are, in fact, greatly enlarged compared to the adjacent limb segments. These findings are not only due to the aforementioned changes in the bones, but also because the soft tissues are thickened and coarse (Figs. 20 and 21). Again, in comparison, the hands and feet of genetic microcephalics are generally small and delicate, the skin thin, as in children, in comparison to the rest of the body.

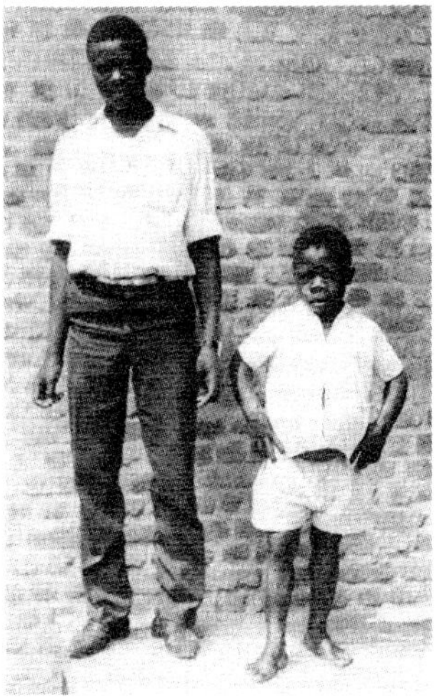

Fig. 20. Adult myxedematous cretin, male, 21 years old, height 1.17 m and unaffected male, 20 years old, height 1.72 m from Idjwi Island, now Democratic Republic of Congo. From Obendorf *et al.*, 2008, with permission of Papua New Guinea Institute of Medical Research.

Fig. 21. Adult myxoedematous endemic cretin approximately 27 years old from Sumatra (Aceh), Indonesia, circa 1930 with unaffected female of the same ethnicity (foster mother). From Obendorf *et al.*, 2008 (thesis of L. B. van Bommel, 1930).

Though none of these soft tissue features can help directly in the examination of fossils, they may well be important, as we shall see later, as evidence relating to the veracity of oral descriptions.

Is Flores Iodine-Deficient? Could Cretins have Occurred There?

These voluminous findings seem to deny all hypotheses save that the fossils are cretins. In order to further test this hypothesis, however, we need to know if the environmental and other conditions for long continued production of cretins exist now, and could have existed for a long time on Flores.

Very large areas of the world have limestone regions where iodine is leached from the hills and with diminished sea-to-hills return of iodine in rain. These include many parts of Europe and North America. In these countries, however, goitre and cretinism have been largely removed from the populations for more than a century by public health measures involving added iodine (iodised salt) in the diet. It is partly for this reason that it is difficult to find skeletons of cretins. The collections of a century or so ago have been largely lost or discarded; for example, the cretin skeleton in the Royal College of Surgeons of England was destroyed during the Second World War leaving only the half skull that was on display.

In many other places: large tracts of South America, Africa, Asia (including both mainland China and island areas of SE Asia), iodine deficiency is also present and hypothyroid goitre and cretinism are still endemic (Figs. 22–24). The incidences of adult hypothyroidism and goitre in some of these areas are often close to 100% often judged simply by assessment of visible or palpable goitres (Fig. 25). With such figures go surprisingly high percentages of cretinism (1%, 4% and even as much as 11%, in different world regions). Moreover, these rates may be minimum because local communities generally hide the most seriously affected individuals during medical examinations and censuses.

The lack of treatment in these regions is not because of lack of knowledge. Most governments are well aware that treatment and prophylaxis are easy and have usually mandated them. The World Health

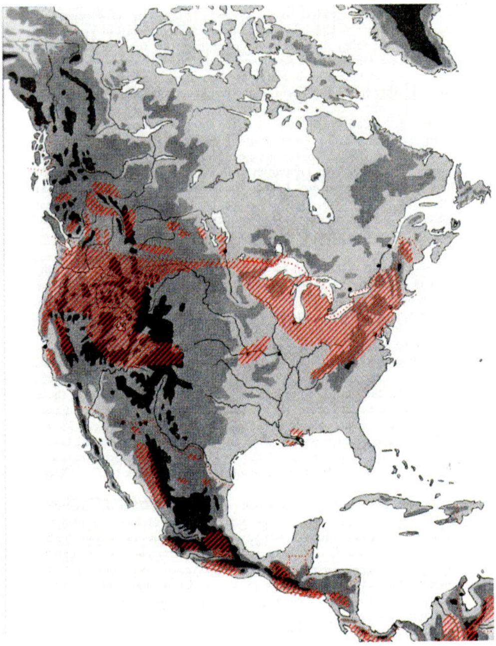

Fig. 22. Map of N. America showing goitre regions.

Organisation has long had excellent reports about worldwide geographic incidences and treatments (e.g. WHO Report, 1960). The problem is that it is just not easy to do it in isolated hilly areas where there is a great deal of poverty, considerable undernutrition, little in the way of health-related services, and no money to pay for treatment. Even yearly visits by health professionals to assess thyroid function and to supply depot injections of iodised compounds reach relatively few people.

In SE Asia in particular, endemic goitre has been recorded throughout the Indonesian archipelago from Sumatra to Timor. Iodised salt was introduced into selected areas from as early as 1927 (Kelly and Snedden, 1960) and has been mandated by law since 1994. Nevertheless, even just recently, the Province of Nusa Tenngara Timur (which includes Flores) had a total goitre rate (as measured by visible or palpable goitre) of 38%. In the Manggarai District of Flores

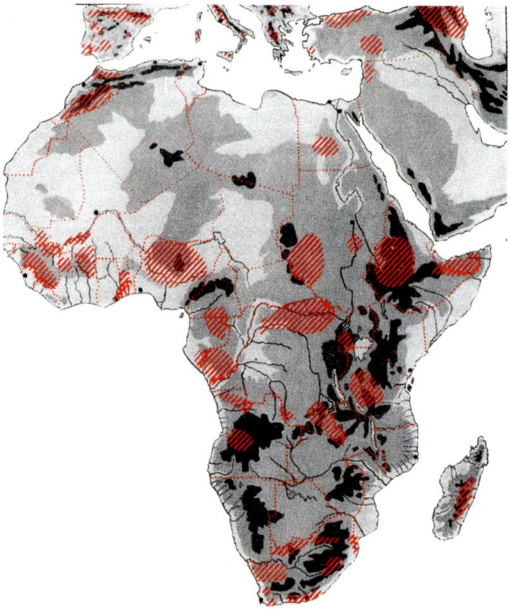

Fig. 23. Map of Africa showing goitre regions.

containing Liang Bua, the total goitre rate was 41% in 1998; this had increased by a further 11% to 52% by 2004.

These high goitre rates do not necessarily imply that cretinism should be encountered on Flores today among agricultural populations. The Manggarai people arrived by sea as farmers 4000 years ago and, as they spread inland, would probably have maintained access to coastal resources, including sea salt (containing iodine) by trading networks. These networks are certainly present in modern times. However, the Manggarai intermixed with the original Nage inhabitants, and, especially in isolated central parts of the island, many prior customs continue. The current goitre rates imply that poor indigenous hill groups in Flores should have produced some cretins (though it must be emphasised that these have not yet been reported). Today's goitre rates imply that prior hunter–gatherer populations in the hills should have been severely iodine-deficient and would regularly have produced cretins.

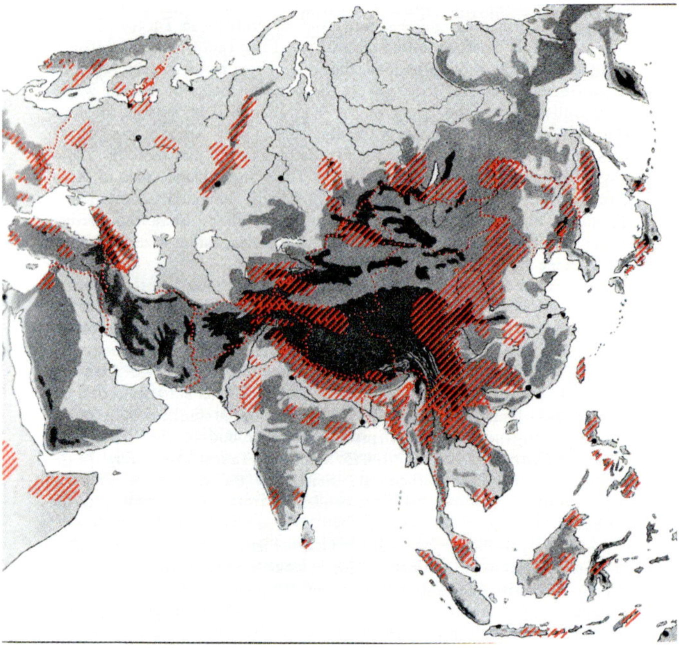

Fig. 24. Map of Asia showing goitre regions.

Fig. 25. An example of the incidence of goitres in children in Sumatra.

Liang Bua itself is a limestone cave and nearby soils are alkaline and probably therefore iodine-deficient. The altitude is 500 m (it would have been higher 18,000 years ago because sea levels were lower). The site is reasonably remote from both the north and south coasts (and again would have been further away with the lower sea levels of 18,000 years ago). Fish bones found at Liang Bua are river fish that would be deficient in iodine. River waters (a river flow nearby) in such regions are low in iodine. All these factors would have precluded access to iodine-rich seafoods.

More Goitrogenic Factors on Flores!

There are a number of additional environmental factors that potentiate iodine deficiency. Experimental studies in rats show that high thiocyanate and low selenium potentiate the increase in thyroid necrosis produced by low iodine. This is a rat model for cretinism (Contempre *et al.*, 2004).

Let us look first at thiocyanate. High dietary thiocyanate is a major component of the iodine deficiency problem in many regions of the world. Thus, in Africa (Uele and Idjwi Island), iodine deficiency hypothyroidism is potentiated by increased serum thiocyanate that stems from cyanogenic glucosides in a specific food item: cassava (Delange, 1974). This is also so in SE Asia where one source is the highly cyanogenic fruit of the durian (*Pangium edule*) described in Borneo as 'the hunter–gatherer's refrigerator' (Zahorka, 2000). On Flores itself cyanogenic plant foods are present today and would have been available to earlier hunter–gatherers. They include native bamboos (*Gigantochloa* spp. and *Bambusa* spp.) which are highly cyanogenic. In Ruteng Park specifically (only about 10–15 km from Liang Bua) are cyanogenic *Dioscorea* tubers, probably the bitter yam (*D. hispida*). These are harvested and used to eke out meagre diets and would be especially important food sources in times of seasonal and cyclical droughts. There is also *Acacia* on Flores, many species of which are cyanogenic. Though their seeds are widely consumed by Australian hunter–gatherers we do not know if they have cyanogenic potential on Flores.

The second precipitating factor, low selenium, is also a widespread component of the iodine deficiency situation in many regions of the world. Selenium levels are very low in the parts of Africa and China where iodine deficiency is rife and where there are high percentages of cretins. For example, there are low serum concentrations of selenium in school children in the cretinism area of Idjwi Island in Africa. In local regions of SE Asia, low selenium was found in eggs collected from free-living chickens in the Malang District of East Java (a region with similar geology to western Flores). There are also low serum levels in school children with palpable goitre in central Java. There is no information about selenium specifically in Flores, but the above is indicative.

Are there any Social and Community Factors of Relevance?

Though the environment exists throughout SE Asia, though iodine deficiency goitre rates are high on Flores, and though cretinism is recorded in Sumatra, Central Java and Borneo, we do not know of any recent cretinism in Flores. Direct Dutch control of Flores was only assumed in 1907, and missionaries first entered western Flores only from 1921 after the aboriginal hunter–gatherers had been displaced or assimilated into the agricultural population.

In societies in the West, however (for example, in small hill communities such as Derbyshire, UK, and the Cantons of Switzerland in the 19th century), cretins were not infrequently found. They were cared for by their mothers and the immediate family. But they were often different enough from the unaffected village people that they were retained at home in sheltered situations. Quite often the more seriously affected individuals would be hidden from the sight of visitors though they would be known to the locals. The 'village idiot' syndrome was well known, and though in those days, it was commonly laid at the door of inbreeding, incest, or family rape, it was probably much more often due to cretinism in communities living in markedly iodine-deficient limestone regions. Such cretins, cared for as any other child in the family at first, later continued to be treated as 'children' even when considerably older.

There is much information available that this picture of cretin children still occurs in relevant parts of the world. Kelly and Snedden (1960) note that in North Borneo: 'the worst cases [of cretinism], subhuman in appearance, are seldom seen, as they are hidden in the jungle at the approach of strangers'. It is not unlikely that this accounts for the non-recording of cretins in the hilly areas of Flores today.

In a hunter–gatherer community in prior times, the initial parts of the above scenario could be replicated. Cretin children would be looked after by their mother and perhaps the immediate family. In seasonally mobile hunter–gatherer groups, such small individuals, as normal children, would be taken along with the group. The time would come, however, when the cretins approached adulthood that they might be forced into separation. This would especially occur with the relatively early loss of parents, likely with shortened lifespans in such societies. They would then be ostracised as adults by the wider community due to their abnormal features and behaviours. Unable to travel easily with a mobile community, especially unable to help build normal temporary dwellings in such a community, adult cretins might well separate and shelter in caves. If there were a reasonable number of them (say, conservatively) 5% of all births, they might indeed shelter together.

Becoming separated, and in any case being different from normal, they would not be accorded the usual death rites of the community such as burial, cremation, body exposure, bone scattering, whatever were the hunter–gatherer norms in that place and time. Thus, cretin remains would not receive such rites but would be left in such separate shelters (e.g. caves) that they utilised. A concatenation of factors: loss of parents, seasonal mobility, use of alternative shelters and lack of social norms for systematic disposal of remains, could all explain the absence of normal individuals from caves such as Liang Bua. Such situations could continue over many generations.

Oral Traditions on Flores

Although the current population of Flores are the Manggarai, they were themselves immigrants arriving by sea about 4000 years ago. They either supplanted, or interbred with, or both, the Nage, the original indigenous people of Flores. There is an extant oral history of the Manggarai recorded by Forth (1998). Some of these stories were collected from the Nage (Keers, 1948). In addition to the victorious stories of the celebration of the arrival of the Manggarai by sea, there are stories about these earlier Nage 'ancestors'. These latter speak of *ebu gogo,* known as 'greedy ancestors' of the Nage.

The stories have, says Forth: 'an apparent historicity and matter-of-fact quality'. They imply: 'some empirical basis in a former component of the human population of Flores that is no longer present'. Thus, the *ebu gogo* lived in caves called *lia ula*, caves of the 'children'. They were short, 'roughly built', 'hairy', 'pot-bellied', 'stupid', and the females had 'pendulous breasts'. This last feature implies that, small though they were, they were not actually children, but adults. Other features of these stories (Forth, 1998) can be easily associated with the behaviours of adult cretins, including their stealing of food, inability to cook and 'an ability to speak the local language though one judged by the Nage to be imperfect and interspersed with an "mmm" sound' (Forth, 2005).

Other stories describe the *ana ula,* 'banished children', and note that they share some characteristics with *ebu gogo,* including living in caves, and 'resembling human beings one meter in height who have no language, but make only squeaking noises' and have living descendants among the Nage (Forth, 1998: 102). One particular Nage story describes in detail the helplessness and grief of a mother following the gratuitous killing of her *ebu gogo* child by a Nage ancestor (Forth, 1998: 108). This becomes intelligible if thought of as within a community that included cretin children and normal mothers! Forth even, subsequently, related them to *H. floresiensis* (Forth, 2005).

The anatomical and behavioural descriptions of European cretins include all these features. Anatomically, they are short, crudely proportioned (rather than gracile as hypopituitary dwarfs), often hairy as children (Perloff, 1955), pot-bellied (Delange, 1974) and may have pendulous breasts (Jackson, 1952). Exactly similar descriptions of cretins were observed by a doctor in the highlands of British North Borneo (Clarke, 1951). 'Briefly, outstanding features are reduction in height, often with disproportionate shortening of limbs, a general physical podginess (*sic*) with thick dry skin, short thick fingers, protuberant abdomen, perhaps umbilical hernia, dull expressionless faces, nose with widely patent nostrils and depressed bridge, exaggerated bossing of the skull, and various degree of mental retardation. Many of them show a specious giggling brightness; not a few are deaf or dumb, or both. The worst cases [are] subnormal in appearance'.

Of course, many groups throughout the world have stories of small individuals (e.g. leprechauns, fairies, gnomes and dwarves (and Tolkien's hobbits!)). Such small individuals are generally not human, often live forever or are very long-lived, or at least live only in their world (e.g. middle-earth), are mischievous but clever and usually capable of magical powers. The *ebu gogo*, in contrast, are clearly human, handicapped, die and have human mothers. These stories of the Nage are consistent with and may derive from real experiences of cretinism.

Further Hypothesis Testing

The cretin hypothesis is tested by a great deal of evidence. Could such a situation have given rise to sufficient cretins to make possible the fossilisation of some individuals? A small population of hunter–gatherers (say $n = 25$–100) utilising resources around Liang Bua, with cretinism prevalence of 5% would, on average, have 1.25–5.0 living cretins, and, therefore, 20–60 adult cretin deaths per kiloyear (assuming birth rate and death rate at 30/1000 with 50% survival to adulthood). This is easy enough to explain the number of discovered remains, certainly far more easily than for the much rarer genetic microcephalics.

Until now the status of the Flores fossils has been approached as a straightforward evolutionary question. The remains could be of a new species, a modification of an earlier species, an adaptation to the small island situation, or a possible combination of two or even more of these.

A more complex concept, also based upon hereditary and evolutionary ideas is that they are not a new species, but an autosomal recessive genetic pathology that had become fixed in a small isolated population (of *Homo*, whether *sapiens* or just possibly a relict *erectus* group).

We are providing a third hypothesis that does not involve genetics and evolution, but environment. The Flores fossils may be endemic cretins resulting from environmental factors present in Flores over an extended period of time (Oxnard and Obendorf, 2006; Obendorf and Oxnard, 2006; Obendorf *et al.*, 2007; Obendorf, *et al.*, 2008). This transforms the task from making an evolutionary assessment from the primitive and derived characters of evolution, to an environmental (medical) diagnosis based upon pathological features of iodine deficiency. It is remarkable that so many features similar to those normally present in great apes, in *Australopithecus* and *Paranthropus*, and in early *Homo* (e.g. *H. erectus* and even to some degree, *H. neanderthalensis*) but not in modern *H. sapiens* are generated in humans by growth deficits due to the absence of thyroid hormone. In other words, many of the pathological features of cretinism mimic the primitive characters of evolution making it easy to mistake pathological features for primitive characters. The differences can be disentangled by understanding the underlying biology of characters.

The extended descriptions above show that the cretin hypothesis has already been tested a number of times, including statistical tests. It can, however, be further tested through prediction.

First, if the Liang Bua fossils are human cretins, we can predict that their DNA will be modern human DNA, or possibly older human DNA. I understand that such DNA has been found to date is only modern DNA and it is proposed that this results from modern contamination.

Second, if the Liang Bua fossils are truly cretins, we can also predict that further skeletal remains will also show characteristics of cretinism. These include many, many additional features. We can expect to find a sternum with separate sternebrae that looks like an ape sternum (though, if this existed, the separate sternebrae might, for that reason, not be found or recognised). We can expect vertically short vertebral bodies due to missing epiphyses. We can expect open epiphyses (or loss of epiphyses) in most girdle bones and long bones. We can expect more complete evidence of mixed dentitions, more evidence of reduced and missing hand and foot bones, and so on.

Third, we can predict that the known evidence of about 50% of goitre on Flores as shown by visual inspection will prove to be even higher when more complete medical examination of the local people is carried out. We can also predict that actual evidence of the existence of cretins today will be uncovered.

Fourth, we can predict that environmental testing will unequivocally determine that the local soil and river water in central Flores are truly iodine- and selenium-deficient, and that the native plants available as food in a state of undernutrition contain excess thiocyanates.

Fifth, we can predict that further examination of the weaker, but nevertheless interesting evidence provided by social, cultural and oral traditions of populations with cretins will add to the richness of the hypothesis.

Of course, no single cretin shows all features; neither can we expect the Liang Bua fossils, if cretins, to display the entire range.

A Final Critical Wrinkle

Whether or not the Liang Bua fossils turn out to be cretins, this discussion remains highly significant for totally different reason. This relates not so much to the evolution of a sibling species (*H. floresiensis*) as to the ecology of modern humans (*H. sapiens*). In other words, by far the most important result, made visible by what is likely to be (in fact already is) the highly controversial and therefore visible nature of these ideas, may be the **reawakening** of modern governmental agencies to the fact that, not only in this part of SE Asia, but widely throughout the world, hypothyroidism in adults and cretinism in infants have not been conquered.

For instance, even in Australia, a country in which one would think iodine deficiency had been conquered, a new problem has been identified. Recent studies on iodine levels of school children have shown that about 50% are classified as mildly or moderately iodine-deficient. In New South Wales and Victoria (these are iodine deficiency regions, as also is Tasmania) as many as 14% and 19% of childen, respectively, are classed as moderately deficient. Of course, this is not cretinism, but even mild iodine deficiency results in a reduction in average IQ and a lessened number of gifted individuals. These new findings in Australia are not clearly understood but may be partly due to decreased consumption of iodised salt and partly to reduced use of iodine-based cleaning products in the dairy industry. This may be a call for action in OECD countries that had long thought the problem was solved.

In the iodine deficient regions of the developing countries of the world (in South America, Africa, Asia especially China, and much of South East Asia) the matter is much more serious. In these regions florid goitre, and especially cretinism in the children of iodine-deficient mothers, remains an enormous but hidden problem.

Though the health professions know what to do, though governments know what to mandate, though both undoubtedly believe they are acting correctly, the prophylactic and therapeutic measures, simple and easy as they are, just are not reaching poor people in many iodine deficient areas of the world. A **reawakening** of the world to this problem (because of the high publicity given to these fossils), and the reminder that it is a problem so cheap to fix (in comparison with the high costs of so much modern medicine), would be by far the most important thing that could come out of our ideas.

References

Andersen HJ, Studies in hypothyroidism in children, *Acta Paed Scand* **50** (Suppl 125): 1–150, 1961.

Argue D, Donlon D, Groves C, Wright R, Homo floresiensis: Microcephalic, pygmoid, Australopithecus or Homo? *J Hum Evol* **51**: 360–374, 2006.

Bellini MA, Neves I, The skull in childhood myxedema: Its Roentgen appearance, *Am J Roentgen* **76**: 495–498, 1956.

Brown P, Sutikna T, Morwood MJ, Soejono RP, Jatmiko SEW, Due RA, A new small-bodied hominin from the Late Pleistocene of Flores, Indonesia. *Nature* **431**: 1055–1061, 2004.

Clarke MC, Some impressions of the Muruts of North Borneo, *Trans R Soc Trop Med Hyg* **44**: 453–464, 1951.

Contempré B, Morreale de Escobar G, Denef J, Dumont JE, Many M, Thiocyanate induces cell necrosis and fibrosis in selenium- and iodine-deficient rat thyroids: A potential experimental model for myxedematous endemic cretinism in central Africa, *Endocrinology* **145**: 994–1002, 2004.

Dart RA, The relationship of brain size and brain pattern to human status, *S Afr J Med Sci* **21**: 23–45, 1956.

De Quervain F, Wegelin C, *Der Endemische Kretinismus*, Pathologie und Klinik in Einzeldarstellungen, Julius Springer, Berlin and Vienna, 1936.

Delange F, *Endemic Goitre and Thyroid Function in Central Africa*, Monographs in Paediatrics, Vol. 2, Karger, Basel, 1974.

DeLong GR, Effects of nutrition on brain development in humans, *Am J Clin Nutr Suppl* **57**: 286S–290S, 1993.

Dodge PR, Prensky AL, Feigin RD, *Nutrition and the Developing Nervous System*, Mosby, Saint Louis, 1975.

Dolega D, Ein Fall von Cretinismus beruhend auf einer primären Hemmung des Knochenwachsthums, *Beit z Pathol Anat* **9**: 488–509, Plate XVIII, 1891.

Flower WH, Description of two skeletons of kkas, a pygmy race from Central Africa, *J Roy Anthrop Inst* **18**: 3–19, 1889.

Forth G, *Beneath the Volcano: Religion, Cosmology and Spirit Classification among the Nage of Eastern Indonesia*, KITLV Press, Leiden, 1988.

Forth G, Hominids, hairy hominoids and the science of humanity, *Anthropol Today* **21**: 13–17, 2005.

He TY, X-ray manifestations of bone in myxedematous endemic cretinism, *Zhonghua Fang She Xue Za Zhi* **18**: 279–282, 1984.

Hill WCO, Two examples of infantilism, *Ceylon J Sci (D)* **IV**: 71–118, Plates II–IX, 1936.

Herschkovitz I, Kornreich L, Laron Z, Comparative skeletal features between *Homo floresiensis* and patients with primary growth hormone insensitivity (Laron syndrome), *Amer J Phys Anthrop*, Accepted 17 April, 2007, Published online.

Howells WW, Howells' craniometric data on the Internet, *Am J Phys Anthropol* **101**: 441–442, 1996.

Jackson WPU, Studies of adult cretins, *S Afr Med J* **26**: 605–607, 631–633, 645–647, 1952.

Jacob T, Indriati E, Soejono RP, Hsü K, Frayer DW, Eckhardt RB, Kuperavage AJ, Thorne A, Henneberg M, Pygmoid Australomelanesian *Homo sapiens* skeletal remains from Liang Bua, Flores: Population affinities and pathological abnormalities, *Proc Natl Acad Sci* **103**: 13421–13426, 2006.

Keers W, *An Anthropological Survey of the Eastern Little Sunda Islands. The Negritos of the Eastern Little Sunda Islands. The Proto-Malays of the Netherlands East-Indies*, Indisch Instituut, Amsterdam, 1948.

Kelly WC, Snedden WW, Prevalence and geographical distribution of endemic goitre, in *Endemic Goitre*, World Health Organization, Geneva, pp. 27–233, 1960.

Kornreich L., Heyev G, Schwarz M, Marmajya B, Laron Z, Pituitary size in patients with Laron Syndrome (Primary GH insensitivity), *Euro J Endocrinol* **148**: 339–341, 2004.

Knaggs RL, Cretinism, *Brit J Surg* **16**: 370–400, 1928.

Laron Z, Kornreich L, Hershkovitz I, Did the small-bodied hominins from Flores (Indonesia) suffer from a molecular defect in the growth hormone receptor gene (Laron Syndrome), *Ped Endocrinol Rev* **3**, 2006.

Larson SG, Evolutionary transformation of the hominin shoulder, *Evol Anthropol* **16**: 172–187, 2007.

Lewis, OJ, Evolutionary change in the primate wrist and inferior radio-ulnar joints, *Anat Rec* **151**: 275–286, 1965.

Martin RD, MacLarnon AM, Phillips JL, Dobyns WB, Flores Hominid: New species or microcephalic dwarf? *Anatom Rec* **288A**: 1123–1145, 2006.

Morwood MJ, Brown P, Jatmiko ST, Saptomo EW, Westaway KE, Due RA, Roberts RG, Maeda T, Wasisto S, Djubiantono T, Further evidence for small-bodied hominins from the Late Pleistocene of Flores, Indonesia, *Nature* **437**: 1012–1017, 2005.

Obendorf P, Oxnard C, A model for Liang Bua hominins based on endemic cretinism, *Australas. Soc Human Biol* **20**: 12, 2006.

Obendorf P, Oxnard CE, Kefford B, Are the small human-like fossils found on Flores human endemic cretins? *Proc Roy Soc B* **1098**: 1–10, 2008.

Obendorf P, Kefford B, Oxnard C, Morphometric analysis of the LB1 cranium, *Australas Soc Human Biol* **21**: 9, 2007a.

Oxnard C, Obendorf P, *Homo floresiensis*: A remarkable evolutionary paradox, *Australas Soc Human Biol* **20**: 11, 2006.

Oxnard C, Obendorf P, Kefford B, The bipartite trapezoid and the primitive wrist morphology of LB1, *Australas Soc Human Biol* **21**: 11, 2007b.

Oxnard C, The anatomy of cretinism: features resembling characters primitive for hominoids, and features unique to cretins, *Australas Soc Human Biol* **22**, in press, 2008.

Ortner DJ, Hotz G, Skeletal manifestations of hypothyroidism from Switzerland, *Am J Phys Anthropol* **127**: 1–6, 2005.

Perloff WH, A manifestation of juvenile hypothyroidism, *J Am Med Assoc* **157**: 651–652, 1955.

Schlaginhaufen O, Anthropologische Reminiszenzen von den Feni-Inseln in Bismarck-Archipel, *Z Morph Anthrop* **46**: 282–287, 1954.

Shellshear JL, Elliot-Smith G, A comparative study of the endocranial cast of *Sinanthropus*, *Phil Trans Roy Soc B* **223**: 469–487, 1934.

Straus W, Cave AJE, Pathology and posture of Neanderthal Man. *Quart Rev Biol* **32**: 348–363, 1957.

Tocheri MW *et al.*, The primitive wrist of homo floresiensis and its implications for Hominin evolution. *Science* **317**: 1743–1745, 2007.

Weber J, Czarnetzki A, Pusch C, Comment on 'The Brain of LB1, Homo floresiensis', *Science* **310**: 236b, 2005.

Zahorka H, *Pangium edule Reinw — der Kühlschrank der Jäger und Sammler auf Borneo*, Palmengarten **63**: 121–124, illus. (In German, Engl. summ.).

Chapter 9

Brains, Babies and Vitamin B_{12}

A Side Issue

My involvement with vitamin B_{12} has the most bizarre origin of any project in this book. One day in 1959 I was approached by a graduate student in psychology (then Mr Malcolm Roberts, now Emeritus Professor Malcolm Roberts) with a peculiar request. Would I look after his monkeys medically? This was before the days of central animal holding facilities with attendant veterinarians and so forth. His qualifications were in psychology; mine were in medicine. What more natural than that he would make such a request! He had just imported a dozen or more green monkeys for Skinner box experiments. Having just arrived cooped up in small crates, and having been for weeks or months in holding facilities both in the United Kingdom on arrival and somewhere in Africa after capture but before shipment, these poor creatures were in a terrible condition. Some had died in transit. The rest had diarrhoea and vomiting; they had upper respiratory tract infections; they seemed to be in a state of malnutrition; they were scruffy and had ectoparasites; they probably had enteroparasites though I could not actually be sure of that. Worst of all, they could not speak and tell me what was wrong. It was just like when I did paediatrics. And when I did paediatrics I knew that if a child looked well it was probably ill, and if it looked ill, it might be going to die. The same seems to apply to small monkeys.

I agreed to treat them. I ordered sulphonamides for the gut infections, penicillin for the respiratory infections, and good food and vitamins because they looked as though they needed them. I was careful not to fall into the trap of treating their parasites; that was to wait until they were strong enough to take those powerful drugs. All this was not particularly clever — just blockbuster treatment — with precious little diagnosis. They got better and my colleague was delighted.

Included in the vitamins that I ordered was folic acid (which instantly tells you I did not know much about monkeys or diet; the one thing they are unlikely to be short of is folic acid, a component of many of the plant foods that they normally eat).

Within a few days I was called to see an old rhesus monkey that had been in our monkey colony for a long time. (This colony was a research facility that had been set up by Professor Zuckerman, later Lord Zuckerman, when he took up the Birmingham chair). The animal had become paralysed in its hindlimbs and tail.

> 'I'm afraid nothing can be done', I said, making the snap diagnosis of 'cage paralysis'. This did seem to occur in captive monkey facilities periodically. Then the technician who called me admitted, very bravely:
>
> 'You know you ordered folic acid for those green monkeys. By mistake I gave a dose of folic acid to this rhesus monkey'.

Of course that immediately brought to my mind the situation in humans where folic acid administration to a patient with pernicious anaemia, a vitamin B_{12} problem, can precipitate a lower limb paralysis called 'sub-acute combined degeneration of the cord', in America 'combined system disease'. The excess folic acid channels what little B_{12} is left towards the blood-forming system leaving the nervous system deficient.

> 'Perhaps the monkey is deficient in vitamin $B_{12,}$' I said. 'Let's give it a shot of B_{12}'.

Over the next 2 or 3 days the animal was remarkably better. But, of course, having given it vitamin B_{12} I could not test for deficiency. We were no wiser.

Shortly after, however, another rhesus monkey spontaneously showed 'cage paralysis'. This time I did the right thing, I took blood for determination of vitamin B_{12} levels. The result came back at about 25 pico-grams/ml of serum. This is an extremely low level for a human (human values are 200–600). I had to acknowledge, however, that I did not know the normal value for rhesus monkeys. So I tested another dozen or so of our 'normal' rhesus monkeys. All had levels between 20 and 100 (Krohn, Oxnard and Chalmers, 1963). Did this mean that all our rhesus monkeys were deficient? Or was this normal for rhesus monkeys and they were just different from humans? Was the level in the paralysed animal just a red herring?

It was clear that I needed to know more about vitamin B_{12} metabolism. So I read about it. Humans and pigs have levels of 200–600, sheep and cows, levels in the 1000s, rabbits, levels in the 10,000s–40,000s. Nothing, but nothing, seems to have levels below 150 except people with untreated pernicious anaemia and its rare complication, subacute combined degeneration of the spinal cord, together with certain other interesting human conditions such as infestation with the fish tapeworm (that selectively resorbs the vitamin B_{12} that should go to the host), certain intestinal conditions, *and* our rhesus monkeys!

Yet, of course, I still could not be certain. There was nothing for it but to visit Shamrock Farms on the South Coast of England where, in those days, imported rhesus monkeys arrived in the United Kingdom. Analyses of blood taken from some 20 recently arrived animals confirmed that rhesus monkeys are similar to humans. They have serum levels of 200–600. Therefore, *all* our animals were deficient, paralysed or not! The Department of Anatomy at the University of Birmingham's research colony had never less than several hundred monkeys.

Many questions arose. What is the course of the developing deficiency? What is its cause? What evidences are there of the deficiency? What happens upon treatment of deficient animals? What about the small number of animals with cage paralysis? What differences are there between these rhesus monkeys and humans? What is the picture in other primates? What are the implications for the many different scientific investigations that have been carried out on these rhesus monkeys over the years? Did this problem also exist in other research monkey colonies and in zoos? Are there implications for monkeys in field situations? Are there implications for evolution? These were among the plethora of questions which I attempted to answer over the next decade.

Metabolism of Vitamin B_{12}

Before, however, I could take this finding further I needed to know more about vitamin B_{12} itself. I was aided here by Lester Smith (1960), a biochemist with Glaxo Laboratories. First I needed to be reminded that vitamin B_{12} is not a regular 'B' vitamin. It is a molecule that belongs to the family of cobalamins and this overall family is involved in many fundamental biochemical reactions in living things. Most of these are in a variety of microorganisms. At least two are fundamental in mammals. One involves support of successful cell division and it is therefore necessary for all rapidly growing systems such as blood, skin and babies where there is rapid multiplication of cells. The other involves myelin formation, the production of the insulating material of many nerve cells throughout the peripheral and central nervous system.

In mammals one source of the vitamin is the diet. Dietary supplements that contain B vitamins do not contain vitamin B_{12} unless it has been specifically added. The B vitamins are generally found in plant and yeast extracts. Vitamin B_{12} is not found in yeast and other vegetable products but in foodstuffs containing animal products. It is also found, as it happens, in faeces, earth, dirt and sewage sludge where, in each case it is produced by the activities of microorganisms.

Thus, in carnivores and omnivores, because they eat animal products, vitamin B_{12} is present in the diet, is rendered suitable for absorption by combination with factors in the stomach, and in this form is then absorbed in the small intestine where, again, there are special cell surface factors involved in its absorption.

In creatures eating mainly plant foods (which do not contain the vitamin), it enters the tissues of the body by one of a number of mechanisms. First, in those animals (e.g. cows, sheep) that have multilobed stomachs it is manufactured by microorganisms in the stomach. It is then available for resorption in the small intestine. Second, vitamin manufactured in the large intestine and appendix can also be made available to the small intestine if, as a result of coprophagy, the faeces are re-introduced into the gut. Animals such as rabbits and guinea pigs produce soft faecal pellets at night that are ingested. Third, it can be made available to the small intestine in those animals with a very large appendix (rabbits) by reverse peristalsis of the contents of the appendix into the caecum and thence back into the small intestine.

These different mechanisms for vitamin B_{12} intake result in very different serum levels. Omnivores and most regular carnivores (e.g. pigs, humans, dogs, cats) tend to have levels of about 200–600 units. Those carnivores that are small (e.g. insectivores), that must eat very large (relative to their size) amounts of animal food have levels of about 1000. Animals with multilobed stomachs (e.g. cows, sheep) tend to have levels of 2000 to 6000. Animals that have coprophagy and very large appendices (e.g. rabbits, guinea pigs) often have levels of 10,000–40,000 or more.

The nutritional requirements of vitamin B_{12}, whatever the animal, are very small. The daily requirements in humans are much smaller than the amounts in the diet. As a result, human vitamin B_{12} levels are more than two orders of magnitude larger than the daily requirements. A simple dietary deficiency is thus difficult to produce in humans. It may take from 3 to 10 years for a dietary deficiency to produce low serum levels of the vitamin. Such dietary deficiencies tend to occur somewhat more frequently in residents in long-stay nursing homes or other institutions where diets may be deficient for long periods of time. Deficiency does not typically occur in those types of vegetarians who, while eschewing meat, take milk, eggs and sundry other animal products. It does, however, occur in people who are true vegetarians. The deficiency can also occur in the progeny of long-term true vegetarians. Though foetuses and babies normally get the vitamin from the mother, they may not do so if the mother, being a long-term true vegetarian, is providing a deficient vitamin environment from the beginning of pregnancy. In all these cases vitamin B_{12} is an essential dietary supplement.

In addition, in humans, despite good levels in the diet, deficiency can be produced in various medical conditions. These include pernicious anaemia whereby autoimmune damage to the stomach causes lack of factors necessary for vitamin B_{12} absorption in the ileum. Likewise, infestation with the fish tapeworm, *Diphyllobothrium latum*, (*Dibothriocephalus latus*) can cause a deficiency. These tapeworms are found in people like Scandinavians and Japanese who traditionally eat large amounts of raw fish. The vitamin deficiency is produced because the worm selectively absorbs vitamin B_{12} thus removing it from the host's gut. In humans various operative procedures can produce vitamin B_{12} deficiency. The older treatments for peptic ulcers, of partial or especially complete gastrectomy, eliminated, presumably, the source of intrinsic factor for absorption of vitamin B_{12}. Removal of large amounts of small intestine as in the older treatments of Krohn's disease removed the site of absorption. Creation of blind loops of bowel in various surgical procedures is also associated with deficiency of the vitamin.

In animals, the situation varies. It seems not to be possible to produce a dietary deficiency in small insectivorous creatures because they will not live unless they are provided with insects or an insect-equivalent diet (that inevitably contains very large amounts of the vitamin). It is also not easy to produce the deficiency in truly herbivorous animals that have various mechanisms for making available vitamin B_{12} produced by microorganisms outlined above. Deficiency in these animals can only be produced by heroic measures. For example, preventing coprophagy by placing the animal in a body splint that stops it getting its mouth against its anus during the night, and at the same time chemically stressing it, say with thyroid hormone that probably increases its requirements for vitamin B_{12}, may render the animal deficient.

What is the situation in those primates such as rhesus monkeys and chimpanzees that are phylogenetically close to humans and that do not have any of the special mechanisms just outlined? What about those primates that are primarily leaf-eaters (such as colobus and langurs, spider and woolly spider monkeys,) that, though almost totally vegetarian, have multilobed stomachs maintaining, therefore, mechanisms comparable to those in cows and sheep (Oxnard, 1969)?

Problems with Captive Primates

Because I was already engaged in other works for my PhD thesis, this vitamin B_{12} work originally was a hobby carried out at the weekends. I would take blood from the animals (with the agreement of Professor Krohn; they were his animals for experimental purposes) on Monday; the samples would be analysed during the week (with the help of Dr Marshall Chalmers and his haematology laboratory in the Queen Elizabeth Hospital, Birmingham — he was a remarkable senior medical man who was willing to help a young physician gratis). The results would be available on the Monday two weeks later. As a result coffee time on Mondays became the time when I reported each new set of findings to my fellow research students. Almost every week, there were indeed new things to report (Krohn, Oxnard and Chalmers, 1963; Oxnard, 1964).

First was the fact that the results for our monkeys (34 animals that had been in our colony for more than a year and a half) truly were a statistically significant order of magnitude less (with levels of 20–70) than that of 43 animals (with levels of 200–680) that had just entered the country. Second, a cross-sectional study of 106 rhesus monkeys that had been in our colony for 0, 6, 12, 18 and 24 months implied that this reduction was logarithmic (Fig. 1). Third, a longitudinal study of 15 monkeys followed regularly from 8 to 24 months confirmed the logarithmic course of the developing deficiency (Fig. 2). Fourth, because it seemed imperative to treat those many animals in the colony that were *not* to be the subject of my subsequent vitamin B_{12} studies, we were able to show that injections of vitamin B_{12} immediately restored blood levels to normal (Oxnard, 1964). We further showed that very small doses given to infants restored the levels immediately but were insufficient to maintain the levels. In other words, we established a minimum treatment dose. Finally, we obtained the pattern of vitamin B_{12} levels in a variety of other primates (Oxnard, 1966). Many of these were from species other than rhesus

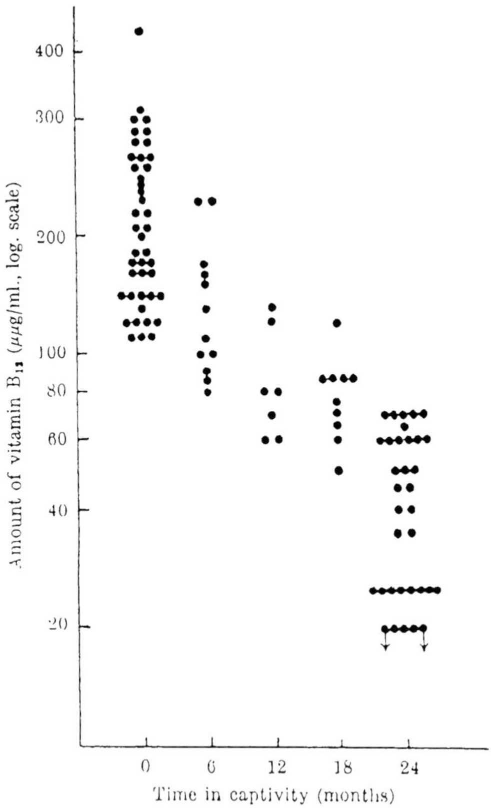

Fig. 1. The relationship between the amount of vitamin B_{12} in the serum and time spent in captivity.

monkeys in our own research primate colony; many more were from animals at the London Zoological Society. There were several different patterns among these primates and these started to make clearer the cause of the deficiency.

Thus, the strepsirrhines (they were called prosimians in those days: pottos, slow and slender lorises and tree shrews, this last because, though not now recognised as a primate, is the non-primate mammal most closely related to primates) had blood levels of the vitamin between 1000 and 4000 — irrespective of the length of time they had been captive (Table 1). These creatures are all completely or

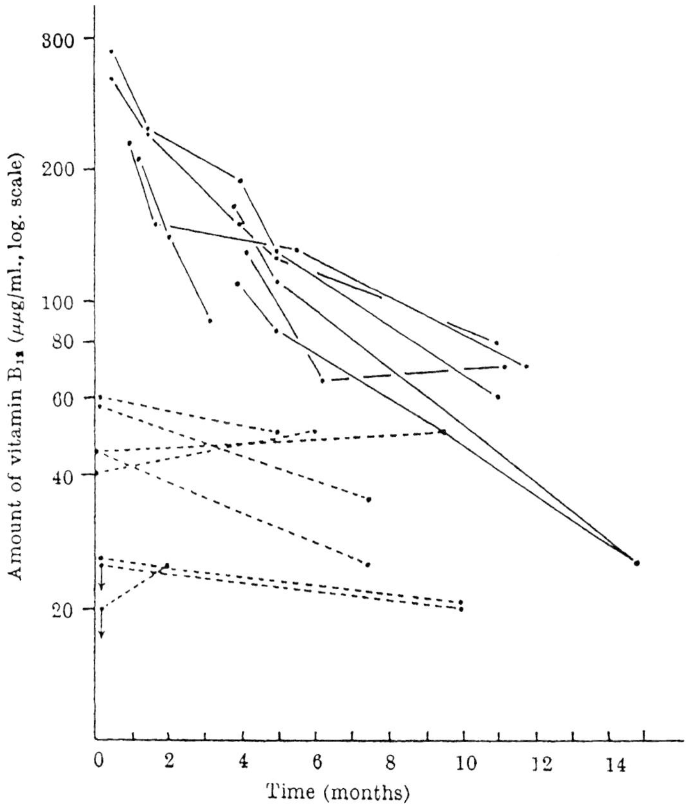

Fig. 2. Change in the amount of vitamin B_{12} in the serum with time spent in captivity. Solid line: changes in recently captive animals, dashed line: changes in animals captive for more than 2 years.

Table 1. Vitamin B_{12} levels in tree shrews and prosimians.

Specimens	Numbers	Vitamin B_{12}, in $\mu\mu$g/ml serum
Tupaia glis	1	> 4000
Galago demidovi	3	1680, 2920, and 1220
Galago crassicaudatus	3	> 4000, > 4000, > 4000
Perodicticus potto	3	4000, 2000, > 4000
Loris tardigradus	3	1360, 1840, > 4000
Nycticebus coucang	1	1210

largely insectivorous in the field. They will not live in captivity unless they received an animal diet (ours had meal worms). They are, thus, similar to the insectivores mentioned above; they seemingly cannot become deficient however long they are in captivity.

The half dozen colobines (Old World leaf-eating monkeys) that we eventually examined had levels from 1000 to 3000. These monkeys have a sacculated stomach (not too dissimilar from the multi-lobed stomach of sheep and cows). The stomach contains microfauna presumably capable of producing vitamin B_{12} that is then available for absorption in the small intestine. We were never able to get blood samples from any of the leaf-eating monkeys of the New World (e.g. woolly monkeys, spider monkeys). I assume that they, too, produce vitamin B_{12} in the stomach that is subsequently absorbed in the small intestine, and can confidently predict that they, too, should have high levels of the vitamin (Oxnard, 1969).

Most of the non-colobine monkeys and apes examined (e.g. a few chimpanzees and gibbons, several patas and talapoins monkeys, a few baboons, a single mandrill, several marmosets and one douroucouli — these latter two species are our only examples of New World monkeys) all had, when recently captive, levels of 200–600 (Table 2). This is the same as in our recently captive rhesus monkeys and normal humans. In specimens of these species that had been captive for more

Table 2. Vitamin B_{12} levels in non-colobine anthropoids.

Recently captive specimens	Vitamin B_{12}, in $\mu\mu$g/ml serum	Captive for more than 2 years	Vitamin B_{12}, in $\mu\mu$g/ml serum
Erythrocebus (2)	180, 410	*Erythrocebus* (2)	100, 64
Papio (1)	424	*Papio* (4)	50, 100, 360*, 480*
Cercopithecus (2)	420, 700	*Cercopithecus* (2)	310*, 280*
Mandrillus (1)	228	*Macaca* (3)	116, 62, 156
Callithrix (1)	450	*Cercocebus*	40
		Aotus	100
		Hylobates (2)	34, 350*
		Pan (2)	120, 1100*

* = long captive animals in special non-deficient situations — see text.

than 2 years, however, blood levels were 20–150, just like our long-stay rhesus monkeys.

There were several most interesting individual exceptions. First, a drop in levels with time in captivity did not occur in the green monkeys (*Cercopithecus* in Table 2) that we were holding. These particular animals, however, had been given vitamin B_{12} on entry and were fed a considerable amount of animal food during captivity (they were the ones used for the Skinner box experiments that started this whole matter). Second, a reduction did not occur in one of the gibbons (Table 2). It turned out that this animal (in the Regents Park Zoological Gardens) received one ounce of meat daily during captivity. Third, two of the long-stay baboons were not deficient (Table 2). They were kept outside for much of the year and, in addition to plant foods provided by the keepers, also ate whatever they could find outside: insects, birds and birds' eggs, small rodents, discarded food such as half-eaten hamburgers, and so on. Finally, unlike the other chimpanzees, one chimpanzee had levels above 1000 every time its blood was examined (Table 2). It turned out, however, that this animal had developed pica (the habit of eating its own faeces; personal communication from D. Morris). This habit may have been psychological in origin due to its small enclosure, lack of contact with conspecifics, and other aspects of captive conditions that prevailed at the time. These exceptions, however, also pointed us towards the cause of the deficiency when it occurred. What else could it be but a dietary deficiency?

Finally, we were able to examine several animals both at the time of capture and some months later (Table 3). This showed the

Table 3. Vitamin B_{12} levels in those individual monkeys that were examined twice, on entry to the colony within the first month and again some months later.

Specimens	No	Months captive	Vitamin B_{12}, in $\mu\mu$g/ml serum	Months captive	Vitamin B_{12}, in $\mu\mu$g/ml serum
Macaca	(3)	1, 1, 1	370, 320, 230	3, 3, 3, 6, 6	60, 80, 70, 70, 70
Erythrocebus	(2)	1, 1	280, 210	7, 7	80, 100
Cercopithecus	(3)	1, 1, 1	1100, 800, 360	11, 11, 11	660, 384, 312

reductions that we had come to accept in captivity (but not the talapoin monkeys that were fed minced meat and meal worms).

Of course, all this was long before the days of a university using commercial monkey chows. A university research monkey colony found it much cheaper, thinking that a monkey is an animal with a banana in its hand, to supply cheap fruits bought late in the day from the market. Our animals were, thus, on a totally vegetarian diet from entry. In those days (the 1950s) it had not been clearly recognised that, in fact, many monkeys eat some animal food items. It was only in the 1960s that Jane Goodall (1965) reported her findings of termite 'fishing' by chimpanzees. The 1930 sighting of precisely the same phenomenon by Fred Merfield, Powell Cotton's 'shooter', had been overlooked (Merfield, 1956). So, too, was much earlier information. Thus a Liberian postage stamp of 1906 figures a chimpanzee 'fishing'. The picture used in the stamp can be traced back to a specific picture of a chimpanzee in Dresden Zoo drawn by Gustav Murtzell in 1887 that was exhibiting this behaviour. Goodall was presumably unaware of these prior sightings, and none of this was readily recognised in 1959 when the vitamin B_{12} work was started.

Of course, such dietary deficiency could not occur today because captive animals are routinely fed monkey chows. These contain offal that, being an animal product, is loaded with vitamin B_{12}. In fact, within 6 weeks of my second paper on vitamin B_{12} appearing in *Nature*, the monkey chow company had changed the label on their product to read 'Vitamin B_{12} Added'. *They did not know* that this was unnecessary because of the offal content of their chows. *I did not know* that this was a patentable idea; otherwise I might now be royalty rich!

Though it seemed we had identified the deficiency as dietary, there were still many many questions to be asked. What evidences might there be of this, up to that time, hidden deficiency?

Sexual Cycles, Pregnancy and Babies

It quickly became apparent that diet was not the sole factor involved in reduced levels of vitamin B_{12}; reproduction was also important.

Table 4. Variations in Vitamin B_{12} levels in pregnancy and the puerperium (in $\mu\mu$g/ml serum).

	Recently captive monkeys			Long-stay monkeys		
Status	No	Mean	Range	No	Mean	Range
Mature	15	271	110–680	19	43	< 20–70
Pregnant	18	105	50–240	8	79	50–120
Puerperal	10	62	< 20–90	12	36	20–60

Studies of reproduction by a large team of investigators lead by Zuckerman: Krohn, Eckstein, Mandl and many others, were one of the main reasons for the existence of the research colony. Pregnancy and the puerperium were quickly found to be important (Oxnard, 1964). Those animals that came direct from the field situation but that happened to be pregnant or puerperal at examination had reduced vitamin B_{12} levels (Table 4).

Thus 15 mature non-pregnant recently captive females had levels of 110–680, and were therefore similar to recently captive males. In contrast, 18 females that were pregnant on capture had somewhat reduced levels (50–240), and 10 that were immediately post-partum at entry had very low levels (20–90). In other words, though levels in monkeys in the field are generally similar to those of humans and far higher than those of long-stay captive monkeys, the advent of pregnancy and the puerperium in the field situation was associated with reductions in vitamin B_{12} levels. Because, however, recently captive non-pregnant females were no different from males, it seemed that, in the field situation, this must be temporary.

In humans it is well known that the foetus, through the placenta, 'parasitises' the mother for vitamin B_{12}. This continues after birth: the baby taking the mother's milk that preferentially contains very high levels of vitamin B_{12}. Nevertheless humans have such large amounts of the vitamin that pregnancy does not reduce human maternal levels to deficiency grades except in the very few cases of macrocytic anaemia of pregnancy. Rhesus monkey foetuses must be likewise parasitic upon their mothers, but the maternal levels seem not to be maintained as in human mothers. The monkey mothers obviously returned to normal quickly; the levels in non-pregnant recently captive females being the same as in males. Rhesus monkey mothers (and indeed also many other non-human primate mothers) regularly eat the placenta. This would immediately provide a big source of vitamin B_{12} that would return maternal levels to normal. Such behaviour sometimes occurs in some human societies. Certainly, individual pregnant women have cravings for various substances, some of which contain vitamin B_{12}.

These pregnancy–puerperium findings were further extended by the determination of vitamin blood levels in the foetuses and babies (this was permitted by access to materials resulting from investigations of other researchers on these animals, Table 5). Thus, the reduced levels (mean of 122) in 10 recently captive pregnant monkeys were nevertheless associated with normal levels (mean of 579) in their 10 foetuses. In contrast, 8 long-stay mothers with low levels (mean of 38) had newborn babies that were just as deficient (mean of 47). In other words, though both sets of mothers had reduced levels, the recently captive mothers rebounded and long-stay mothers did not. Likewise, though the babies of the recently captive mothers had normal vitamin levels, the babies of long-stay mothers had very much reduced levels.

Presumably, then, the reduction in blood levels in recently captive monkeys was related at least in part to the maintenance of normal levels in their foetuses. The reduced levels in the long-stay mothers were so very much reduced (due to long time in captivity on a vegetarian diet) that their babies suffered. Further, the fact that pregnancy could

Table 5. Vitamin B_{12} levels in $\mu\mu$g/ml serum in mothers and near-term foetuses or neonatal infants.

Recently captive monkeys			Long-stay monkeys		
No. of pairs	Mean vitamin level		No. of pairs	Mean vitamin level	
10	Mothers 122	Foetus 579	8	Mothers 38	Babies 47

easily produce a temporary deficiency in recently captive mothers implies that rhesus monkeys in the field may be dithering along on the verge of deficiency. This may be because they are primarily vegetarians and only eat marginal amounts of animal food. In such a situation, pregnancy may be enough to reduce the levels. Indeed, levels may also vary at other times in the year when an abundance of vegetarian foods (fruits) means that the animals consume less in the way of animal products. All this could also explain why a solely vegetarian diet during captivity starts producing the deficiency immediately.

The last piece of evidence here comes from the effects of injecting vitamin B_{12} as a therapeutic measure during pregnancy as compared with the non-pregnant state. Thus, non-pregnant animals showed a rise in levels after injection. Pregnant animals did not, but their babies showed a very marked rise. This further implies that the baby parasitises the mother at the mother's expense in deficiency situations (Table 6).

None of this occurs in humans. A truly vegetarian diet in humans may take many years (up to 10) to produce a deficiency. This is presumably because of the very large amounts of the vitamin in humans resulting from the fact that humans, unlike apes and monkeys, are primarily omnivores and therefore take in large amounts of the vitamin relative to daily requirements. It is nevertheless known that the new born babies of true vegetarian mothers (some on vegetarian diets for many years) may display deficiency symptoms: redness of the palms and soles, a neurological flap of the

Table 6. Effects of injection of vitamin in $\mu\mu$g/ml serum on amounts in the serum.

Non-pregnant long-stay monkeys		Pregnant long-stay monkeys		
Before injection	After injection	Before injection while pregnant	Mother after injection	Baby after injection of mother
25	310	70	80	800
25	330	25	100	640
60	880	60	130	570

hands, and electrical changes in the brain, that are all associated with the deficiency!

Was this the whole reproductive story? Though one of the primary reasons for the existence of the large rhesus monkey colony in Birmingham was to carry out many different researches into reproduction (studies of the efficacy of the early contraceptive devices and drugs, fundamental investigations of the function of the ovary, studies of various reproductive hormone interactions, etc.), there had always been some difficulties because the fertility of the rhesus monkey colony seemed to be somewhat low. There were just not as many babies as might have been expected. Of course, with today's knowledge of animal colonies there could have been many reasons why this was so. The animals were not kept in the social conditions known today to be required. They were usually kept singly in small cages, and were brought together in pairs for copulation in specially constructed larger cages! The environmental, physical and psychosocial conditions were just not what is known is required today. Babies were born; it is true, but not in any great numbers given the large number of mating opportunities that were part of the colony protocol. However, once the vitamin B_{12} deficiency had been discovered, and especially since the discovery of the effects on reproduction just noted, it seemed prudent to look into possible further relationships with reproduction.

For example, at that time it was already known that the vitamin B_{12} content of the blastocyst fluid in other animals (e.g. rabbits, cows, pigs) was high. It was also already known that the vitamin B_{12} content of semen was high (and especially that vitamin B_{12} had to be added to bull semen when stored for subsequent artificial insemination). Similar information was available for humans. There was, at the time, anecdotal evidence that reduced vitamin B_{12} levels in semen in humans was associated with reduced male fertility. It was well known that human females with pernicious anaemia (pernicious anaemia is more common in females than males) had to be warned, on treatment of their anaemia with vitamin B_{12}, that they should, in the language of those days, 'take precautions' if they did not wish to become pregnant soon after treatment.

Further, in our colony, it had been known that the reproductive cycles of the rhesus monkeys were not very regular. A new search of the old records showed that the cycles were not simply irregular; they were regularly irregular in the sense that the irregularity seemed as though it were due to single or double or even on occasion triple missed cycles. Could it be that the animals (we are talking here only of the ones that 'were given the opportunity') did indeed become pregnant but that the conceptus products were resorbed at early stages in development with no external signs of abortion and a delayed return to cycling?

We also obtained semen samples (using electro-ejaculation carried out by Professor Peter Krohn) from a number of the deficient males in attempts to determine semen levels of vitamin B_{12}. The results were equivocal. This was mainly because of the difficulties of carrying out the vitamin assays on such materials, and also perhaps because the sample of animals was not large enough. Thus, the overall upshot of these studies was that the idea was truly possible, but that the evidence at the time was not sufficient to be sure.

All of this produced a further question however. 'Why are rhesus monkeys different, in these various reproductive matters, from humans?' Before we could attempt an answer it seemed necessary to look for other manifestations of the deficiency.

Blood, Skin and Growth

In humans, the commonest manifestation of vitamin B_{12} deficiency is in the blood: pernicious anaemia. This is a condition that was untreatable before the discovery of vitamin B_{12}. It was accordingly natural that we should look to blood values in our rhesus monkeys for the stigmata of that condition. However, none of our colony animals displayed reduced haemoglobin levels (anaemia) or the enlarged red blood cells and their precursor cells (macrocytes or megaloblasts) that can be found in microscopic analysis of blood films in pernicious anaemia. Given the above history, however, it seemed prudent to take matters further. We thus repeated these tests and included more

Table 7. Blood differences between normal and vitamin B_{12}-deficient monkeys.

Measure	Normal monkeys	Deficient monkeys	Significant
HB%	85.1	82.9	No
PCV%	42.7	41.0	No
RBC (m/mm^3)	4.93	4.55	No
MCHC%	29	29.1	No
MCV cμ.μ	*86.7*	*92.1*	No

detailed haematological measures. The results confirmed that in our colony animals, haemoglobin levels, red cell counts and various other measures: mean corpuscular haemoglobin concentration, mean corpuscular volume and mean corpuscular haemoglobin, were all statistically similar to the values both in the recently captive monkeys and in normal humans. In other words, anaemia like that of pernicious anaemia in humans (a macrocytic megaloblastic anaemia) seems not to occur in rhesus monkeys (Table 7). Of course I published this finding (Oxnard, 1964).

Almost immediately I had to eat my words. It was not that I my original cross-sectional observations were wrong. It was rather that, when I used a longitudinal design, that is, comparing each deficient animal with itself before and after treatment, I found that indeed there were significant changes. Haemoglobin levels and mean corpuscular haemoglobin concentrations were increased. Volumes of the extra-large red cells (mean corpuscular volume) were reduced to normal. This work was carried out through a further collaboration with a clinical colleague, of Dr Eric Spicer. These changes were so small, however, that they could scarcely be called anaemia, and they could only be statistically discerned in the 'before and after treatment' comparisons.

Mean corpuscular volume is the best measure of the macrocytosis of pernicious anaemia and other related anaemias. In humans with vitamin B_{12} deficiency the mean corpuscular volume is strikingly elevated (between 110 and 140 cubic microns) as compared with normal (78 to 95). Statistics are not needed. In contrast, in all our rhesus monkeys, both deficient and normal, the mean corpuscular volume ranged between 86 and 92; this is similar in both deficient and normal animals, and entirely within the normal human range. However, when animals were examined before and after treatment the mean corpuscular volume fell by an average of 6 cubic microns in each animal, and this difference was statistically significant. In other words, the cells are larger in the deficient monkeys but to such a small degree as to be only recognisable through before and after comparisons. The huge macrocytosis of deficient humans does not occur in rhesus monkeys. Of course, this meant I had to write another paper (Spicer and Oxnard, 1967)!

The blood picture in vitamin B_{12} deficiency in humans is essentially one of interference with the normal mechanisms of blood cell production. This also occurs in all other tissues where cells proliferate. Thus, in vitamin B_{12} deficiency all epithelial cells show megaloblastosis. This includes those lining the entire gastrointestinal tract — buccal mucosa, tongue, stomach, small intestine — together with many others — especially in reproductive organs such as cervix, vagina and uterus.

In fact, the human condition of vitamin B_{12} deficiency is often first noticed through the development of mucocutaneous lesions. A doctor almost always looks at and in the mouth. As a result, fissuring at the corners of the mouth (angular cheilosis), a smooth tongue lacking papillae (glossitis) and white patchy gums and cheeks (stomatitis) are readily apparent. Microscopic examination of buccal smears shows that the epithelial cells have increased nuclear size and greater variation in nuclear shape. Though similar changes in humans have also been found in iron deficiency anaemia, they are especially well recognised in association with low levels of vitamin B_{12} deficiency in pernicious anaemia (and also in deficiency of folic acid, something that we later examined). These epithelial changes are reversible with vitamin B_{12} therapy. We therefore looked for such lesions in the deficient long-stay rhesus monkeys and for improvement after treatment.

A similar glossitis and stomatitis were readily apparent in 19 deficient rhesus monkeys as compared with 25 recently captive non-deficient monkeys (Fig. 3). There was a macroscopic glossitis (a smooth tongue, especially on the sides and tip) with, especially, a general

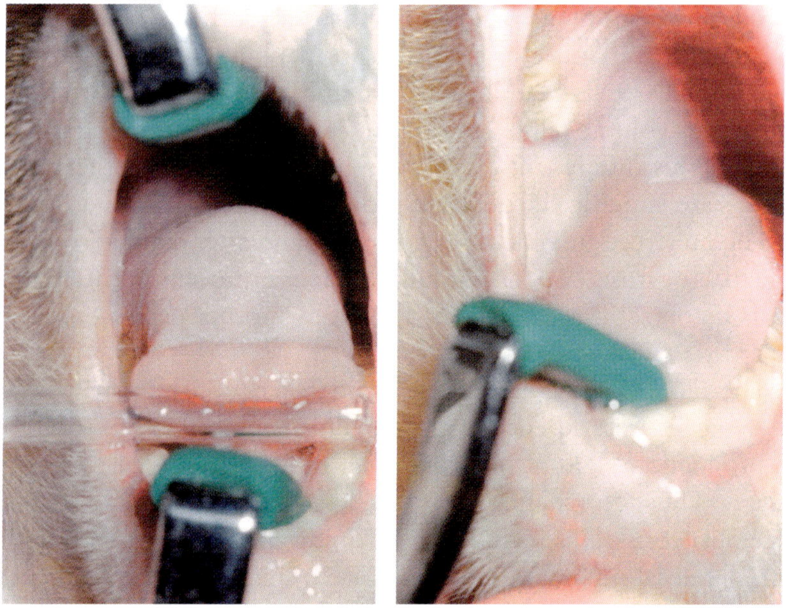

(a) (b)

Fig. 3. Papillae on a normal rhesus monkey tongue, compared with the smooth shiny tongue of a deficient monkey.

reduction in numbers and size of the tongue papillae. White atrophic lesions of the oral mucosa were evident both on the buccal and gingival mucous membranes. Angular cheilosis seemed to be present, but we could not be certain because these lesions at the corners of the mouth are difficult to distinguish from the trauma of animal handling.

A dental colleague, Dr Margaret Rose, and I undertook more detailed microscopic examinations. Buccal smears showed various lesions in the deficient animals. With the aforementioned enlarged cells went enlarged cellular nuclei, increased numbers of binucleate and multinucleate cells, greater irregularity in shape of nuclei and increased numbers of bilobed nuclei. The staining (for chromatin) of the cell nuclei was lighter and more irregular than normal. Cell divisions were greatly reduced. All of this was in comparison to the situation in recently captive animals. Later with another colleague, graduate student (and later Professor) Ildemaro Torres (Torres,

1968), these findings in the oral mucosa were confirmed. In addition we noted that several of the deficient animals had lesions of gastric mucosa (atrophy of the gastric mucosa — three cases). All the non-deficient animals had a normal gastric mucosa. These epithelial lesions resemble completely what is found in vitamin B_{12} deficiency in humans. It is likely that similar changes are also present in the cervix, vagina and uterus and it is not impossible that they could relate to the problems in reproduction (see last section). However, we did not examine cells from such regions because I was not then aware that they could occur there.

Those animals that were given parenteral vitamin B_{12} also showed improvements in epithelia. All animals improved, some showing a reduction in the smooth shiny tongue in as short a time as 3 weeks. After 2 months of treatment all animals had fully papillated tongues and apparently normal mucous membranes. Microscopic examination of buccal smears showed that the various lesions characteristic of the deficient animals had disappeared. There were no gastric lesions in those of the treated animals that came to post-mortem, but, of course, there was no pre-treatment evidence with which to compare them.

Following the situation in blood and skin we could logically expect that there would also be interference in growth generally. Studies in many animals have shown that a deficiency of vitamin B_{12} leads to gross abnormalities of development and growth. Shortly after conception the deficiency may result in resorption of the conceptus products. After birth it is manifest as decreased growth. It is well known in the meat industry that the vitamin is useful in promoting maximal growth of many animals maintained on otherwise normal diets.

It was not, however, such information that lead me to look at growth. It was, rather, a series of accidental observations made during attempts to discover the minimum dose of vitamin B_{12} needed to maintain blood levels. I had done this by giving small parenteral doses of the vitamin to three infants on a regular basis over some months with monitoring of serum vitamin B_{12} levels. Three other infants born at approximately the same time acted as controls. The doses given were so small that after an initial surge in blood levels of the vitamin in the days following injection, they fell back to their previous deficiency levels. I did this three times in infant rhesus monkeys. Because the babies were routinely weighed weekly, this meant that I could plot changes in weight with changes in vitamin B_{12} levels. The plots showed that their weights surged shortly following injection and then, shortly after that, fell back. However, it so happened that three other rhesus babies not given the vitamin were weighed at the same time. These were a very small control group, of course. To my chagrin, they showed precisely the same saw-toothed curve of weight on age as did the treated animals. In other words, the original finding is little more than an anecdote; the samples are too small; the control results imply that other secular factors in the colony could have been involved.

Yet, with the help of yet another colleague, Dr Roger Flynn, a properly controlled and statistically examined study of growth in infants over a period of some 700 days was carried out (Flinn and Oxnard, 1966). The results were startling. In both males and females separately, growth after the period of administration of the vitamin was increased (Fig. 4). This was especially marked in males (the regression of weight on age in treated males averaged 6.8 as compared with 2.6 in untreated males). The figures for females were a little less but all differences were very highly significant statistically ($p = 0.005$). At one point, indeed, we became worried that we might have to order bigger cages!

This longitudinal study of rhesus monkey infants shows clearly what is already known for many other deficient animals, that treatment with vitamin B_{12} is often associated with markedly increased growth. In general, growth rates in late infancy after treatment show considerably less variation than before, and are of the order of 50%–100% greater than before treatment. Such changes are large enough to have major biological significance in the rearing of healthy rhesus monkeys. As low levels are found in captivity in many other primates (see above), it may well be that for optimum growth all apes and monkeys in captivity (except leaf-eating monkeys) should receive diets (such as monkeys chows) that contain animal products.

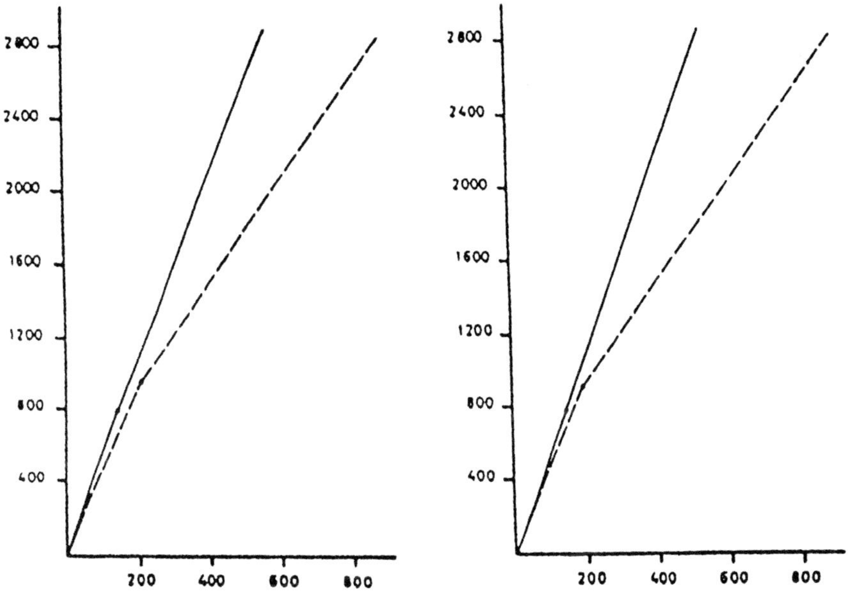

Fig. 4. Comparisons of growth regressions in rhesus monkeys. Dashed line is growth in untreated animals, solid line in treated animals. Left plot: females; right plot: males. *x*-axis: age in days; *y*-axis: weight in grams. The inflexions in the curves are the points at which treatment was instituted.

Even in the field this may be important. Though levels of the vitamin in animals in the field must be thought of as normal, the vagaries of field life may well mean that blood levels fluctuate with the seasons. Fruits form the major dietary component at certain times in the year and this may reduce animal food consumption to almost zero. At other times reduction in fruits may mean that small amounts of animal products may be consumed. For instance some primates seen 'eating' fallen fruit were actually tearing the fruit open and consuming contained bugs. Sometimes meat-eating is an accidental occurrence — a field primate may kill a conspecific or a member of another species, or some other small creature; a puerperal mother may (commonly does) eat the placenta; birds' eggs may be consumed; and so on. As a result of such factors, monkeys in the field show, as judged by the variation in levels in recently captive animals, considerable fluctuations in vitamin B_{12} levels; they may be on the margin, much closer to deficiency than we might expect.

Yet a fourth time, then, we might ask about differences between these various non-human primates and humans.

Nerves and Brain

Much to our surprise, by far the most interesting signs of the deficiency in rhesus monkeys were in the nervous system (e.g. Oxnard and Smith, 1966; Oxnard *et al.*, 1969, 1970; Torres *et al.*, 1971). What we found needs to be viewed against the extensive background of what is known in humans.

In humans vitamin B_{12} deficiency syndromes themselves are not that common (for pernicious anaemia, of the order of 25 new cases per 100,000 people per year). When the deficiency occurs as a result of dietary problems the manifestations take very much longer to develop (if they develop at all). It may take as much as 10 years before a dietary deficiency manifests itself as the macrocytic megaloblastic anaemia; this may never occur in individual cases. Of the small number of individuals who show such an anaemia, only a very few advance to the neurological complication of subacute combined degeneration of the cord. The neurological complications, when they occur, are dire medical emergencies. Patients complain of pins and needles in the feet, have changed reflexes, and become paralysed in the legs. This worsens over just days and becomes permanent unless treatment with parenteral vitamin B_{12} is instituted immediately. Such treatment of course, precludes further study of the condition. Invasive examination of the nervous system in a condition that can be cured by vitamin administration is not an option.

In great contrast was the situation in our long-stay rhesus monkeys. Over a two-year period we noted a total of five animals (in a colony of several hundred) that developed obvious paralysis. The duration of their captivity in our colony varied from 1.5 to 26 years. This is an incidence of paralysis that is much greater than in humans. (In parenthesis, it is an incidence that was not associated with the macrocytic anaemia; see above). The paralysis followed a consistent pattern. Undoubtedly these animals, could they speak, would have complained of 'pins and needles' in their hind limbs and tails. They usually sat with their feet in inverted positions so that the plantar surfaces were not in contact with the cage floor; some had self-inflicted wounds on their feet and tails; both of these possibly evidence of paraesthesia. The worst cases could only drag themselves around the cages seemingly from hindlimb paralysis. In the less-affected animals mobility was impaired so that either they did not climb about the cages at all, or they climbed clumsily and with apparent difficulty, avoiding contact with the substrate on their soles and bearing the body weight mainly on their heels. Clumsiness in movement, though apparently partly due to the weak grasp of the foot, seemed to be less evident if the animals watched their limbs. Sometimes the limbs showed a tremor. Sometimes they seemed to have difficulty handling a small object, such as a nut; was this loss of manual dexterity or evidence of a visual problem?

Neurological examinations of monkeys are not easy, though they are a bit easier if the monkeys are paralysed. Yet signs were there. There seemed to be disturbed sensory and motor function in the paralysed animals. Touching the feet elicited distress. Weakness of the lower limbs was obvious. There were changed tendon and skin reflexes; initially these were exaggerated and later reduced. It proved impossible to test power and tone save in the most paralysed animals. Of course, the paralysed animals were the first to be examined post-mortem.

Materials obtained at post-mortem, peripheral nerves, spinal cord and brains, were examined with the help of expert neuropathologist Professor Walter Smith. Much the most obvious findings were lesions of the peripheral nerves of the lower limb, and of the dorsal and lateral columns of the spinal cord.

Lower limb nerves showed loss of the insulating cover (demyelination) that appeared much worse distally (Fig. 5). There were similar lesions in the posterior (sensory) nerves entering the spinal cord (Fig. 6). There was an equivalent demyelination in the spinal cord (in the posterior and, to a lesser extent, the lateral columns, Fig. 7). In individual animals demyelinating lesions were found elsewhere in the

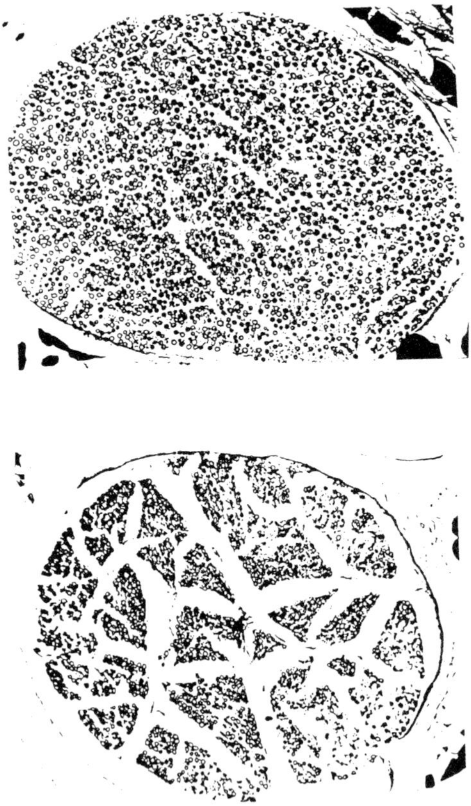

Fig. 5. Comparisons of numbers of myelinated nerve fibres (open circles) in proximal (above) as compared with distal sections of the sciatic/tibial nerve.

nervous system, in the white matter of the cerebral cortex (Fig. 8) and in the optic nerves and optic chiasma (Fig. 9). Later we found a whole series of lesions in the optic system, and Dr Valerie Hinds, in her doctoral work with Professor Walter Smith and me, took our initial findings very much further (Hind, 1969). Her ophthalmoscopic examinations showed flattening of the disc. Her histology of the retina showed thinning with histopathological changes (Fig. 10). The original demyelination of the optic nerve and chiasma were mirrored by equivalent lesions in the optic tracts and the related parts of the brain (the lateral geniculate bodies).

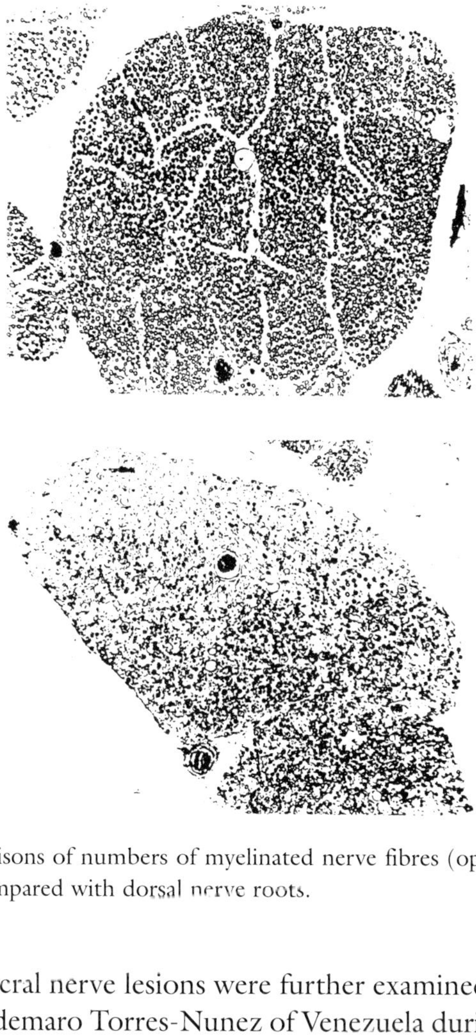

Fig. 6. Comparisons of numbers of myelinated nerve fibres (open circles) in ventral (above) as compared with dorsal nerve roots.

The peripheral nerve lesions were further examined quantitatively by Professor Ildemaro Torres-Nunez of Venezuela during his doctoral work with Professor Walter Smith and me. Torres-Nunez not only confirmed the lesions already discovered but rendered much of the information quantitatively and examined teased individual nerve fibres from different portions of the peripheral nerves. These findings confirmed that the initial lesions showed segmental demyelination in the peripheral nerves (Fig. 11). The die-back phenomenon that we had previously observed (based on the numbers of degenerate fibres per cross section at different points on the peroneal and sciatic nerves)

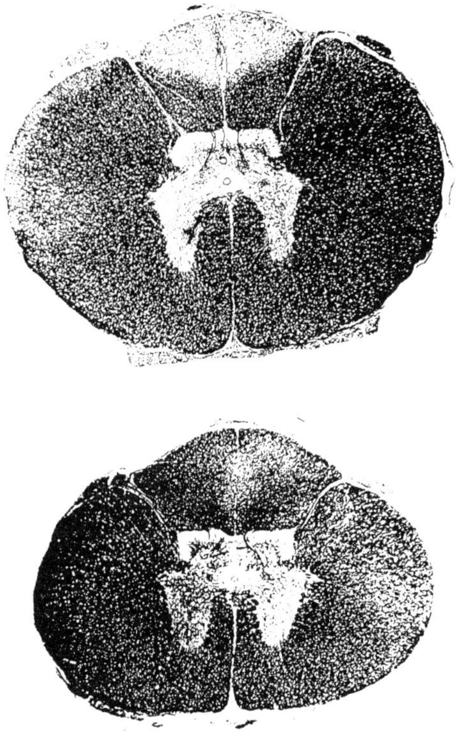

Fig. 7. Loss of myelinated fibres in the dorsal and to a lesser degree lateral columns of the spinal cord in a deficient paralysed animal above, compared with lesser demyelination in a deficient but unparalysed animal.

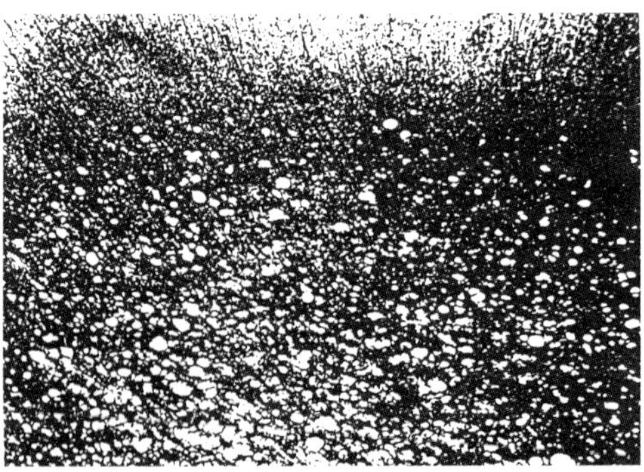

Fig. 8. Spongiform degeneration (loss of myelin) in the cerebral white matter in a deficient paralysed animal.

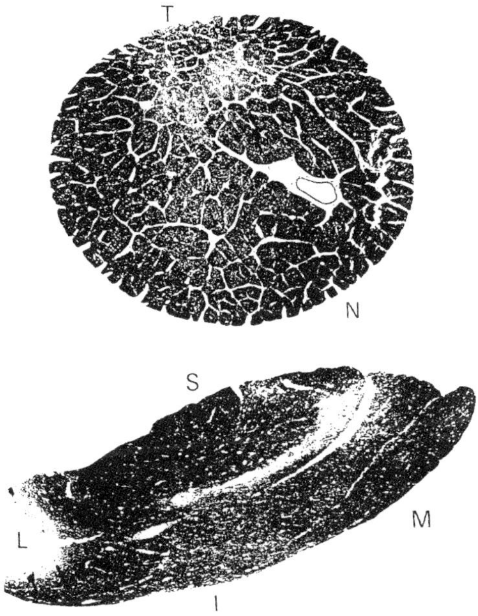

Fig. 9. Loss of myelin in (above) the optic nerve and (below) the optic chiasma in a deficient animal (from Hind, 1969).

was a secondary phenomenon following the initial segmental demyelination. We had first thought that the deficiency of the vitamin affected the neuronal cell body so that it could not support a complete length of the axon. This might have meant that the axons would die back to a length that could be supported by the cell body. Though we published this speculation (and it was indeed a correct finding (Oxnard and Smith, 1966), Torres-Nunez' studies of individual teased fibres showed unequivocally that the initial lesions were segmental demyelination (Torres *et al.*, 1968). The initial seat of the problem was a deficit in the Schwann cells. This resulted in the loss of myelin from individual segments. In the central nervous system, i.e. in the white matter of the cerebral cortex, the demyelination was patchy because neurodendroglial cells, unlike Schwann cells, serve myelinating functions in a local patch around several neurons. However, because we examined the worst-affected cases first (a natural thing to do), what we found, that gave us the impression

Fig. 10. Comparison of the retina and disc in a normal animal (above) and a paralysed animal (below); disc atrophy is evident in the latter. The central artery of the retina is indicated.

of a die-back phenomenon, was an axonal degeneration (Wallerian degeneration). This was clearly secondary to the initial supporting cell demyelination. Again, therefore, in publication, we had to retrace our steps (Torres *et al.*, 1971).

Obviously, however, it was necessary not only to compare the neurological lesions in the paralysed animals with the neurological condition of the normal (non-deficient) recently captive animals, but also to compare them with the non-paralysed (and apparently neurologically normal) yet deficient long-stay animals. Initially, seven long-stay non-paralysed monkeys came to post-mortem and could be examined for neurological lesions. Their duration of captivity varied

Fig. 11. Three teased peripheral nerve fibres showing normal axon above and segmental demyelination below. The middle fibre is from an animal that had been treated for a short period of time and shows beginning of remyelination (from Torres-Nunez, 1968).

from 3.5–16.6 years. Their vitamin B_{12} levels were low and therefore no different from those of the paralysed animals. They showed no clear neurological signs on examination before death, no trophic lesions and no overt evidence of paralysis (though I did think that tendon and plantar reflexes might be reduced, it was hard to be sure in the examination of the restrained animal).

To my surprise, every one of these seven apparently normal animals displayed some demyelination in peripheral nerves of the lower limbs, and mild but definite spongiform degeneration of the posterior and lateral columns of the cord qualitatively similar but of lesser degree than those in the paralysed monkeys. Later we were to discover lesions like these in every long-stay rhesus monkey on a vegetarian diet in our colony (Table 8) on which I was able to conduct a post-mortem. These data provide an incidence for the neurological lesions in the vitamin B_{12}-deficient animals of virtually 100%. Yet, few of the non-paralysed animals showed obvious neurological deficits on examination while alive. Also, of these 33 animals, not a single one

Table 8. Number of rhesus monkeys with nervous system lesions.

Status	Total no.	No. with overt clinical signs	Site of nervous lesion	No.
Non-deficient	17	1	Nerve	1
			Spinal cord	1
			Brain	0
Deficient	12	12	Nerve	12
			Spinal cord	12
			Brain	3
Deficient but treated	14	10	Nerve	10
			Spinal cord	10
			Brain	2

had a macrocytic anaemia though some were demonstrated to have a minor degree of macrocytosis (as estimated by mean corpuscular volume that became slightly but statistically reduced after treatment with the vitamin; see above).

The relationship between the incidences of lesions of blood forming tissues and the nervous system in humans (high and low respectively) is thus the complete reverse in rhesus monkeys (low and high, respectively).

Lateral Thinking: Implications for Evolution

We can summarise these findings. The rhesus monkeys that formed the major part of this study had been kept in our animal colony on vegetarian diets for long periods of time. Save for the occasional animal with cage paralysis (a handful only in a colony that remained at several hundred animals for many years) they were apparently healthy. The various lesions defined above were totally unsuspected until the accidental finding of vitamin B_{12} deficiency. Except for overt paralysis, the various lesions regressed following long-term treatment with parenteral vitamin B_{12}. Once, however, we knew what was happening, the lesions never developed in animals subsequently introduced into the colony because they were then fed a monkey chow diet containing large amounts of vitamin B_{12}. Likewise in other institutions holding monkeys in captivity, all are fed monkey chow or other foods containing vitamin B_{12}.

Though we do not have documentation at this level of detail, such information as we can garner from other primates implies that (except for the leaf-eating species, colobus and langurs in the Old World, and the spider and howler monkey groups in the New World) monkeys and apes are all similarly liable to vitamin B_{12} deficiency in long-term captivity on a vegetarian diet. We established that this was true of animals in one zoological garden. It was inevitable, therefore, that we should come to consider the differences between non-human and human primates in vitamin B_{12} deficiency.

All this is completely different from the situation in humans. In humans, the blood-forming system is the obvious target of the deficiency with a reasonable incidence of macrocytic megaloblastic anaemia. Although less obvious clinically, all epithelial systems are probably equally at risk and epithelial lesions may well have a similar origin as the blood-forming organs, that is, pathology resulting from interference in abnormalities of cell division decreasing growth. The lesions of the nervous system in humans, involving especially demyelination of peripheral and central axons, are likely to be due to deficiencies in supporting cells (Schwann cells in the peripheral nervous system, oligodendroglia in the central nervous system) as well as, in all likelihood, metabolic changes in the neuronal cell bodies themselves. The pathophysiology of the nerve cell is different from that of growing cells.

In rhesus monkeys specifically, but in all apes, in most Old World monkeys (excepting the leaf-eating monkeys), and possibly in many New World monkeys (excepting the non-prehensile tailed monkeys, also leaf-eaters) the situation is the opposite of that in humans. In these non-human species, the neurological lesions are dominant, the blood-forming and epithelial tissue lesions, though present, are minor.

Part of the explanation of the difference may be given by the findings in pregnancy and the puerperium together with knowledge of the natural diets of these species. In the field situation all these primates eat some quantity of animal food. The taking of birds' eggs, the catching of insects, the occasional killing and eating of small animals, sometimes the killing and eating of larger prey, even the eating of killed con-specifics, all occur in these animals. This has now been well documented by Craig Stanford (1999). However, it is also easily documented that, the above notwithstanding, the primary diets of all these species are a variety of plant foods. It is only occasionally that animal products are obtained and when they are, they are often shared very unequally, a dominant male, a favorite consort and her baby, get the lion's (*sic*) share. The remainder of the troupe gets just scraps. It is thus likely that most of these animals receive only small amounts of animal foodstuffs, and that only occasionally. As a result, the normal situation for most animals is one of primarily vegetarian food. This implies that only small amounts of vitamin B_{12} are available most of the time. It is entirely likely that this access to the vitamin varies enormously across the seasons, perhaps less animal products being eaten during seasons when fruits are at their most abundant. Vitamin B_{12} levels are likely marginal and this is why the animals immediately start to become deficient upon a totally vegetarian diet in captivity.

This marginality in vitamin B_{12} sustenance is further emphasized by the findings in pregnant animals in the field. In pregnant animals, as indicated above, the fact that vitamin levels are low indicates that a deficiency in the mother is very rapidly produced when the foetus parasitises the mother. This does not occur in animals treated with vitamin B_{12} in captivity (nor of course, in normal humans).

Finally, folic acid availability must also be considered. Because, in the field, the food is primarily vegetarian, folic acid will be present in great abundance. The blood-forming and epithelial systems require both vitamin B_{12} and folic acid, whereas the nervous system needs only vitamin B_{12}. Thus, in the presence of excess folic acid, the small amount of vitamin B_{12} there may be channeled towards the maintenance of the blood-forming and epithelial systems. As a result, the nervous system is starved of the vitamin resulting in a very high incidence of the neurological lesions in captivity on a vegetarian diet. Certainly this occurs to degrees that do not exist in omnivorous humans.

All these factors imply that even in the field these animals are on the verge of deficiency most of the time. Of course, this does not matter in the field situation because changes in conditions (obtaining a small amount of animal food occasionally, eating the placenta, even eating earth and earth-contaminated roots and tubers) would restore reduced values to normal.

In contrast humans are omnivorous creatures, eating considerable amounts of animal products. They probably have been doing this for a million years or more. Though the australopithecines were likely vegetarian (see their large teeth), members of the genus *Homo*, which seem to go back even further than australopithecines, were presumably not. The taking of animal foods may have started with scavenging by groups of humans long before precision hunting evolved. As a result, humans may have had much larger dietary amounts of vitamin B_{12} than australopithecines. This increased amount may be the reason for a vegetarian diet not producing a deficiency for many years. In some cases a deficiency may never be produced. Not all true vegetarians are deficient in the vitamin. This could account for the difference in timing of the deficiency in the apes and monkeys as compared with humans.

These same factors could also account for differences in the nature of the lesions in vitamin B_{12} deficiency in apes and monkeys compared with humans. Both humans and the particular non-human primates being discussed here need folic acid as well as vitamin B_{12}. Folic acid is involved with vitamin B_{12} in the normal function of the blood-forming and epithelial systems. It is not involved in the normal function of the nervous system element of vitamin B_{12} action. As a result, in the non-human primates, the huge amounts of folic acid (from the large amount of plant matter in the diet) channel vitamin B_{12} to protect the blood and epithelial systems. When vitamin B_{12} is reduced in these animals, the nervous system is what suffers first and most. This would account for why almost 100% of the animals, when deficient, have nervous system lesions (even though, in most animals, clinical evidence of such lesions is not easily observed before death).

In humans, in contrast, the relatively smaller amounts of folic acid together with the much larger amounts of vitamin B_{12} that are normally present protect the nervous system. As a result, when deficiency occurs it is the blood-forming and epithelial systems that are most affected. This presumably explains, at least in part, why the blood system lesions predominate in humans, the nervous system lesions in apes and monkeys.

This information has an important implication. The brains of modern apes are at the upper end of a trend towards increasing brain size in primates. Humans, in turn, are enormously further evolved in this direction than even the apes (see Chapter 10). However, notwithstanding increased size and complexity in the brains of apes and monkeys, their vitamin B_{12} status with consequent easy damage to their brains, suggests that they have been operating at the margin in this regard. If this were true, it would also have been true for other primarily vegetarian fossil relatives. Certainly, the structure of the teeth, jaws and skulls of creatures like *Paranthropus* and *Australopithecus* imply primarily vegetarian diets. It may well have been the unique (among primates) onset of a fully omnivorous diet at some point in a non-ape, non-australopithecine, pre-human lineage that might just have supplied the vitamin B_{12} conditions that removed this neural constraint allowing a 'Cerebral Rubicon' to be crossed. Could this new metabolic background, a background making it easier to maintain and increase axonal myelination, have allowed many other changes in the brain (e.g. development of much greater size and complexity)? Is it possible that it was this dietary marginalisation in apes and monkeys that prevented it occurring in them? We will examine the fit of this idea with the material of the next chapter.

Lateral Thinking Again: Medical Possibilities

There is another, completely different story that stems from these studies. This is to do with folic acid, Vitamin B_{12} and clinical effects on the brain. The beginnings of the brain and spinal cord form in the very early embryo where the epithelium on the back folds in upon itself forming a groove: the eventual nervous system. This groove deepens and becomes a tube lying under the skin, the neural tube. This eventually becomes the spinal cord and brain. This process requires folic acid. When folic acid is deficient the neural tube remains open at certain points and this gives rise in humans to severe malformations like spina bifida (known as neural tube defects). It is inconceivable that this could occur in primarily vegetarian monkeys, their dietary intakes of folic acid being so high. In human mothers, however, folic acid supplementation can prevent spina bifida and related defects.

There is a well-known danger here. If, by chance, mothers are also already deficient in vitamin B_{12} (because they have pernicious anaemia or any of the other causes of vitamin B_{12} deficiency), then added folic acid (to prevent neural tube defects in pregnancy) may precipitate neurological lesions. The added folic acid channels what little vitamin B_{12} there is towards resolution of the blood-forming and epithelial conditions. As a result the developing foetal nervous system is starved of vitamin B_{12} and neural tube defects ensue. This is parallel to what happens all the time in the non-human primates. Marginal levels of vitamin B_{12} together with large amounts of folic acid render almost all the animals susceptible to nervous system lesions when they enter captivity under the conditions of a vitamin B_{12}-deficient vegetarian diet.

The folic acid/spina bifida/neural tube defect link was observed in the 1990s in the Research Institute for Child Health in Western Australia. Supplementation with folic acid has become a health policy in Western Australia in this new century. The aim, to reduce the incidence of such defects in human babies, has been successful.

In addition, as early as 1966 (Oxnard and Smith, 1967) it was evident that Vitamin B_{12} deficiency might be important in the elderly (International Herald Tribune, Health and Science, 2008).

Consider the following scenario. In the 1960s I was studying vitamin B_{12} in pregnant monkeys. I was looking at the levels in foetuses and neonates. I was measuring transplacental passage of the vitamin. I was involved in nervous system histopathology of the deficiency. I was trying to assay folic acid in order to look at the relationship with vitamin B_{12} (though I had great difficulty here, the folic acid assays were much more difficult to perform at that time than the vitamin B_{12} assays). I knew about the relationship between vitamin B_{12} and folic acid.

Why did I not discover the spina bifida link in the 1960s? Why did we not discover then, that we could prevent this problem by giving folic acid in pregnancy? Why not indeed? Of course, it is entirely possible that in the 1960s I would not have recognised the problem. I might well not have had the perspicacity. But then again I just might. And if I did not, perhaps one of my colleagues or students might. One just cannot say what might have happened. But we did not have the chance to make the discovery; the work stopped. Why?

The vitamin B_{12} work ceased for several reasons. Rhesus monkeys that originally cost about £10 each became 100s of $US each. The cost of their food increased from that of buying it oneself in the market to that of using a central feeding system. The cost of shelter for the animals increased from keeping them in an inexpensive, non-profit making, departmental colony, to keeping them in an expensive central animal holding facility. The cost of treating them medically by a young physician at no charge changed to the cost of care by extremely expensive veterinarians. The charges for clinical treatment went from courtesy service by a colleague to costs in a large veterinary unit. Even the costs of drugs went from gifted drugs from our pharmacy, to officially ordered and paid-for drugs. There were a host of similar factors as the times took us from the collegial collaboration of my younger days to the user pays and bottom line considerations of later times. All this meant inevitably that there was not enough money.

This is not a criticism of the centralised facility *per se*; goodness knows improvements in animal holding facilities were long overdue. But it does point to what can be lost if the managerial and financial bottom line is pursued at the expense of the collegial and knowledge bottom line.

It is further possible that the work might have been undertaken by the Agricultural Research Council (ARC) of the United Kingdom (the original awarder of some of my research funds). The ARC (UK) wanted to do this and spent a large amount of money attempting to create the vitamin deficiency in a new colony of baboons (not rhesus monkeys) over a three-year period. Unfortunately, they did this without contacting me. I had left the United Kingdom for the USA and did not know what they were trying to do. I was, perhaps, the only person in the world who already knew that this could not be easily done with baboons. The conditions under which baboons are kept in captivity (see above) usually preclude the possibility of them becoming deficient. As a result even this follow-up work was abandoned.

My own further participation in this work was abandoned for other reasons: the increasing activities of the animal rights lobby. One effect of the 'gentle' wing of that lobby was the creation of ethics and animal rights committees. These were important steps in improving the conditions of the animals and their surroundings. However, they also lead to increasing bureaucracy and delays with resulting very great difficulties in getting permission for work on animals particularly primates. This finally culminated in an inability to work on primates at all. Again, this is not to say that such committees were not important, and certainly they have produced marked improvements in the situation of experimental animals. But it does point to how protections against one set of circumstances may unwittingly damage a second.

Another effect stemmed from the 'violent' wing of animal rights groups. It became just too dangerous to do this kind of work. The Wisconsin Animal Facility (this was just up the road from the University of Chicago where I was at the time) was bombed by protesters. Experimental animals were 'freed' into the Wisconsin winter. Five years worth of data for a doctoral thesis was destroyed. Someone was killed. Other animal, especially primate, facilities have been similarly treated. I have many friends whose addresses and telephone numbers are incommunicado because of threats from such groups.

These things all stopped the vitamin B_{12} work at that point. They were precluded by the financial bottom line, user-pays, bureaucratic, administrative and interference mechanisms described above. This is not a grumble about these mechanisms *per se*. Clearly, improvements in animal handling, in the ethics of experimentation, in the running of animal facilities, and the protections of both animals and staff were necessary and have occurred.

However, because of the way that they occurred, because the work had to stop, we may just have had 30 years worth of spina bifida babies that might, that just might, have been avoided! A lot of families might just have been spared enormous anguish! A lot of normal babies might have been born instead!

In addition, as many as 30% of the elderly have Vitamin B_{12} deficiency and do not know it. This, too, might have been avoided.

Lateral Thinking a Third Time: What About Research?

A finally different thought intrudes. Our rhesus monkey colony had been set up specifically to tackle a range of applied problems of great significance to humans. It was the lineal descendant of the original group of monkeys that Professor Zuckerman had in Oxford before the Second World War. The colony was enormously increased, however, by his abilities to obtain funding. It was eventually used by a very great variety of investigators in his department in Birmingham. These included studies of breeding and hormones, development of contraceptive drugs, structure and function of the nervous system, examination of growth and aging, vitamin B_{12} deficiency, etc.). What other medical anatomy department of those days (or even indeed of today when anatomy has generally been downsized to teaching units scarcely undertaking any research at all) employed, in addition to anatomists: zoologists, physiologists, chemists, physicists, engineers, psychologists, veterinary scientists, a public schoolmaster, three fellows of the Royal Society and three peers of the realm? Such a diverse group of investigators produced a great deal of work using this animal colony.

It is of interest that the results of many studies carried out on what were thought to be normal experimental subjects, have never been re-examined in the light of such a widespread vitamin B_{12} deficiency with such extensive pathology! Indeed, I once heard it said that it did not matter. 'After all, a monkey was just a walking womb'!

References

Flinn RM, Oxnard CE, The relationship between growth and the administration of cyanocobalamin in the rhesus monkey, *Folia Primatologica* **4**: 432–437, 1966.

Goodall AG, Chimpanzees of the Gombe Stream Reserve, in DeVore I, (ed.), *Primate Behavior*, Holt, Rinehart and Winston, New York, pp. 425–437, 1965.

Hind VMD, Degeneration in the peripheral visual pathway of captive monkeys, Doctoral thesis, University of Birmingham, UK, 1969.

Krohn PL, Oxnard CE, Chalmers JNM, Vitamin B_{12} in the serum of the rhesus monkey, *Nature* **197**: 186, 1963.

Merfield FG, *Gorillas Were My Neighbours*, Longmans Green, London, 1956.

Oxnard CE, Some variations in the amount of vitamin B_{12} in the serum of the rhesus monkey, *Nature* **201**: 1188–1191, 1964.

Oxnard CE, Vitamine B_{12} nutrition in some primates in captivity, *Folia Primatologica* **4**: 424–431, 1966.

Oxnard CE, A note on the ruminant-like digestion of langurs, *Lab Primate Newsletter* **8**: 24–26, 1969.

Oxnard CE, Smith WT, Neurological degeneration and reduced serum vitamin B_{12} levels in captive monkeys, *Nature* **210**: 507–509, 1966.

Oxnard CE, Smith WT, Vitamin B_{12} and psychiatry, *Lancet* **1**: 161, 1967.

Oxnard CE, Smith WT, Torres-Nunez I, Peripheral neuropathy and hypovitaminosis B_{12} in captive monkeys, *Proc Sec Int Cong Primatol* **3**: 162–168, Karger, Basel, 1969.

Oxnard CE, Smith WT, Torres-Nunez I, Vitamin B_{12} deficiency in captive monkeys and its effect on the nervous system and the blood, *Lab Animals* **4**: 1–12, 1970.

Smith EL, *Vitamin B_{12}*, Methuen, London, 1960.

Spicer EJF, Oxnard CE, Some haematological changes during pregnancy in the rhesus monkey, *Folia Primatologica* **6**: 236–242, 1967.

Torres-Nunez I, Degeneration of the nervous system associated with vitamin B_{12} deficiency in captive primates, Doctoral thesis, University of Birmingham, UK, 1968.

Torres-Nunez I, Smith WT, Oxnard CE, Degeneration of the peripheral and central nervous system in vitamin B_{12} deficient monkeys, *Experientia* **25**: 273–275, 1968.

Torres-Nunez I, Smith WT, Oxnard CE, Peripheral neuropathy associated with Vitamin-B_{12} deficiency in captive monkeys, *J Path* **105**: 125–146, 1971.

Stanford CB, *The Hunting Apes, Meat Eating and the Origins of Human Behavior*, Princeton University Press, Princeton, 1999.

Chapter 10

New Wrinkles on Old Brains

From Bones to Brains

Even just a glimpse at my scientific work makes it clear that, over the years, I have measured a very large number of bones. Bones are hard. Measurements of bones are 'hard'. Hard measurements are good meat for statistical analysis. The emphasis on bones is not just mine; it is true of most biological anthropologists.

Of course, much of the information content of bone form and architecture relates to function. It depends not only on bone form, but on soft tissues (ligaments and muscles), upon postures and movements (often locomotion), upon the contexts in which the movements occur (often behaviour and lifestyle), and upon the milieu within which the behaviours exist (often, even, the entire environment within which the lifestyles occur). As the years have passed I have therefore been lead to apply the statistical methods that are primarily used for bones (Oxnard, 1983/1984) to these non-bony contexts (Oxnard *et al.*, 1990; Oxnard, 2000).

The non-bony components are more difficult to study. There are thousands of skeletons in the museums and medical schools of the world resulting in the possibility of tens of thousands of measurements. There are far fewer cadavers with soft tissues, and especially few carcasses of non-human species. Even when there are good samples of bodies, obtaining measurements about soft-tissues: from dissection and micro-dissection, light and electron microscopy, physiological and biochemical experiments, and so on, are major logistic burdens. As a result, such studies are usually restricted to a small sample of specimens of just one or a very small number of species. This is inadequate for studying animal diversity and evolution. Measuring other non-structural aspects of animals (for example, movements and locomotion, behaviours and lifestyles, and the environments within which these occur in evolutionary contexts) is even more difficult. Neverthless I have attempted studies using statistical methods ('metrics') in all of these areas (see other chapters).

One particular soft tissue structure that could be studied using measurements and analyses in an evolutionary context is the brain. But where could I get the data? This question was answered in a most unexpected way. As a result of a series of unfortunate incidents relating to the termination and later death of his previous supervisor (Dr David Rindos), a senior graduate student, Willem de Winter, came to me for help. Would I be willing to be his new doctoral supervisor?

The project which de Winter had been originally tackling with Rindos, involved a synthesis of opposing theories of natural selection. But I felt uncomfortable supervising a doctoral project on theory alone; I am more of a 'real data analysis' person. In any case de Winter had largely finished this part of his thesis and it later appeared as a paper of monographic dimensions in *Biology and Philosophy* (de Winter, 1997a). However, de Winter was also fascinated by the flexibility of human behaviour and we had many discussions on this topic. These discussions and his interests in my book, 'Animal Lifestyles and Evolution' applying statistical analyses to non-bony contexts, lead us towards a conjoint interest in the evolutionary flexibility of the brain (de Winter, 1997b).

Obtaining a lot of data about the brain, from many specimens, in many different species, representing many and varied taxonomic groups, is no trivial matter. It would be impossible within a doctoral project. But, of course, I knew there was already a very large set of data on the sizes of brains and brain parts in many mammals collected by the very careful work of Dr Heinz Stephan over a vigorous research lifetime. Analyses of these data have been frequently published. Most of the studies are univariate and bivariate though some early multivariate statistical analyses were carried out in the 1960s and 1970s by workers such as George Sacher

(1970) and Ralph Holloway (1979). De Winter and I set out to reanalyse these data using more extended modern multivariate techniques.

Our abilities to undertake this work were especially enhanced by the gracious gift from Professor Heinz Stephan of his full set of measurements. We were, thus, not limited by the use of summary information from his published work, all that was available to other workers. This was, incidentally the fourth remarkable *gift* of complete raw data that I have received. Adolph Schultz had much earlier provided me with his measurements of primate cadavers, Francoise Jouffroy her measurements of primate limb bones, and Wu Rukang his data on fossil teeth from China.

Stephan's full data are remarkable: the absolute volumes of 11 major brain parts from the medulla to the neocortex, on each of 921 specimens of 363 species of various insectivores, bats, primates, tree shrews and elephant shrews (thus representing three large and two very small orders of mammals). Stephan's summary data have been investigated many times. They have generally shown the importance of the relationship between brain size and body size. Could we really expect our study of the full data to give new information about brain evolution?

Brains and Complexity

New advances in the study of the brain, its structure, its function, its development, especially its structure while functioning and developing, and its relationships with behaviour, cognition and mind, are providing whole new views about this most central of organs. Structural studies have moved from the coarse macroscopic recognition of individual parts of the brain, through standard microscopy, cyto-architectonics, nuclear connections, ultrastructural microscopy, cellular synapses and sub-cellular organelles to ultimate brain molecules. Functional investigations have progressed from studies of interference with function by crude macroscopic brain lesions, through studies of carefully placed microlesions, intracellular recordings, and interactions along brain cell axons and across individual synapses between cells, to underlying chemical events associated with function. Developmental mechanisms have come to be understood at finer and finer levels, from overall organogenesis, through mechanisms of movements of cell layers, single populations of cells and cell death, to the genetic basis of brain development and the many complex gene expression factors that govern it. Perhaps most of all, these separate approaches: structural, functional and developmental, have started to coalesce. Nowadays brain structures can be seen developing and functioning through many new non-invasive imaging techniques. All this allows better understanding not only of how the brain works, for example, in terms of movement and sensation, but also of how it functions during sleep, during preparation for action, during thinking, during emotions, and so on.

Such new lines of investigation, however exciting, and with such major implications for normal human brain function and in disease, employ more and more complex methods and reveal the workings of smaller and smaller brain components. As a result, the logistical problems of carrying out such studies on whole brains in an evolutionary perspective loom ever larger. It is scarcely possible for a single investigator to be able to use the entire battery of techniques in even just a single functional brain system in a single species. How much more difficult would it be, then, to replicate all of this, not just in one specific brain system (such as visual or olfactory) but in most or all major brain systems, and in the integrated brain overall? How enormously more difficult would it be to carry out such studies, not just in one experimental sample of animals, not just in a few exemplar species such as the rat, the monkey (usually a rhesus or squirrel monkey), the ape (usually a chimpanzee), and the human (usually a white male), but in large numbers of species that encompass a very broad diversity of brain organisations in an evolutionary setting?

Yet such animal diversity is necessary to illuminate brain evolution (Preuss, 2000). As a result, studies of brain evolution in large numbers of specimens, species, families, orders and classes of vertebrates have been confined to the size of the brain, or at best, the sizes of its major parts.

Brains and Size

For this reason, most broad evolutionary studies of brains have been aimed at investigations of brain and brain part size, including especially brain/body size relationships (Fig. 1). Such investigations have been carried out throughout the whole of the latter half of last century and into the present (e.g. Stephan and Andy, 1969; Sacher, 1970; Jerison, 1973; Holloway, 1979; Martin, 1983;

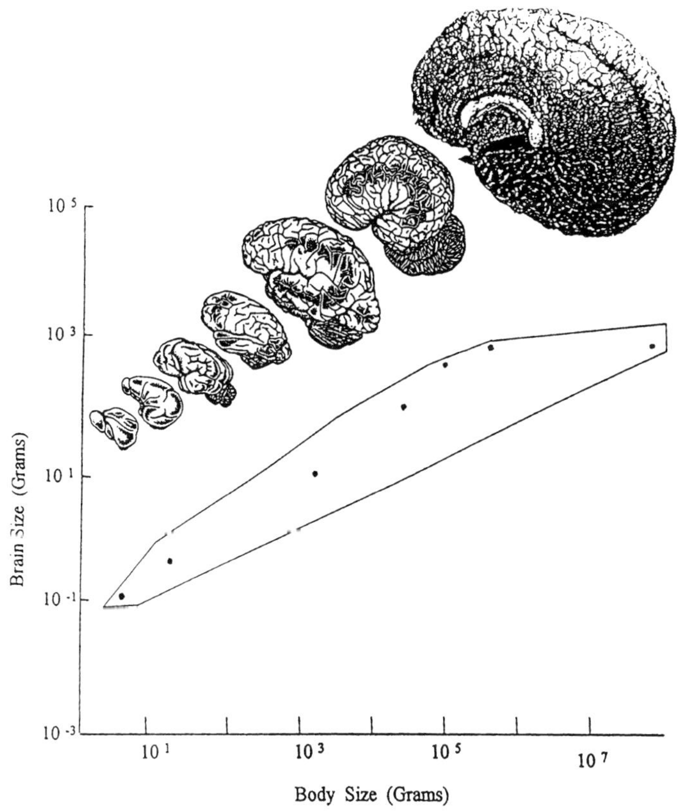

Fig. 1. The relationship between body size and brain size (both logged) in a total of 50 species of mammals is shown by the polygon that encloses all species. The positions (dots) and sketches of six representative species (armadillo, hare, dog, chimpanzee, human, dolphin and blue whale) are provided. Combined from and modified after Jerison (2001) and Roth (2001).

Roth and Wulliman, 2001). Brain and body sizes in mammals (when plotted on double logarithmic paper) show a straight-line relationship from the smallest bats to the largest whales (a range of about 8 orders of magnitude). This contrasts with different straight-line relationships for other vertebrates (reviewed in Jerison, 2001).

In some ways this is very exciting, suggesting as it does, a constancy of brain organisation in mammals compared with other vertebrates that is quite remarkable. Yet this result is also rather boring. Does overall brain size not have any other meaning? Holloway (2001) puts it another way: 'it is the allometric constraints which might be deemed "trivial"'.

This linear relationship between brain and body size seems especially to hold for most primates. Almost every investigator has attempted to explain this. Linear indices of 'brain complexity' have been erected. Linear concepts of increasing 'hominisation' or 'intelligence' of the primate brain in a 'progression' towards humans have been put forward.

All such ideas suggest that brain organisation is really rather similar in most mammals, especially primates. This implies a conservative and very ancient pattern of brain development. How likely is it that such a simple linear concept is truly definitive of brains as complex as we now know them to be, possessed of myriad different intelligences, and the result of evolutionary processes that must be very far from linear?

The Tyranny of Size

The enormous differences of mammals in other aspects of their anatomy suggest to me that perhaps the brain data are also more different than just size. Perhaps, these differences are 'hidden' by what I have called elsewhere 'the tyranny of size'. Is it possible that the size range of the brains (and indeed of the bodies) of mammals is so great that it is actually hiding other information in the data? Is it possible that the data, if examined in a different way, might yield a different picture?

With the hindsight of our prior morphometric studies, it seemed obvious what could be done. If we are interested in understanding brain function, the proportion between two brain parts rather than the absolute size of one brain part, may be more useful. Such a proportion might reflect, at least in part, the functional cross-talk between the parts. Of course, this is crude compared to the proportions (say) of nerve cell numbers or connections. As explained before, however, such data are not available in a major evolutionary context.

Armed with this idea, therefore, and after consulting with neurobiologists, de Winter and I devised a series of brain proportions that we thought might reflect, at least in part, functional relationships between major brain regions. Recognising, first, that the medulla forms the gateway of most projections linking brain and body, we thought that the proportion of the main entry port of the brain, the medulla, to other major brain parts, might reflect, to some considerable degree, the amount of neural traffic between the body and the other major brain components (Passingham, 1975). Accordingly then we reorganised the sizes of the other brain parts as proportions of the size of the medulla. Recognising, second, that all brain parts exchange many links with the highest brain level, the neocortex, we reorganised the various brain part sizes as proportions of the neocortex.

This idea is supported not only by these functional considerations, but also because these variables are spread across different developmental growth fields of the brain (Deacon, 1990). Sacher, as long ago as 1970 had a similar idea and in his hands seemed to give greater information than raw volumes alone. Thanks are particularly due to Lutz Slomianka, Alan Harvey, Alan Pettigrew and John Haight, career neurobiologists, who guided us in the invention of these proportions.

This process gave a total of 19 derived variables for analysis. Each new variable directly describes an aspect of the relative form of each individual brain specimen. This crucial property sets them apart from variables obtained from allometric or other statistical estimates. Statistically scaled proportions reflect abstracted properties of a sample. Direct proportions are genuine descriptors of each individual (Harvey and Krebs, 1990). Further, for many purposes in biology, absolute sizes, the length of a muscle lever arm, the diameter of a bone and the size of a hand, are functionally unimportant. What may be functionally important is the proportion between two lever arms (a biomechanical variable), the ratio of the diameter of a bone to its wall thickness (a stress analysis variable), and the comparison of the size of the hand to the size of the branch it grips (a locomotor variable). In this case what might be functionally important is the ratio of the cerebellar size to neocortical size. It is better to use such variables because their proportions would not lose the ability to study the biological form in a functional context. Thus we hoped that the brain proportions calculated here would reflect, at least in part, relationships that were biologically important in understanding brain part interactions. In addition, because proportions scale each variable on each specimen to unity, the resulting data set is no longer dominated by very large size differences, though elements of size, of course, are still present within the data.

Brain Components: One at a Time

We first examined this new data set one variable at a time. Figure 2 provides an overview of the distributions of the data using what is called 'main effects ordering' (Cleveland, 1993). In this, the values of the variables in each sub-plot are ordered from bottom to top according to their increasing values within one of the groups. In this case we chose insectivores, the group commonly regarded as the mammalian order (amongst those studied here) with the most conservative brains (Stephan *et al.*, 1991). Thus, in the plot for insectivores (bottom right hand frame of Fig. 2), the box plots are arranged from the smallest brain part (the septum) through the next (the schizocortex) up to the largest (in insectivores — the palaeocortex). This provides a plot showing a smooth upward trajectory of values of brain parts.

If the data for some other mammalian group were organised in the same way as in insectivores, then, notwithstanding different brain sizes in that other order, the main effects plot would show the same smooth upward trajectory. This did not occur in these Orders.

In fact, figure 2 shows that, within each of the other taxonomic groups, the pattern loses its smoothness and becomes quite jagged,

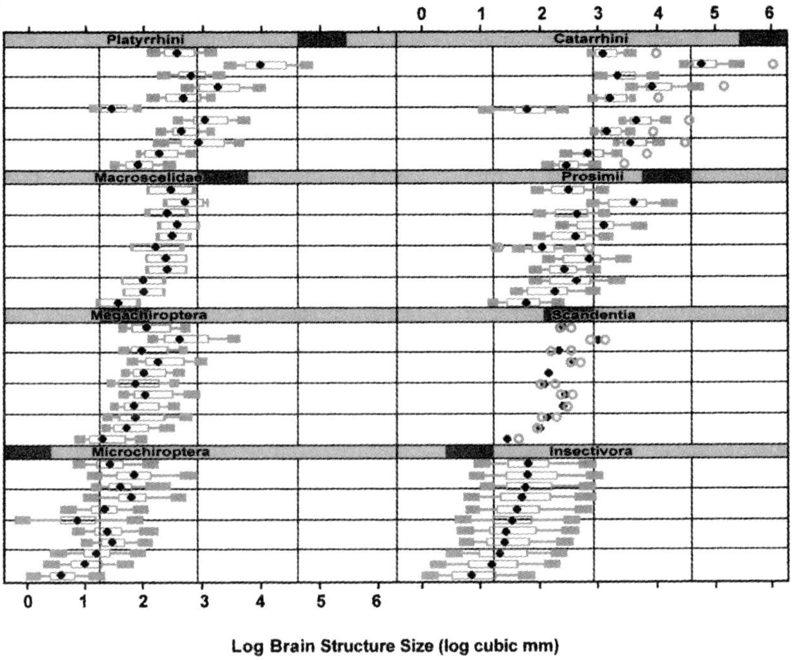

Fig. 2. The sizes of brain parts taken one-at-a-time displayed by what is called 'main effects ordering' in each of the major groups of species as indicated. The plot at lower right gives the average sizes of each brain part in insectivores when ordered from the smallest part (septum) to the largest (palaeocortex). The remaining plots are for small bats, large bats, tree shrews, elephant shrews, prosimian primates, New World primates and Old World primates as indicated. In each case the order of brain parts is kept the same as in insectivores. The disruption in the smooth curve of the plots for the other groups shows how they differ from one another. The disruption is at its maximum in the Old World primates indicating that the greatest difference from insectivores is in the brain organisation of Old World primates.

increasingly so with increasing evolutionary distance from insectivores. For example, among the bats the olfactory bulb is relatively small in the Microchiroptera, but relatively much bigger in the Megachiroptera. This produces an abrupt difference between these two plots, and in their comparison with insectivores. For example again, in Old World primates, the neocortex is relatively the largest. This also interrupts the smoothness of the primate curve and this curve also shows marked departures for every variable level compared with insectivores. Such interruptions in pattern are found in every taxonomic group and demonstrate that evolution has produced many patterns of brain organisation.

Brain Components: All Together

Though these data have been available for many years, and though the statistical methods for looking at large numbers of variables taken together have been available for even longer, the neurobiology and the statistics have not often been brought together. However, Sacher (1970) and Holloway (1979) both applied early multivariable statistical methods.

It is only recently that such work has been taken up again. Thus Finlay and Darlington (1995), Barton and Harvey (2000) and Clark *et al.* (2001) re-examined the sizes of 10 different brain structures using similar multivariate statistical techniques. They all used Stephan's published data on mean sizes for each species. Their different techniques agreed in showing that the size relationships in the data were so strong that their main statistical descriptors arranged the species means by overall size to the extent of 97%, 92% and 90% of the total information, respectively. In other words, other information was limited to 10% or less. This seemed to imply enormous conservatism. As a result, Finlay *et al.* (2001) ask: 'How can we reconcile this evidence for cross-species conservatism in patterns of brain enlargement with the strong intuition that species-specific brain adaptations must exist?' This conservatism comes at the very time that developmental and environmental lability and flexibility in most other body features are being recognised as very important in evolution.

These workers, nevertheless, all recognised that, though the primary finding suggested marked developmental constraints upon brain organisation, there were some degrees of adaptive mosaicism at the level of the highest phylogenetic groups (Orders). It is of interest that this was also evident in the much earlier studies of Sacher (1970) and Holloway (1979).

Some very recent work makes further advances using new types of data. Thus, Sherwood *et al.* (2004) studied samples of specimens rather than species means and drew their data from magnetic resonance images of brains. They avoided the problems of using raw or allometrically corrected volumes by using, as we did, ratios, but in their case they were of each brain part volume to the volume of the whole brain. The difficulties of doing such more complex studies of the brain are clearly shown by the fact that, in the time available, they could examine only four great ape species and only five brain parts. They did, however, make new findings related to locomotor abilities and ecological milieux.

Phylogeny and Function in Mammal Brains

During all this time, we (de Winter and I) had also been applying multivariate statistical methods to Stephan's data but with major differences to prior workers. We had triple the number of species (from the 131 of Finlay and Darlington to 363) and more than five times as many specimens (from 131 to over 921) in the five different orders. In addition, we had the original unpublished variables, not the published transformed variables. Because, for most groups we had more than single specimens, we could take account of intragroup variation. As a result, in addition to using the standard Principal Components Analysis for examining specimens in a single packet, we were able to use a method called Canonical Variates Analysis for examining the separations of specimens as several known groups.

This work started in 1993 and, of course, we first applied the same method, principal component analyses, to the raw data, happily, the same technique used by Finlay and Darlington. This technique (principal component analysis) allows differences among specimens to be revealed without any *a priori* assessment of groups. In some studies, such analysis shows that there are no subgroups within the data; the specimens are just linearly or elliptically arranged. However, when subgroups are truly present, PCA may reveal them to a degree, although, because the data are examined as a single packet, such subgroups generally show much overlap. The method is, however, an excellent first data reduction technique.

Though we had a much larger data set that could possibly have changed the results, we were delighted to see, when the paper by Finlay and Darlington (1995) appeared, that our results analysing the raw variables (as they did) with the same technique were identical to theirs even to the first decimal place. This was so despite the much larger data set that we had. Such replication is most valuable; it is a *sine qua non* for good science; it is so often neglected. Our main aim, however, was the analysis of the 19 new proportional variables described earlier. It was our hope that this analysis would provide much more information about groups of species and clusters of variables.

Groups of species in the new analyses

With these new data our results were remarkably different from those of the prior workers. We first tested their relationships with size. We agreed that there is indeed a single multivariate direction that contains 100% of the size information. However, no one order is directly aligned along it. As a result each individual order shows a lesser relationship with size than the data as a whole (which was 95+%). The separations of the primates reflected size by only 78%; the insectivores and bats reflected even less (29% and 24%, respectively). This implies first, that the trends of organisation of brain components are not constrained almost completely by size, and second, that they are different in each Order.

Much more important than size, however, was the fact that the three major orders (insectivores, bats and primates) lay almost at right angles to one another in three dimensions (Fig. 3). This implies that the pattern of organisation of the brain components is completely different in each major order. Our next thought, therefore, was to look for phylogenetic separations within each order. Some exist. Among bats, the fruit bats (Megachiroptera) are fairly clearly (and statistically) distinguished from the other bats (Microchiroptera). Among primates, the strepsirrhines (creatures like lorises, lemurs and bush babies) are partially (and significantly) separated from the haplorrhines (tarsiers, monkeys, and apes and humans). These two findings

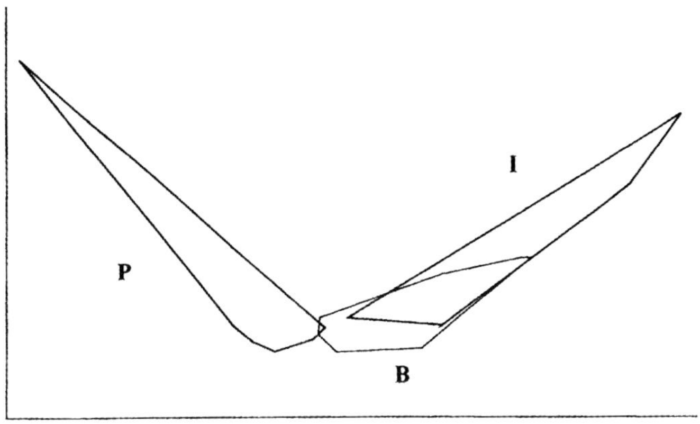

Fig. 3. Examination of 19 brain variables in 921 specimens of mammals taken altogether using principal component analysis. The first principal component is horizontal and the second vertical. The polygon marked P contains all the primate specimens; the polygon marked I contains all the insectivore specimens. These lie generally within the space of the first two principal components (i.e. in the plane of the paper). The polygon marked B contains all the bats, and, though it appears to be superimposed upon the insectivores, it actually lies at approximately right angles to them out of the plane of the paper mainly in the third principal component axis.

also supported most prior studies. No lower phylogenetic groups (e.g. sub-orders, superfamilies, families or subfamilies) were distinguishable in any study.

If not size and phylogeny, what else? Much to our surprise, other unsuspected separations were revealed. First, various insectivores were clearly separated into species groups that have in common, semi-aquatic, fossorial and terrestrial lifestyles (Fig. 4). These lifestyle associations are further strengthened by the fact that each contains species of phylogenetic groups that have been separated for very long periods of time. These new species groups seem to be truly lifestyle convergences that cut across phylogeny.

For example, one cluster of species implying that they have strikingly similar brain organisations, includes the potomogalid otter shrews from the African mainland, the *Limnogale* of Madagascar, the talpid desmans (*Galemys* from the Pyrenees and *Desmana* from Russia) together with the Eurasian water shrews (of the genus

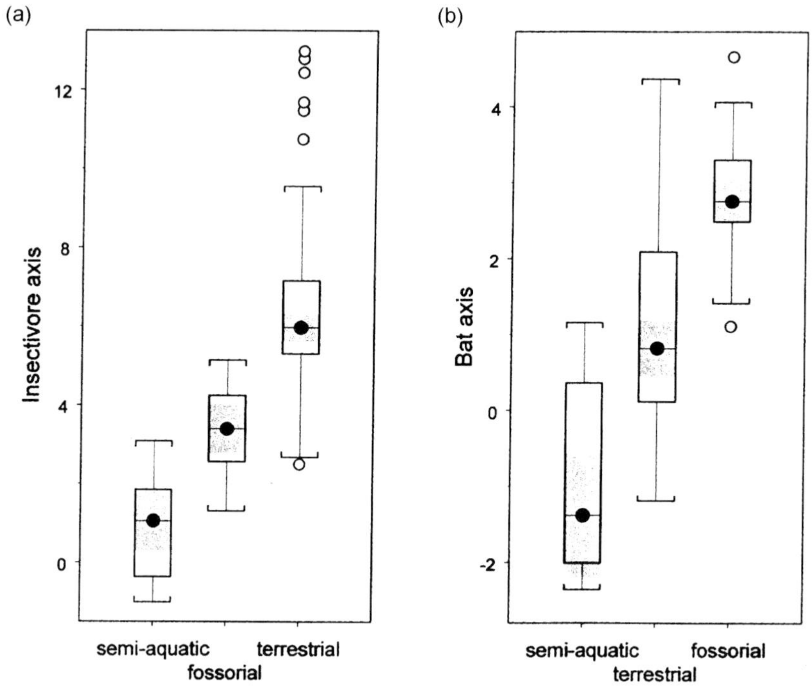

Fig. 4. Box plots showing differentiations of lifestyle groups among insectivores. The first plot indicates statistically significant separations of semiaquatic, fossorial and terrestrial insectivores along the first principal component axis. However, there are also significant insectivore separations along the direction of the third principal component axis (the one effecting primary separations of bats), and this axis shows additional separation of semiaquatic and terrestrial insectivores.

Neomys), and the erinaceoid moonrats (Echinosorinae) of Asia. These species are from five different families, three superfamilies and five widely different regions of three continents. As a group, they are all semiaquatic.

Again, there is a cluster showing similar brain organisation that comprises the golden moles (Chrysochloridae) of Africa and the phylogenetically far distant true (talpine) moles of Eurasia. They are all fossorial in lifestyle.

Finally there was a cluster of species with similar brain organisation comprising the phylogenetically separate tenrecs and shrews.

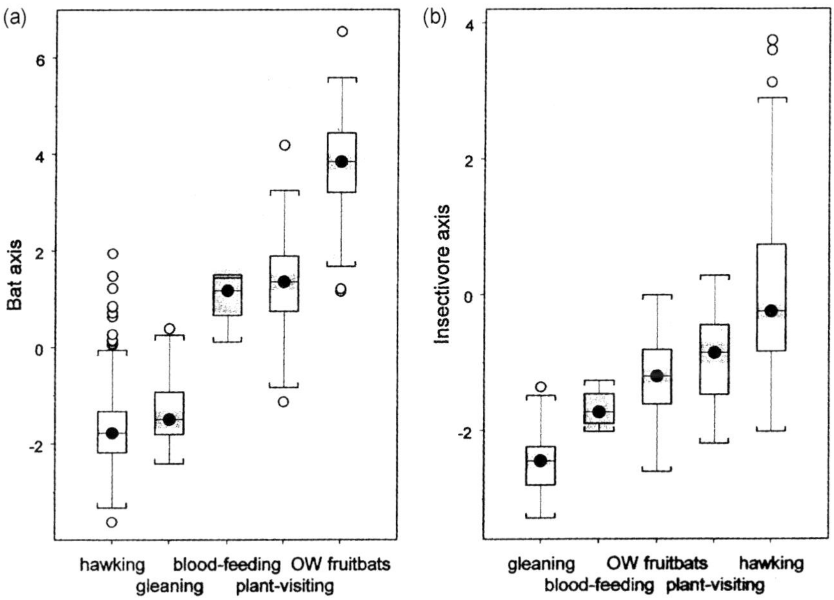

Fig. 5. Box plots showing differentiations of lifestyle groups among bats. The first plot (along the third principal component axis) indicates significant separations of hawking, gleaning, blood-feeding, plant-visiting and fruit-eating bats. In this plot there is no significant separation of hawking and gleaning bats, or of blood-feeding and plant-visiting bats, respectively. The second plot, however, shows that the second principal component clearly and significantly separates gleaning from hawking, and blood-feeding from plant-visiting, respectively. Again, therefore, like the insectivores, this component performs an additional separation of species not shown in the first plot.

These are all terrestrial and, again, their similarities cross the phylogenetic boundaries.

Similarly, among bats, there were several separate clusters of species that shared diets in their lifestyles (Fig. 5). As in insectivores, these lifestyle associations are strengthened by the fact that each contains species of phylogenetic groups that have been separated for very long periods of time. They seem to be truly lifestyle convergences.

Thus, there were nine species that share the habit of carnivory. They are from five different continents and long-separated phylogenetically. A particularly convincing subgroup of this carnivorous lifestyle cluster is the fish-eating bats. The fish-eating bats from the New World (e.g. *Noctilio*, the fisherman bat) are totally unrelated to the Old World fish-eating bats (e.g. *Megaderma*, the Asian false vampire bat). They share other special anatomical adaptations not found in any other bats: e.g. the rather large feet that are used in their 'fishing' activities. In contrast is a separate group of the various leaf-nosed bats of the New World together with certain microbats of the Old World. These are all fruit-eaters. Sometimes, the degree of similarity is amazing. One of the most convincing propinquities within this group involves fig-eating. The New World fig-eating stenodermine microbats (*Artibeus*) are completely intermixed with the Old World fig-eating bats (epomorphines) of the genera *Hysignatus*, *Epomorphorus*, *Epomops* and *Micropterus*. There is even a group of bat species that are omnivorous, taking insects and some small vertebrates as well as fruits and nectar. These species are also spread across both the New and Old Worlds. They are grouped approximately half way between the two groups described above. Thus among the bats there are clear separations of species with hawking, blood-feeding, gleaning (of insects in foliage), nectar-taking and fruit-eating lifestyles. In each case the concept that these are genuine lifestyle groupings is reinforced by the fact that, except for the fruit-eating megabats that are entirely Old World and almost certainly monophyletic, they contain both New and Old world forms separate for many millions of years.

Finally, there are clear separations among primates (Fig. 6). Within the monkeys and apes alone, there are distinctions between the powerful springing and leaping monkeys (e.g. marmosets, titi monkeys, langurs and colobus monkeys from both New and Old Worlds) which lie at one extreme, the generally quadrupedal monkeys (e.g. capuchins, macaques, baboons, also from both the New and Old Worlds) which lie centrally located, and the climbing acrobatic forms (e.g. spider monkeys, gibbons, orang utans, yet again from both New and Old Worlds) which lie at the other extreme. The very close proximity of the New World spider and woolly spider monkeys to the lesser and greater apes is especially interesting because, in simple brain size terms, the spider and woolly spider monkeys have considerably smaller brains than the Old World monkey genera such as macaques

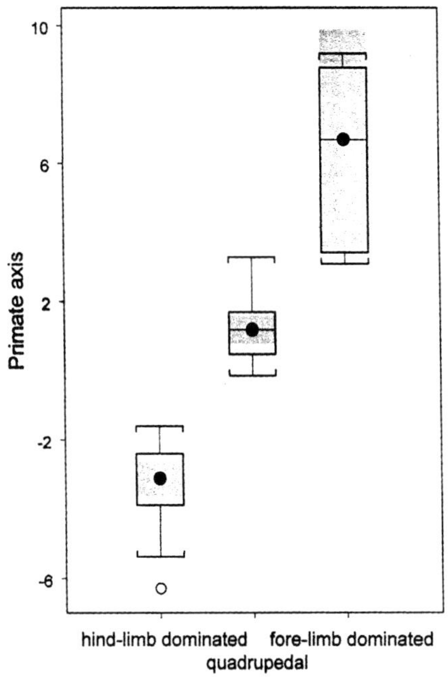

Fig. 6. The single box plot showing the differentiation of lifestyle groups among Old World primates. Hind-limb-dominated, quadrupedal and forelimb-dominated locomotor modes are statistically significantly separated. The tarsiers are the single lowermost open circle in the hindlimb-dominated cluster.

and baboons (see also Sherwood *et al.*, 2004). Yet this overall size difference is overcome in the placement of these upper-limb dominant New World monkeys with the apes. In each case, as in insectivores and bats, these associations seem to reflect genuine lifestyle convergences because each group contains species from different taxonomic groups at very high levels.

Though I have used single terms to refer to the lifestyle groups in each of these Orders (environmental terms for insectivores, dietary terms for bats, locomotor terms for primates), I do not mean to imply that these brain convergences relate to such narrow descriptors. In each case the one-word descriptors are surrogates for a wide array of functional, behavioural and environmental features that in total add up to the lifestyles of the various creatures. For example, my use of the terms aquatic, fossorial and terrestrial in insectivores includes all the other lifestyle factors (e.g. locomotion, predation and escaping, and foraging and feeding) that are associated with aquatic, fossorial and terrestrial ground niches.

Similarly, my use of dietary terms in bats includes a whole host of lifestyle factors: foraging styles, type of flight pattern, use of different sensory and motor abilities in both flying and feeding, and so on. Thus, bats in the carnivore cluster, even the exclusively insectivorous ones, have in common that they glean prey from the ground, vegetation or air, and hence must be capable, in contrast to fruit-eating species, of distinguishing their quarry from an acoustically cluttered background. The bats in the fruit-eating and nectar-eating groups depend not only upon fruit and nectar but also on the need to remember spatial and temporal distributions of plant food sources, and to make use of olfactory cues.

Similar caveats apply to the locomotor terms for the primate clusters. Hindlimb- and forelimb-dominated locomotor descriptors, respectively imply wide-ranging sets of lifestyle factors (not just locomotion but also eating, escaping, playing, even various social and sexual arrangements) that are important in primates and undoubtedly much more involved in brain organisation than simply my summary one-word descriptors.

The repetition of brain size-species lifestyle associations three times seems to require that we accept the hypothesis of similar brain organisations in relation to convergences in lifestyle. Does this also require that we reject direct phylogenetic relationship as an associated factor? It would seem at first sight that we should. There is so much cross-phylogenetic propinquity. However, these findings do not mean that there is no phylogenetic information in these data and analyses. Lifestyle arrangements must indeed contain phylogenetic information. After all, functional adaptation is a part of phylogeny. It is just that the convergences of lifestyles (and the implications for brain organisation that they imply) cut across and are more powerful than the phylogenetic arrangements. This may be why the phylogenetic groups do not stand out (see Oxnard, 2000). If the sum total of the lifestyle arrangements is greater than the sum total of phylogenetic information, then the phylogenetic information may be hidden. Partitioning the lifestyle information into its phylogenetic and non-phylogenetic portions might allow the phylogeny to shine through. But that is a matter for another chapter. These brain/lifestyle associations are convincing in their own right. They are replicated so many times in so many different groups. Yet they would be even more convincing if the clusters of variables producing them also made neurobiological sense in the light of lifestyles (Brown, 2001).

Clusters of variables in the new analyses

The separations of the various species just described depend upon the differential contributions of the brain variables. These contributions are not easily recognised by inspection of the original statistical (principal component) axes because each of these contains portions of the separations of each of the Orders (for example, both primates and insectivores are largely separated by both components 1 and 2, and bats by components 1 and 3 (Fig. 3, earlier). Because, however, the separations between the main orders are largely orthogonal, de Winter was able to rotate the principal component universe until the separations in each order were primarily along a new set of axes. When this was done, primates were largely parallel to a new first component, insectivores to a new second component and bats to a new third. The contributions of the variables to these new components are, then, those that contribute to the separations in each respective order, and are easily determinable.

The nature of the variables is perhaps best understood in diagrammatic form in Fig. 7. This shows a hypothetical section through a brain (hypothetical because it is not actually possible to place all the brain structures in a single planar cut). The various brain components that are figured are those of Stephan because these are what are represented in his data. Since then, some of the names have been reorganised and coalesced in different ways by neurobiologists. The various components range from the medulla (at the base of the diagram) to the two neocortices (at the top). Above the medulla is, in the midline, a cartoon of the pons, and laterally of the two cerebellar

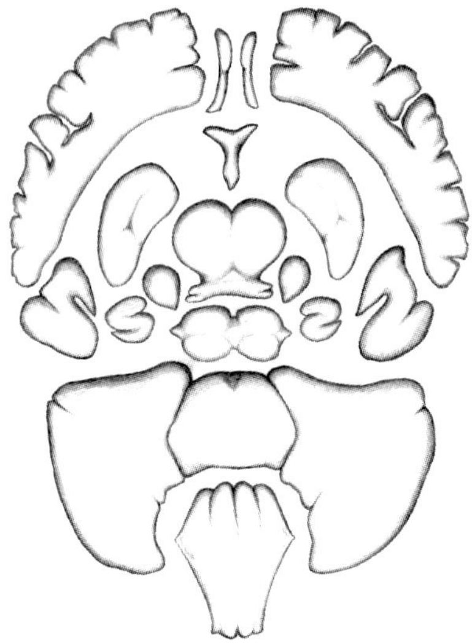

Fig. 7. A simplified diagram of the brain structures examined in this study. Each structure is drawn with a shape that is recognisable for that structure in a human brain. However, because of the three-dimensional complexity of the brain, it is not actually possible to place all these structures on the same anatomical section of a brain. The diagram does however provide a summary picture for visualising the structures.

hemispheres. In the centre of the diagram further midline structures lead upwards through the midbrain, diencephalon and septum to the two olfactory bulbs placed between the neocortices at the top. On each side, between the neocortex above and the cerebellum below, lie the palaeocortex and hippocampus medially, and the schizocortex and striatum laterally. Each of these structures is drawn in a cartoon manner that mimics their actual shapes.

The separations of insectivore species along a trend from terrestrial tenrecs and shrews to fossorial and aquatic insectivores (new axis 2) involved mainly differences in brain organisation related to expansion of the following variables: septohippocampal, mesencephalic and basal forebrain systems relative to the neocortex. This particular conjunction of brain parts seems to make sense in relation to the lifestyles of insectivores. For example, it could be related to brain functions

Table 1. Trend from terrestrial tenrecs and shrews through fossorial and aquatic species.

Brain organisation	Brain function	Species behaviour
Trend towards expansion of septohippocampal and mesotelencephalic systems and expansion of basal forebrain relative to neocortex	This combination of systems may mirror a trend towards a 'goal attainment' mechanism and elaboration of ability for spatiotemporal location	Trend from creatures that exploit two-dimensional niches to those capable of using distance information to exploit three-dimensional niches of increased complexity

involving increasing elaboration of abilities for spatiotemporal location and a goal-attainment system as one passes from the simpler two-dimensional lives of terrestrial tenrecs and shrews to the more complex three-dimensional lives of the aquatic and fossorial species (Table 1).

Likewise, the separations of bats along a trend from insectivorous bats at the one extreme to fruit bats at the other (new axis 3) involve mainly differences in brain organisation related to expansion of septohippocampal and limbic structures relative to medulla, and expansion of olfactory bulb and palaeocortex relative to the neocortex. These anatomical trends could be related to brain function differences subserving gradually increasing enhancement of both spatiotemporal memory and olfactory acuity. This, in turn, mirrors a sequence in animal behaviours from carnivorous bats taking small insects locally on the wing using echolocation, to frugivorous bats flying long distances to fruit-bearing trees and thus having increased need to remember spatial and temporal distributions of highly patchy, ephemeral and odoriferous food sources (Table 2).

Likewise again, the cluster of variables in the new axis 1 producing the trend across the primates (e.g. from bush babies to gibbons) involves a trend of increasing expansion of the neocortex, striatum, cerebellum and diencephalon relative to medulla. This particular combination could relate to the expansion of higher and higher levels

Table 2. Trend from insectivorous micro bats to large fruit bats.

Brain organisation	Brain function	Species behaviour
Trend towards expansion of septohippocampal limbic structures associated with expansion of olfactory bulb and palaeocortex	This combination of systems could be a trend towards enhancement of both spatiotemporal memory and olfactory acuity	Trend from creatures taking insects to large fruit-eaters with the need to remember spatial and temporal distribution of highly patchy, ephemeral and odoriferous foods

Table 3. Trend from hindlimb-dominated prosimians to forelimb-dominated monkeys and apes. Humans are placed far beyond the monkeys and apes.

Brain organisation	Brain function	Species behaviour
Expansion of neocortex, striatum, cerebellum and diencephalon relative to medulla	Expansion of the highest levels of the hierarchy of voluntary motor and sensory control	Trend towards increasingly complex capacity to strategically plan and control complex motor and sensory interactions

of voluntary sensory and motor control, a trend, in other words, towards greater and greater degrees of complex voluntary behaviour and a more complex capacity to strategically plan and control complex motor actions. Interestingly, this trend is continued furthest in the additional extension of humans from the other primates. Humans also have behavioural attributes that relate to the expansion of higher levels of sensorimotor control and increasing ability to strategically plan and control actions (Table 3).

It is worth recording that, if we had found these combinations of brain variables in axes relating to other orders, we would have known immediately that they made no neurobiological sense. For example, expansions of neocortex versus medulla, if found in the insectivore axis, would be functional nonsense. The same would have been true if we had found expansion of septohippocampal, limbic and olfactory structures in the primate axis.

In summary, then, the combinations of variables differ enormously among these three major orders. They make functional sense for each order in terms of their known lifestyle variations. Whatever may be the genetic and developmental constraints on the brain, they must be sufficiently relaxed as to produce the different divergences between and convergences within the respective Orders.

Phylogeny and Function in Primate Brains

Though the aforementioned studies are aimed for five orders of mammals, these data also allow us to look in detail at primates. The previous methods (principal component analysis) are designed to allow differences among the specimens to be revealed without any *a priori* estimation of what groups may be present in the data. However, there is a great deal of information about the primates that is independent of our study. As a result, we can test how clearly *a priori* groups exist within the data. This requires a statistical approach known as canonical variates analysis (Ashton *et al.*, 1957; Reyment *et al.*, 1984).

Some *a priori* groups are phylogenetic clusters. These are independent of our data being obtained from other anatomical and molecular data. Their presence (or otherwise) tests hypotheses about the relationship of brain organisation to phylogeny. Can we also test the reality of the lifestyle clusters? If these were solely obtained from the findings of our analyses, testing their reality in this way would be circular. But again, like the phylogenetic groups, there is a great deal of external information that indicates their reality.

Canonical variates analyses have another very useful property. The separations that are achieved are in terms of group standard deviation units. This allows degrees of similarity and difference to be compared across the study. It is true that in theory this requires that patterns of variances and covariances between and among the groups be generally similar. In practice however, in a very large number of investigations in which I have been involved, this is usually true enough (and tests showed this was the case here) as not to vitiate the analyses. The caveat here is that certain species seem often to defy that statistical requirement. Thus, among primates the orang utan is often far more variable than any other species; among bears it is the giant panda that stands out. The very existence of such 'grotesque' species may be pointing to something biologically important about them; perhaps, predation on them is very much relaxed.

Accordingly, then, canonical variates analysis was carried out upon the primate data with the species assigned to higher level phylogenetic groups. Of course, no specific system of phylogeny is totally fixed. There are always controversies at even the highest levels. However, compared with decisions at subspecific, specific and generic levels (about which there are often major differences of opinion, see Groves 2001), separations at family and superfamily levels are somewhat less controversial. Figure 8 shows the separations in the first two axes (encompassing, together, 67% of the total variance) that are achieved when the species are grouped into their families and superfamilies. The only separation at this level that is statistically significant is that between strepsirrhines and haplorrhines. This has been found by many other workers (e.g. Sacher, 1970; Holloway, 1979; Barton and Harvey, 2000; Clark *et al.*, 2001).

When we test the lifestyle associations, the differences are far more significant. Thus, Fig. 9 is the same plot as Fig. 8 but with polygons drawn around particular species groups. Thus, species such as bush-babies, tarsiers and dwarf lemurs with the indri and sifaka lie close together in the polygon on the right. Though grouped phylogenetically in different families, superfamilies and even infra-orders, these are all species with lifestyles dominated by powerful hindlimb leaping activities. They are completely separate from a second cluster of species lying in the polygon on the left: gibbons, spider and woolly spider monkeys, chimpanzees and orang utans. These species are also in different higher level phylogenetic groups: superfamilies and families yet share lifestyle elements that depend largely upon forelimb activities such as hanging, climbing and swinging.

Both of these lifestyle groups are very tightly constrained, very well separated from one another, and contrast markedly with the

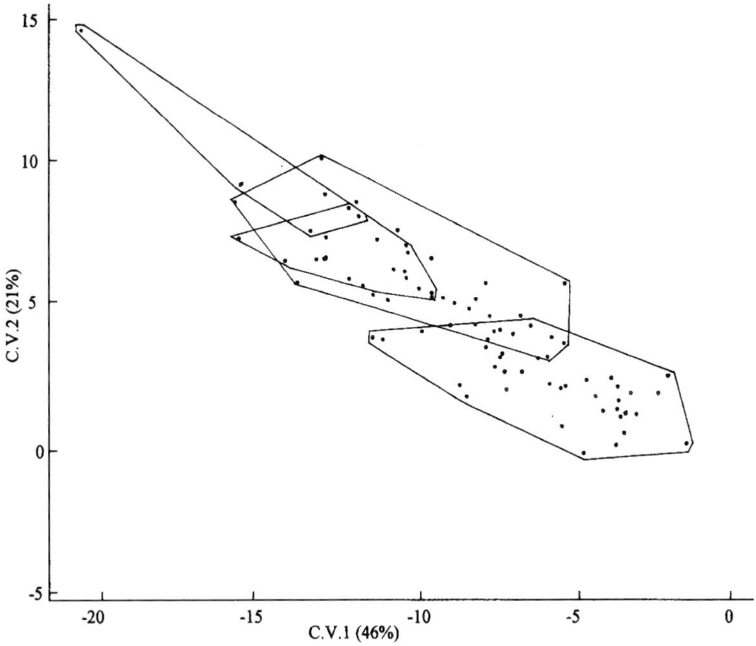

Fig. 8. A canonical variate plot of the first two axes of the brain data for primates. The many small dots are the locations of the means of each species. Polygons have been drawn around the major taxonomic groups. The polygon to the far left of the diagram contains all the prosimian (strepsirrhine) species. The large central polygon contains all the anthropoid (haplorrhine) species (except humans). These two polygons are statistically distinct. There is also a small polygon within the central large polygon that contains Old World monkeys alone. And there is a long thin polygon reaching up to the top left of the picture that contains only great apes and humans. These two are not statistically significantly different from the totality of apes and monkeys, though it is true that humans alone are a significantly long way from any other species (see later this chapter).

poorly defined, strongly overlapping phylogenetic groups of Fig. 8 (see also similar results from the restricted data, fewer variables, in apes alone, Sherwood *et al.*, 2004). These lifestyle groups are extremely tight, being less than 3 standard deviation units across. Of course, this is the same information that is hinted at by the earlier principal components analysis (de Winter, 2004) but because of the group-separating function of canonical variates, the separations are

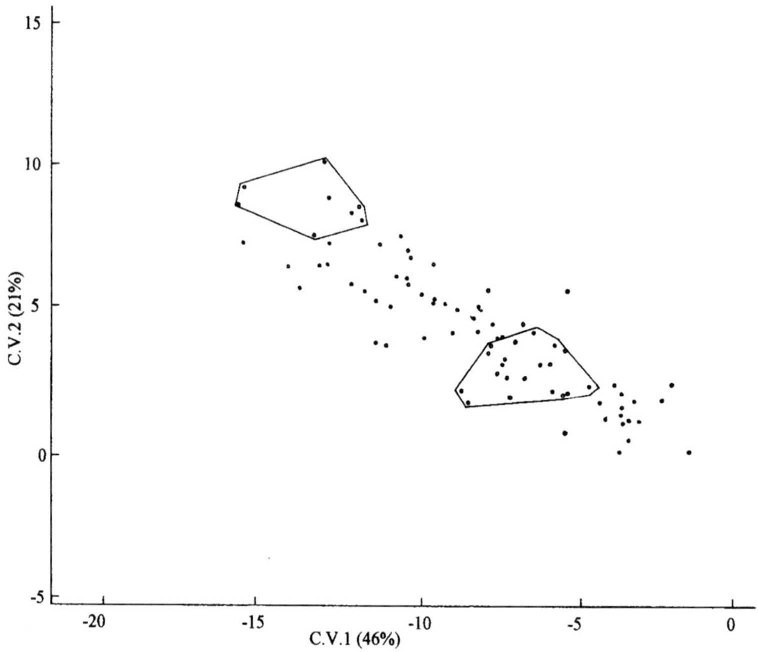

Fig. 9. This shows the species means in the same canonical variates plot as Fig. 8 but with polygons drawn around the lower-limb-dominated species (on the right) and the upper-limb-dominated species (on the left). The tightness of these two polygons shows just how similar is brain organisation in these functional groups of species.

much greater and their degree of separation can be visualised through the standard deviation unit scales of the axes.

Let us look finally at humans. Thus, the entire spread of the non-human primate species (the large polygon, Fig. 10) is about 14 standard deviation units. The spread of the Old World non-human primates (the smaller polygon, same figure) is about 8 units. On this scale (and in the same figure), we can see that humans (separate small circle) are 8 units away from the closest other primate, chimpanzees (small circle within the polygon of non-human species). That is, in these two axes alone, humans are separated from chimpanzees to a degree similar to that encompassing all Old World monkeys and apes. This is much greater than that

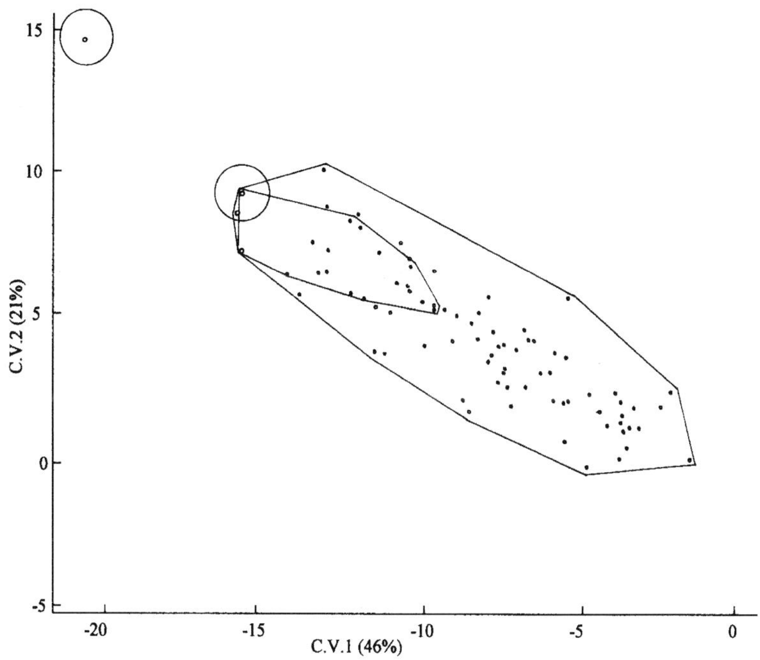

Fig. 10. Again, the same canonical variates plot as Fig. 8 but with the large polygon drawn around all non-human species and a smaller polygon around Old World monkeys and apes. There are circles around chimpanzees (the circle within both polygons) and humans (the circle on the far left). The scales on the axes are in standard deviation units and they show that humans are 8 standard deviation units separate from any other primate including chimpanzees. This difference is to be compared with the spread across all Old World monkeys and apes (also about 8 units) and across all non-human primates (14 standard deviation units).

might be expected given the oft-cited DNA similarity of humans and chimpanzees. Yet large though it is, it is only an extension of the direction of separation already existing within all non-human primates.

This is not, however, the whole picture. There are also three other statistically significant axes containing only relatively small separations: Axes 4, 5 and 6 with 4.4%, 3.6% and 2.3% of the variance, respectively. This adds up in total to only 10% of the overall variance among the primates. But this entire 10% is aimed at separating only one species, humans, from all others. This separation represents a further very large 22 units of separation of humans from everything else (including chimpanzees). This difference is independent of and additional to that in axes 1 and 2 (Fig. 11). The total difference of humans from chimpanzees is thus an enormous 30 standard deviation units. This implies an enormously different level of brain organisation. Let us now examine how this human difference is produced.

Brains in Chimpanzees and Humans

This huge brain difference between humans and chimpanzees leads to two questions. Which brain variables produce the 8 standard deviation units of separation between humans and chimpanzees in an extension of the trend common to all non-human primates? Which brain variables produce the new 22 units of difference that is unique to humans?

We already know the answer to the first of these questions. The pattern of brain variables separating humans from chimpanzees reflected in canonical variates 1 and 2 together is the same as that found in the principal components analysis (see above and de Winter and Oxnard, 2001). To repeat, it includes primarily increases of neocortex, striatum, cerebellum and diencephalon relative to medulla. These are increases in higher centres in relation to their links with the body through the medulla as gateway well as the medulla *per se*. This is diagrammed in Fig. 12.

The pattern of brain variables separating humans in the 4th, 5th, and 6th axes is quite different. It involves, first, decrease of septum associated with an independent increase of palaeocortex both relative to neocortex. It includes, second, increase of septum (but independent of the above), decrease of palaeocortex and increase of midbrain, all three relative to neocortex. It comprises, third, increases of septum and midbrain relative to neocortex as in the second (but that is independent of the second because they are in an additional axis, perhaps because they are not associated with a decrease in palaeocortex as in the second). These contributions are

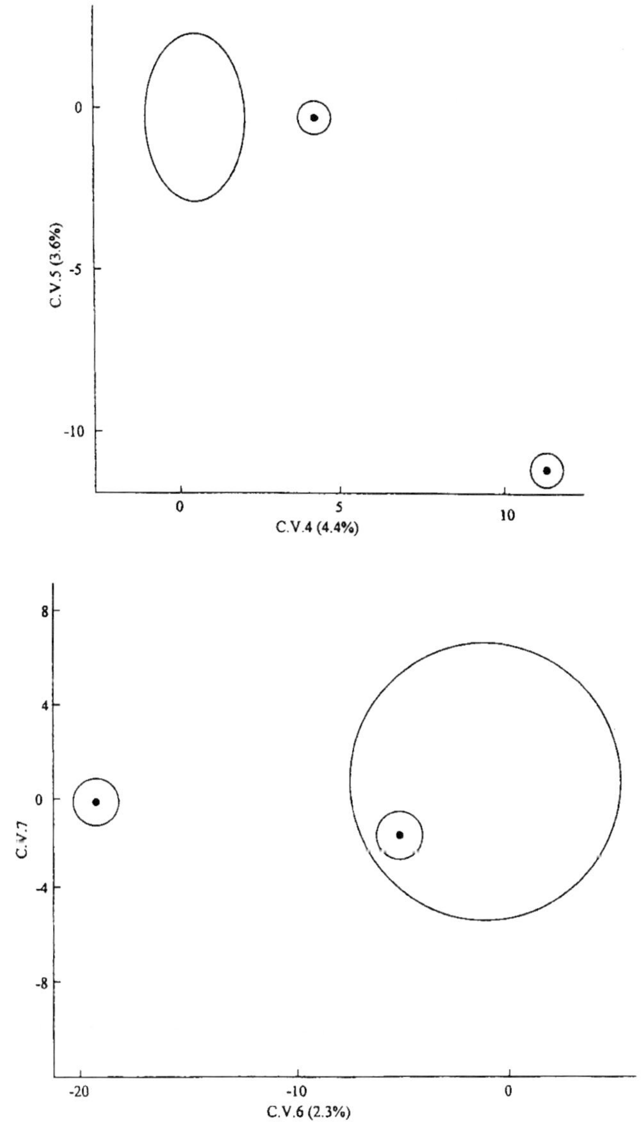

Fig. 11. Plots of higher canonical axes (4 against 5, and 6 against 7) from the same study as Fig. 8. The distant small circle in each plot indicates the position of humans. The large ellipse (upper plot) and large circle (lower plot) contain all other primates. The small circle close to the ellipse and within the circle, respectively, indicates the position of chimpanzees. Again, the scales are in standard deviation units with an additional 22 units of difference from chimpanzees giving a total of 30 units (axes 1 and 2) from chimpanzees.

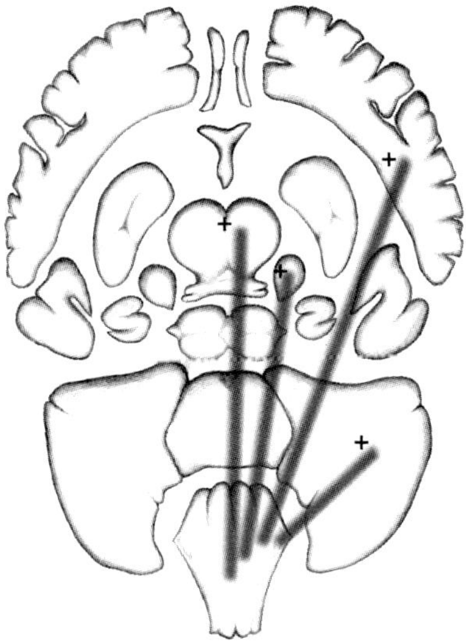

Fig. 12. This is the same diagram as Fig. 7 showing the various brain components but with lines joining those components that contribute most to the separations in canonical variates plots 1 and 2 in Fig. 10. That is, these are the variables contributing to the linear trend among all non-human primates and leading into the 8 standard deviation units extension of that same trend towards humans.

complex and not due simply to the neocortical size increase alone. They are shown in Fig. 13.

What one can clearly see in Fig. 13 is, first, that these various clusters of variables involve structures that all lie within the brain. None includes inputs or outputs through the medulla to the body. Second they all involve interactions (or loops or modules) between three or more brain parts. These include components of midbrain, diencephalon, septum, olfactory bulb, palaeocortex and neocortex. The interactions (or loops or modules) cross in a reverse manner because they involve septum and palaeocortex twice in opposite ways with neocortex, and septum alone with neocortex. Some make positive contributions, some negative. In addition, all these links are independent of the straightforward contributions to canonical variates 1 and 2 (Fig. 12). They are confined to humans!

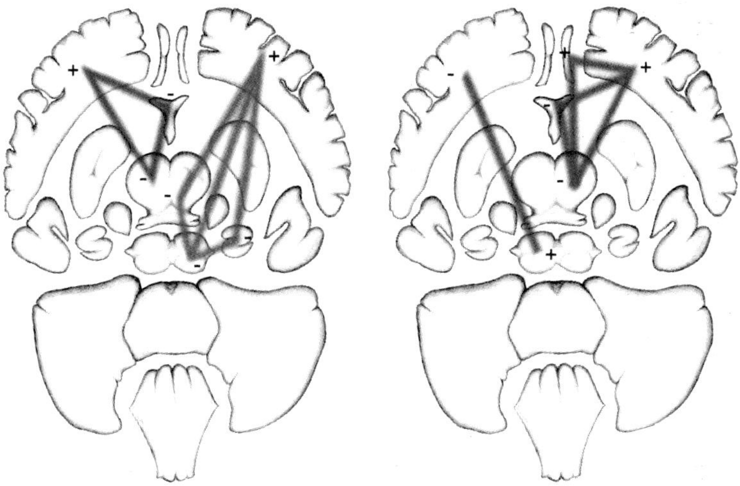

Fig. 13. Again, the same diagram as Fig. 7 showing the various brain components but with lines joining those components that contribute most to the separations of humans alone in canonical variates plots 4, 5, 6 and 7 in Fig. 11. Two frames are necessary because of the complexity of the links. In comparison with Fig. 12, the links are much more complex. They involve three links in looped form, and both positive and negative contributions. These arrangements are unique to humans and produce the additional 22 standard deviation units of difference from the chimpanzee.

Of course, there are additional possibilities. If there are new interactions *between* the neocortex and these other extra-neocortical components in humans, how many more new interactions may there be *within* different brain parts themselves, especially within that part that has enlarged most: the neocortex? Undoubtedly these exist. Indeed they may actually be far more important functionally than the extra-neocortical relationships we have found.

First, for example, the primary visual cortex contributes largely to the volume of the neocortex. This makes sense. Humans are very 'visual' creatures. Second, for example, there are human-specific organisations of the visual cortex associated with immunological reactivity in layer 4A (Preuss and Coleman, 2002). Preuss and Coleman believe this implies particularly that motion and luminance contrasts were modified in human evolution. Third, there are more than three-fold increases in prefrontal cortical volume compared to the general neocortex in humans (Schoenemann, 1999, 2001). Schoenemann thinks that this may relate to higher order cognitive tasks such as planning and strategy. These and other internal neocortical effects cannot be identified from our study because our data do not 'dissect' the neocortex. Internal neocortical interactions like these are undoubtedly present within the single neocortical volume that is the only neocortical measure that we have. They may well be responsible in part to the unique value for the whole neocortex that we see in humans.

What could our findings mean for function? At the level of brain structure they could relate to enormously increased complexity, as compared with chimpanzees, of the interactions of the neocortex with those other extra-neocortical brain parts. If only we had data that could also tell us what was happening within the neocortex! At the level of brain function, they could relate to any one or more of those abilities in which humans far outstrip other primates. These include increased level of communication, of abilities to process information, of capacity to plan and control complex strategies, and of increased propensity for abstract thought. Of course, this is little more than a wish list. We require further structural data and further understanding of function to define more sharply the implications of these differences between chimpanzees and humans. In addition we need more thinking and interpretation based upon the already known functional relationships within the brain.

Glial cell volume

It is also possible that these findings on brain part volumes have another, quite different, functional implication. The data analysed here all relate to the volumes of brain parts with the idea that it is neuronal bulk that contributes most. We can, however, ask the question: is there any other component of the brain that contributes largely to brain and brain part volume? The answer is, of course, yes. The volume of the brain and its parts is not only due to neurons and their processes, but also due to the volume of the various glial cells. This is a volume that occupies more of the brain in humans, as much as 40%, than in chimpanzees and the other apes (Sherwood *et al.*, 2006).

How important could this be? The view of glial cells that I was taught years ago was that they were primarily nerve-cell-supportive. This is almost certainly true. They are involved in such activities as: bringing nutrients from the blood to the nerve cells, maintaining appropriate ion balances in the nerve-cell microenvironment, protecting nerve cells against pathogens, and specifically, for oligodendroglia centrally and Schwann cells peripherally, forming and maintaining myelin sheaths of axons of nerve cells. However, some recent studies (e.g. Fields and Stevens, 2002; Fields, 2004) are beginning to imply additional new ideas about glial function.

Thus, glial cells appear to engage with nerve cells in a continuing two-way dialogue from the embryo to old age. This dialogue influences the formation of new synapses at well-used nerve cell contacts and the elimination of old unused ones. It thus helps determine which nerve–nerve cell contacts get stronger and which weaker. The glial cells also seem to communicate among themselves (through chemical rather than electrical mechanisms) as a separate but parallel communication network to the electrical nerve network. They talk with different local pools of nerve cell contacts over considerable distances. As a result, through them, the synaptic activity of one nerve cell pool can influence development of synapses in another without there being a direct electrical link between them. In effect, the glial cells act as a new kind of chemical-communicating mechanism both among themselves and between nerve cell populations.

It has long been known that there are relatively more glial cells per volume of brain tissue in larger as compared with smaller brains. When first discovered, this fact was laid at the feet of brain geometry. The density of nerve cell bodies decreases as a function of brain size. The extra space due to that is occupied by more and longer nerve cell processes. But cell size increases only slightly with body size. There must therefore be more glial cells per nerve cell in larger brains. At a time when the only functions of glial cells were supportive of nerve cells this did not seem to especially affect nerve cell function. However, if glial cells also function in changing nerve cell connectivity, then this is a different matter. Especially, if this effect continues throughout life it could have very important ramifications in aging brains.

However, even if this new view of glial function should turn out to be incorrect, the metabolic implications for brain function must still be extremely important. A new brain might be expected to require a new metabolic support (Sherwood *et al.*, 2006).

These glial possibilities are rendered even more interesting by new studies of the genes and molecular systems involved in brain size. Some of these are outlined in more detail below. But one specific gene form (SIGLEC 11 — a particular sialic acid binding receptor) is known to be unique to human brains (unique, that is, in being absent in chimpanzee, bonobo and orang utan brains). This factor is especially evident in human microglial cells (Hayakawa *et al.*, 2005). Could such a factor be responsible for producing the relative and absolute increase of glial cells in humans? Would this affect brain connectivity and function through adding and subtracting nerve cell links over physiological time? Would this, in turn, produce the flexibilities and changes that are now known, in humans, to extend into maturity and even old age? This could help explain how surprisingly well old brains can cope with some of the neurological insults of aging. More interesting for our purposes: could the relative and absolute increase in glial cells, through adding and subtracting nerve cell linkages over evolutionary time, produce the flexibilities and adaptations that are clearly greater in humans than in other primates? Could this account for part of our finding of the unique separation of human brains from other primate brains?

Is it possible that all this glial involvement in modifying, supporting and extending nerve cell connectivity be part of the reason why such crude data as relative brain part volumes are able to provide the rich lifestyle associations demonstrated here?

Brain volume overall

Let us return again to the question prefacing the discussion of glial cells. Are there any other components of the brain that contribute largely to brain and brain part volume? Congenital malformations of the brain may provide useful clues here. Mutated gene systems that cause brain reductions imply the possibility of other mutations that produce brain enlargement. These might be part of the business of growing a brain during development. Could such effects spillover into evolution? Certainly, within the last few years many genes have been mapped that seem to be involved in building a bigger brain.

There are, thus, at least eight genes that help create the very large human frontal neocortex. Deleterious mutations in these produce a smaller and thinner neocortex than normal, and the children with the condition walk slowly, and are inarticulate and clumsy. Other mutations in such a system might cause particular brain regions, like the frontal lobes, to enlarge during evolution.

Further, some families have microcephaly genes, and their offspring are born with tiny heads and tiny brains and sometimes suffer from mental retardation. Their brains look normal but are about the size of a chimpanzee or gorilla. Could other changes in microcephaly gene systems have been involved in the evolution of the large brain of humans?

Much more specifically, a particular gene (ASPM) is active in the proliferative regions where nerve cells which eventually migrate to other areas of the brain are born. One region or domain of this gene is repeated more often as brains grow larger. Thus, a simple worm has two copies of the region, a fruitfly 24, a mouse 61 and a human 74. One might think that the changes in the ASPM gene could be responsible for the very large human cortex. However, in this case, Geoffrey Woods' laboratory (Jackson *et al.*, 2002; Bond *et al.*, 2002) has shown that macaques, chimpanzees and gorillas, with smaller brains, have the same number of repeated ASPM genes as humans. It remains to be seen, however, whether these domains are equally as active in apes as in much larger brained humans.

More recent work on ASPM shows, however, that variants of it are specific regulators of brain size, and that their evolution in the lineage leading to *Homo sapiens* has been driven by very strong positive selection. This applies particularly to one genetic variant (haplogroup D haplotype 63) that may have arisen as recently as only 6000 years ago (Mekel-Bobrov *et al.*, 2005). Another variant, microcephalin (MCPH1) that also regulates brain size, seems to have evolved under very positive selection, and arose in the human lineage about 37,000 years ago. Undoubtedly, there have been others over the period since humans separated from chimpanzees. Perhaps, like the myosin heavy chain in terms of muscle reductions (the MYH 16 of a prior chapter), uniquely human molecules for brain increases started 2 million or more years ago.

It seems as though we are close to understanding these various processes. It may soon be possible to tell which genes, which molecules, which mechanisms are crucial to the development of brains. In particular, we are close to understanding the development of mammalian brains as distinct from vertebrate brains generally, of primate brains as distinct from mammalian brains generally, and, this will be critical, of human brains as distinct from other primate brains, even from those of our closest living relatives, the chimpanzees. We seem so close to the apes in body, so different in brain.

Skull volume: A red herring?

Again, let us ask are there any other components of the brain that contribute largely to brain and brain part volume? The question of brain expansion has been implied in studies of masticatory muscles (Stedman *et al.*, 2004). These workers have found that human masticatory muscles are considerably smaller than those of closely related primates. This seems to have been due to the inactivation in a gene complex (myosin heavy chain — MYH 16 — in humans but not in apes and monkeys; see an earlier chapter). Stedman *et al.* hypothesised that it was this reduction in these muscles uniquely in the human lineage that permitted increase in the human skull and therefore allowed brain increase through whatever mechanisms. It is certainly true that reduced musculature (especially the temporal muscle) can result in larger (and thinner) lateral dimensions of the skull. This would seem to be due to the mechanisms involving mechanical adaptation (see a prior chapter). It seems to me very unlikely that this alone would have produced an enlarged brain during evolution (though brain enlargement might have occurred coincidentally with other mechanisms).

Another question on volumes

All these ideas imply that the matter of the size of brains and brain parts is not just a crude measure. Perhaps, increase in size *per se* is a primer that permits various kinds of complexity to occur. Perhaps size and especially relative size are more important than we think. Brain size and brain part size, after years in the doldrums, may once again take a more prominent part in brain development and evolution. Certainly, the new surprise in these results is the evidence that they offer of major differences between humans and chimpanzees. How do these differences stand up in the light of generally recognised similarities in humans and chimpanzees in so many other anatomical parts? Before we enter that discussion however, it may be worthwhile touching on another.

Brains in 'Honorary' Primates

What are we to say about those animals that have brains that are much larger even than the 1.3 kg (approximately) of humans? Dolphins have brains about 2.5 kg. Elephants have even larger brains, up to 5.7 kg. Whales have enormous brains, up to 10 kg (Fig. 14).

Given the relationships of brain size and body size apparent in many studies (e.g. Roth, 2001) it would seem that these huge sizes of their brains are due to the huge sizes of their bodies. Indeed, when one looks at body/brain ratios, the very large sizes of the bodies of these two kinds of creatures does make their brains look relatively smaller. However, this is somewhat of red herring.

Whales, being aquatic and buoyed by the water, have very much larger body sizes than terrestrial creatures because this is permitted by the much reduced biomechanical constraints of their aquatic environment. When they are out of water, the largest whales cannot maintain their body form. Like a human brain out of its surrounding fluid on the pathologist's bench, the whale body gradually flattens like a blancmange because of huge internal body loads. The body weights of whales are, thus, somewhat irrelevant to the sizes of their brains.

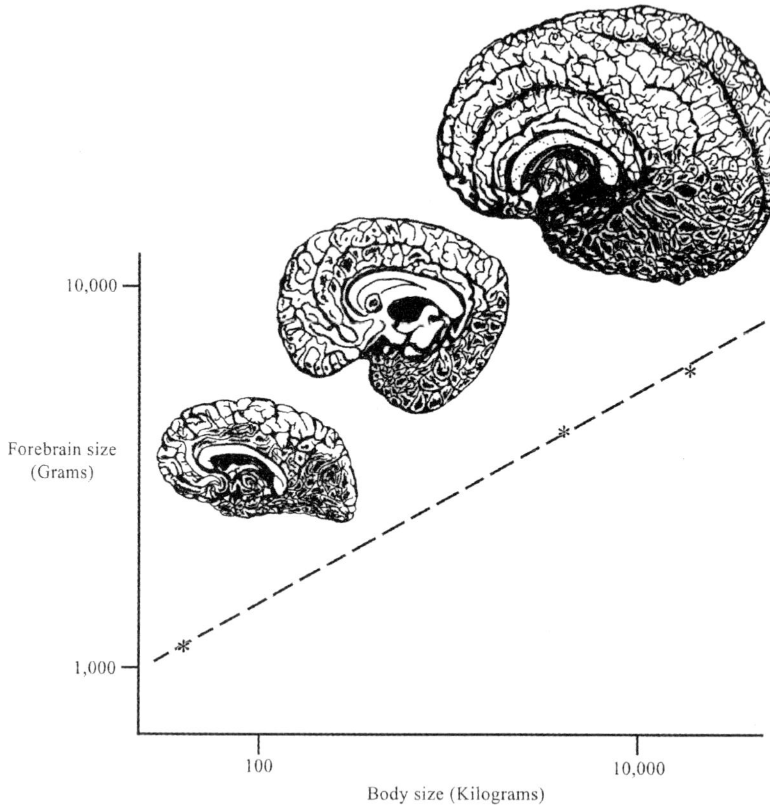

Fig. 14. Brains in 'honorary primates', humans, elephants and whales, and their relationships with body size.

Elephants also can be as large as they (and some ancient elephants were even larger than those of today) exactly are because they, too, have markedly different biomechanical constraints from most other mammals. Because of this they do not possess the marked bent limb stance of most terrestrial animals. If they did, then, with no other change (including no change in brain size) they would be much smaller than they are and the brain body ratio would be much bigger. Again, therefore, the body weights of elephants may also be less relevant to the sizes of their brains, though to a degree less than in whales.

In other words, a relationship between brain size and body size across species makes sense as long as there are no major differences that are related to factors (such as biomechanics) separate from brain size. Here we have two more cases of the 'tyranny of size'. It is much more likely that the absolute enormous sizes of elephant and whale brains (and other creatures like them), together with the internal relative sizes of different brain components, are causally related to the complexities of elephant and whale brain functions, and therefore to elephant and whale behaviours, each, respectively. This is not to say that body size has no effect; of course it does; in part it determines neurological inputs into brains. But these other factors may mean that body size has different and probably much lesser effects on brain sizes in these particular kinds of animals.

Of course, elephant brains do have major similarities to human brains (and to those of other primates). However, there are also major differences. These include: an unusually large and convoluted hippocampus, a very large and distinct temporal lobe for each cerebral hemisphere, and a very, very large cerebellum. These relative differences may be critical. For example, hippocampal size has been found to be extremely variable and poorly correlated with the sizes of other brain structures and with such external factors as lifespan. The genetic regulation of hippocampal nerve cell numbers is likely decoupled from the regulation of other brain structures. These increases are over and beyond those that might be expected simply on the basis of the increased size of the brain overall.

Although we do not have the specific information that would allow us to enter elephants into our statistical analyses of insectivores, bats and primates, we can, I think, be certain that the proportions of the aforementioned parts of their brains alone would place them in yet other statistical axes far outside of those studied here. The organisation of their brains is just totally different. Yet, it is true that there are other features within these brain regions that are similar across humans and elephants (e.g. the white matter to gray matter ratios are the same in elephants and humans for both the neocortex and cerebellum), that is, they are in keeping with the expected values for brains (not bodies) of this size.

All this may relate to a variety of unusual features of elephant lifestyles. Thus, they have remarkable abilities to handle information about sounds including infrasound and vibrations that are well outside (below) the human spectrum. With infraorbital (sensory to a trunk that can pick up a pea) and facial (motor to the very complex muscles of the flexible and powerful trunk) nerves of extreme size (one can almost put one's arm inside the infraorbital foramen of the largest elephant skulls), there must be enormous inputs and outputs in relation to cranial sensory and motor capacity of the trunk. They also seem to have very long-term social memories. The matriarch, entrusted with the survival of her multigenerational family group, seems to remember geographic information such as the location of seasonal water sources, and social information such as the calls not only of family members but of other familiar elephant groups. They seem to have long-term chemical memories of their individuals' urine, especially their mothers', over several decades. They seem to have some kind of understanding of death and to mourn dead companions. Hakeem *et al.* (2005) write 'it is tempting to posit a relationship between the elephant's long memory and large hippocampus'. Further study, however, is required to confirm such a suggestion.

Great care is needed here; it is easy to anthropomorphise elephantine abilities.

In contrast to the brains of both elephants and humans, whale brains, also of great size and complexity, are completely different. In particular, their neocortex is much thinner, much less highly laminated, and contains a much lower density of neocortical neurons (Roth, 2001). Again, however, though it is not known how whales evolved this very different neocortex, we can surmise that the separation from other mammals occurred some 60 million years ago or even more. We can also be fairly certain that this different neocortex was not a simple 'automatic' consequence of a change in absolute brain size. The toothed whales (dolphins) have smaller brain than elephants, and the great whales much larger.

Cetaceans have indeed evolved a very large absolute brain size, but possibly of much more significance are differences in relative proportions of brain regions from those of other mammals. As is also the case for elephants, but even more so, these relative increases are over and beyond those that might be expected simply on the basis of the increased size of the brain overall. In particular, it has been calculated that the prefrontal parts of the neocortex in elephants and whales are much larger in absolute terms even than the large (for a primate) prefrontal cortex of humans. 'What they do with this most massive "highest" brain centre remains a mystery so far' (Roth, 2001).

The lack of information on whale brains (due to the difficulties in obtaining, preparing and examining such large brains) must be set against much better information about what those brains are for. Whales are involved in a whole series of remarkable activities. Presumably the organisation of their brains relates to the recently discovered complexities of their learned 'songs', their clever fish corralling strategies, their long-distance navigating abilities, their highly specialised melon organs with enormously complex facial nerves supplying the facial musculature that deforms the melon into a complex acoustic 'lens', their highly sophisticated ranging and distribution patterns that depend heavily upon spatial memory skills, their complex communicative abilities given their highly complex social structures, and so on.

Undoubtedly in all three of the creatures, humans, elephants and whales, larger brains are not simply a consequence of larger body size. It is clear that elements of the organisation of the human brain are unique among primates. Likewise the organisations of elephant and whale brains seem unique among mammals. In each of these creatures, separately, their larger brains in their different formats evolved in relation to the animals navigating more easily in their complex social worlds, in each case differently from their closest evolutionary relatives (elephants versus hyraxes, whales versus hippos, humans versus chimpanzees). Primates, of course, elephants and whales are not (even if popular assumption wishes to draw correlates with the large brains of humans). But they must be doing something interesting and different with those enormous brains, just as humans are doing something interesting and different with theirs.

One of these days we will have much more data about brain parts in these and other mammals. In the light of what has been found just by looking at insectivores, bats and primates, we can expect major surprises. Today's science is unlikely to permit those studies to be carried out though it may well be that other far more powerful neurobiological techniques may eventually be applied to these creatures.

Special Implications for Chimpanzee and Human Relationships

The especial brain comparison for us remains that between chimpanzees and humans. Humans have brains more than four times bigger than chimpanzees. Despite this, the overall similarity of chimpanzee and human brains has been said to be very great.

This similarity is evident in the well-known and frequently cited DNA relationship. Humans share 98.6% of their DNA with chimpanzees. They also share 98.2% with gorillas, 98% with orang utans and 97% with gibbons. Recent studies of the mouse genome imply that humans and mice share 90% of their DNA. Even the banana is said to share 50% DNA with us!

Is it possible, therefore, that our minds are captured by DNA similarities that are expressed in the high 90s of percentages? When, however, we realise that among all modern humans DNA differences are of the order of only hundredths of 1% only, these 2% and 3% differences start to loom rather large. Is it further possible that most of the small DNA differences between humans and chimpanzees are concentrated in genes that relate specifically to the brain? Is it even possible that some elements of these differences are located in the so-called junk DNA which new research shows as being important in the regulation of brain production by switching brain coding genes on and off? These questions follow from the large standard deviation brain differences that we have uncovered here. Could the apparent 98.6% similarity of DNA hide much greater differences in other brain determinants: gene expressions, chromosome inversions, molecular switches, developmental processes, environmental effectors? What are the effects of differences between chimpanzees and humans in the 4% of non-coding DNA (junk DNA), or of the nearly 1000 inversion differences, or of the very large numbers of multiple copies of some genes (such as DUF1220), or even of specific differences of such gene structures known as HARs, human accelerated regions. Thus, HAR1 is active in foetal tissue in changing the human cortical structure.

Thus, Paabo (Normile, 2001) has shown the DNA of liver, kidney and blood molecular factors are markedly similar in humans and chimpanzees. They are remarkably different to the DNA of these tissues in rhesus monkeys. Paabo finds, however, that the DNA of brains is similar in chimpanzees and rhesus monkeys. Humans are the ones that are markedly different. This is exactly what we have found in our brain metrics; it is chimpanzees and rhesus monkeys that are similar, humans that are remarkably different. Paabo believes that his chimpanzee/human brain differences are so great that they imply that the human brain has evolved between four and six times the rate of the chimpanzee brain in the period since their separation. The brains of chimpanzees and rhesus monkeys, in contrast, have changed relatively little since their evolutionary separation (a much longer time ago).

Varki (also Normile, 2000) has also contributed to this discussion. He has shown that the human brain differs uniquely from that of all other primates because it is lacking a particular surface sugar on the brain cells. It would be interesting to know which fossil brains contained the sugar and which did not. Of course, fossil brains do not exist. However, this sugar, in addition to being absent in human brain, is also absent in human bone, again, alone among living primates. It turns out that the sugar is absent in Neanderthal bones. Could it also have been absent in Neanderthal brains? If this were so, it may also have been absent in the common ancestor of Neanderthals and modern humans. Given new DNA dates for that ancestor, this date could well be 600,000–900,000 years ago or even beyond. Tavare *et al.* (2002) even imply that it could be as far back as 1.5 million years. With that possibility, it is not impossible that this sugar may have also been absent in the bones (and therefore brains?) of older humans such as *Homo erectus* and other *Homo* species? Will the biochemical studies ever be able to tell us this?

Another new type of DNA study (Britten, 2002) is also starting to imply that the 98.6% similarity between humans and chimpanzees is truly not the whole story. Thus, in addition to the 1.4% divergence due to different base substitutions (98.6% similarity), Britten has found that there is an additional 3.4% difference due to the presence of insertion/deletion events (indels). This reduces the similarity of humans and chimpanzees to 95%. The difference could even be greater. Britten writes: 'One interesting observation is that sequence divergence between chimpanzee and human is quite large, in excess of 20% for a few regions' ... [however these are] ... 'complex processes, presumably involving repeated sequences and possible convergence events [that] will require detailed study to understand'. There is undoubtedly, a much bigger difference, but its meaning is yet to be explicated.

Even newer studies are starting to throw up many genes or gene products that, compared with apes, are unique to human systems. These seem to have involved several generations of increases at a number of different times, at least at 5000, 37,000 and 200,000 years ago and possibly even longer. These changes may, indeed, still be going on. For example, recent studies (Cheng *et al.*, 2005) imply that when compared to single-base-pair differences (at the well-known 1.2% figure), segmental duplication effects have a greater impact (2.7%) in altering the genomes in these two species. Further, Linardopolou *et al.* (2005) estimate that, in the studies of another genetic system, subtelomeric transfers, some 49% of differences in such sequences were generated after humans and chimpanzees diverged. It seems that shifting our attention from molecular similarities of chimpanzees and humans to molecular differences is likely to provide a whole new way of looking at the relationships between them and especially what has happened in the period since their common ancestry. All this may apply in spades to the brain.

Finally, evidence is available based on both morphological and molecular data relating to the supposed time of separation of humans and chimpanzees. This used to be placed (e.g. in the 1960s and 1970s) at about 3 million years ago. In the 1980s and 1990s it was thought to be 5 million years. More recently still the figure of 7 million years has been offered (Palmer, 2005). Yet even this is not the end; it may even be earlier. New fossils from Africa imply possibilities of 8, perhaps even 10 million years ago (e.g. Senut *et al.*, 2001,

Brunet *et al.*, 2002). New molecular studies involving a large array of mammals and using three independent sets of fossils for calibration suggest it could be as much as 13 million years ago (Arnason *et al.*, 2000). Modelling of evolutionary processes, though not giving specific times, suggests that common ancestors are likely to be at least twice or more times as old as implied by conventional morphological and molecular studies (Tavare *et al.*, 2002; Oxnard and Wessen, 2001; Wessen, 2002, 2005; and see another chapter). If there is any truth in any of these new and conceptually independent findings, there is an even longer time span for unique human differentiation. The evolution of the human brain may well be continuing, even today.

Evolution of the Human Brain: New Mechanisms?

The very large difference between human and chimpanzee brains and the increased time estimated since their separation leads to yet another question. Could it be that in humans, changes in brain organisation are partly due to a new form of evolutionary process? It would need to be the one that implies very fast brain change, great elaboration of function and behaviour, and totally new lifestyle possibilities.

The new changes in brain organisation may stem in part from elaborations of the brain attendant upon mechanisms for forming new connections not available to our closest living relatives, the great apes. Thus, it is well known, that increased levels of brain inputs and outputs increase the numbers of dendrites and synapses within the rat brain. This kind of change is available to all species. Special effects in humans, not evident in other primates, however, could have much greater effects upon the human brain.

Thus, the level of external inputs and outputs in developing humans is enormously greater than that in other species. For example, a caressing, singing and speaking, pregnant woman can have influences upon the baby in her womb as evidenced by changed patterns of movements of the late foetus to touch and sound. Equivalent maternal (and paternal) influences on the child once born, and for a long period after birth, and likewise the slightly further distant influences of grandparents, siblings and others close to the family, may also have important effects on brain development. These cannot easily occur, or occur to much lesser degrees, in other species, even the great apes. Such influences extend to all infant and childhood experiences, and therefore to 'education' of all types. This covers not just formal education as we generally recognise it, but the 'education' that stems from all influences after birth, even to an infant that cannot, at that point speak, but can certainly feel, see, hear and communicate in its own way.

Similarly, new changes in brain organisation may come through internal input/output effects on development, such as the molecular factors in the cascade of developmental processes that flow from genes during development. It is now well understood that such factors not only modify the instructions of direct genetic factors, but also, through upstream and downstream effects, actually change the genetic instructions themselves. Ideas such as these were proposed theoretically long ago by Waddington (and his epigenetic landscape, 1962, 1977). They have been re-introduced recently by authors such as Moore (*The Dependent Gene*, 2001) and Ridley (*Nature Via Nurture*, 2003).

These new input/output changes in brain organisation may even relate to the new findings about glial cells (e.g. Fields, 2004, and see earlier). These, through their much greater volume in humans, may be capable of determining great changes in individual brain cell connections, and therefore in the interactions between functionally related pools of brain cells.

Finally, it is not impossible that the soil for these changes lies in the removal of constraints against improved brain structure and increased brain function provided by the relatively new evolution in humans of omnivory and the removal therefore of limiting effects of reduced animal protein and vitamin B12 on the brain (see prior chapter).

The new findings described here, the 22 standard deviation units of brain organisation difference between humans and all other living

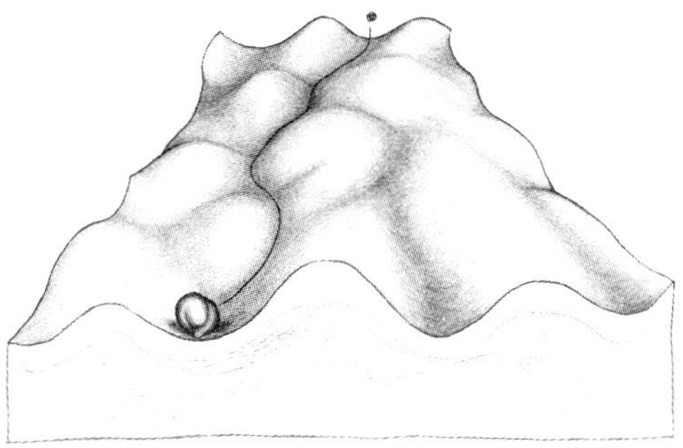

Fig. 15. Waddington's original idea: The adult produced by the genome and epigenome as a ball that has rolled from the genome (above) down a contoured hillside (epigenome) to the adult represented by its final position at the bottom of the (developmental) landscape.

primates, may, thus, also be related to mechanisms like these. Presumably such possibilities are available to non-human primates but do not occur in them because the various effects are much so smaller.

Let us put these concepts into pictures. Waddington's metaphor helps give clarity. Waddington described development from the zygote to the adult as the path that a ball takes rolling down a hill (Fig. 15). The original position of the ball near the top of the hill represents the beginning genetic materials of the zygote. The path the ball takes as it rolls down the hill describes the pathway of development and growth. This pathway depends upon both the starting position (beginning genetic information) and the contours of the hillside (the various downstream genetic and environmental factors, epigenetic factors, as Waddington called them). The final position of the ball at the bottom of the hill represents adult form and function. Of course, Waddington was mainly foreseeing the developmental effects mentioned earlier. However, his idea could well be expanded to include the environmental effects just discussed above. With his metaphor in mind, and this environmental extension of it, let us examine some new implications for the human brain and for evolution.

First, we will retreat from the full landscape as figured by Waddington and move to simpler diagrams that nevertheless

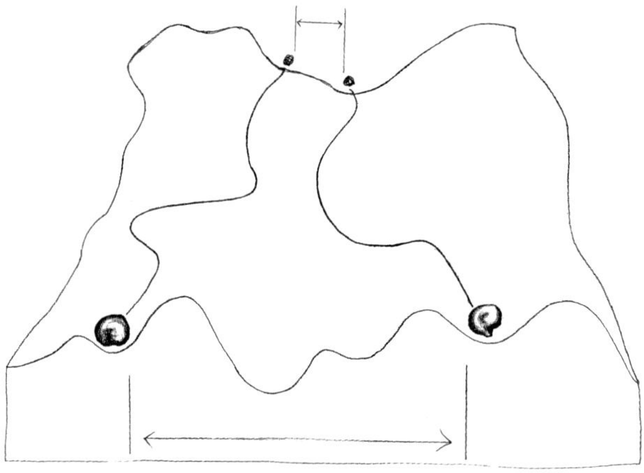

Fig. 16. A small difference in the genome (starting positions of the balls close together) entering different parts of the developmental landscape can produce large differences in final position (the adult).

incorporate all the same features. Figure 16 shows how small differences at the top of the hill (in the genome) together with intermediate differences in the developmental landscape can be expected to produce larger differences in position at the bottom of the hill (adult structure). These larger differences in the adult are what are selected in evolution, the result of both gene and epigene. At this point, our metaphor applies equally to other animals as to humans.

Let us next include the concept of population. Thus, Fig. 17 shows the situation where there are populations of genomes. It diagrams how small population differences in genomes (different starting positions of the ball) and small differences in the various epigenetic factors (differences in the contours of the hillside at different points) could produce considerably larger differences in populations of adults (larger differences in the final position of the balls). Again, this does not imply any fundamental difference between humans and other forms.

Let us now include the effect of changes in the developmental landscape that are due to those contours that are added by the organism because it can (because it is us: humans, upper frame

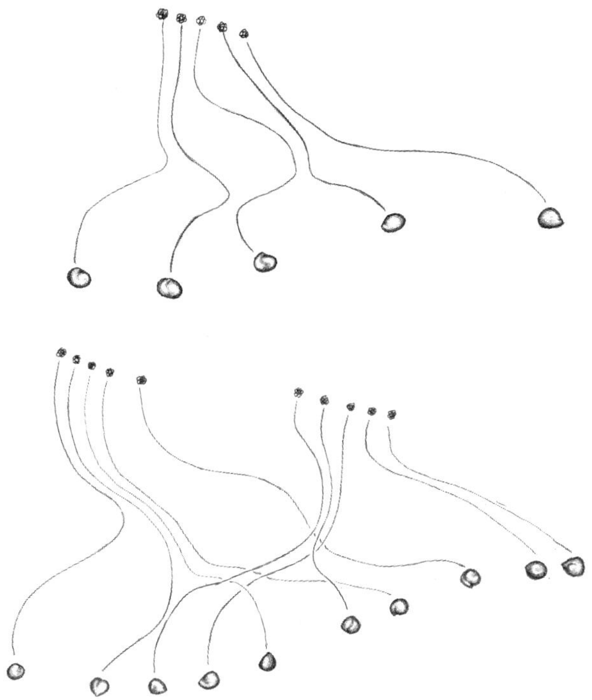

Fig. 17. A population of small genome differences rolling down the developmental landscape can produce large differences in the populations of adults.

of Fig. 18). Such changes might be those we described above: (a) the (now standard) effects of the interactions between the various internal upstream and downstream molecular factors and the genetic factors themselves; (b) the effects of external factors such as sensory inputs, education, etc. (if we think we are clever enough to know what aspects of education actually improve learning); (c) the complexity increasing effects of a much greater volume of glia (if that concept is eventually proven); (d) the removal of vitamin B12 and animal protein constraining limitations, but perhaps most of all: (e) the interactions that may occur between all these different influencing modalities. Figure 18 implies that the effect of all these factors might not only produce new forms for some adults but also eliminate some adults that would otherwise have been produced.

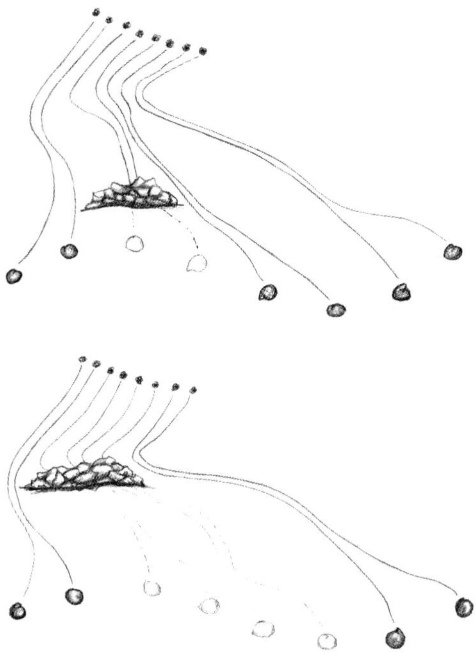

Fig. 18. What if a population of genome differences were influenced by an artificial interference, a barrier, in the developmental landscapes? In humans this could mean an educational barrier. Later education could result in a small separation in the population of adults (upper frame). Early education could result in a large separation in the population of adults (lower frame). Could this lead to the creation of a class and an underclass, and eventually, reproductive isolation between the two?

Could this last, if it were applied only to a portion of the population, give rise to a division, the creation of a class and an underclass?

Of course, the genomes remain the same, and so one would not expect this to produce evolution directly. It could result however in a well-known situation that might engender evolutionary change. That is, such a system might give rise to reproductive isolation and it is well known that this greatly speeds evolutionary separation. Given the normal breeding propensities of humans this would seem, on the face of it, to be unlikely. However, let us go even further.

That is, this artificial interference in the contours of the developmental landscape might be carried out earlier and earlier (higher up the hillside if you like as in the lower frame of Fig. 18). The resulting differences in the different adults might be greater and greater. The figure shows that it might result in a very much increased gap between adult groups, thus greatly increasing the chances of reproductive isolation. Earlier and earlier interpolation of education into the developmental process might do this. This is the reverse of some theories current at the present day (i.e. education starting later and later, often not until the ages of 6 or even 7 in parts of the United States). But, it fits with the educational systems of the past, e.g. education in Scotland started as early as 3–4 years old in my childhood. The Scottish teacher in my day was alarmed if the first task was to teach the child to read; it was expected that the child had learnt to read at its mother's knee; the teacher's first job was to teach the child to write! Even so, the ordinary forms of education with which we are all familiar, are unlikely to produce sexual isolation.

But a new form of education, the use of the computer and information technology, applied at increasingly young ages, to increasingly limited portions of the population, just might do this. Who amongst us has not been embarrassed by the 'teen, even the child, even the infant, who knows how to "work the machine" and who is scathing that nanny and grandad can't'. It is just possible that information technology may do something very interesting to the brains of the information generations of *Homo sapiens.* Would that we could cut up a few infants' brains to find out? Of course not! But the new non-invasive imaging methods might tell us.

Of course, this might be strongly negative. We already hear voices warning about the dangers of computers in the very young (especially obesity). Certainly *Homo nerdensis* would seem to be an evolutionary liability. Uncontrolled access to computers and related gadgets by young children might well be strongly deleterious.

Other changes, however, might be extremely powerfully positive, resulting in *Homo sapientissimus.* I doubt that *Homo nerdensis* would very often breed with non-*nerdensis* individuals; *nerdensis* seems so often to suffer from such major behavioural problems in getting on with its peers that breeding *in toto* might well be reduced or might even not occur (tongue in cheek here). But *Homo sapientissimus* may well seek out only its own! Already improvements in women's education seem to be associated with women tending to seek out only men who are educated at their own level or above. Are 'lesser' males being removed from the breeding pool of these 'super women'?

Are there really a number of new possibilities for the evolution of the human brain? Was Aldous Huxley (1946) correct in imagining his Alpha-intellectuals (given extra stimulation) and his Delta-minuses (today's progeny of the drinking pregnant mum)? Was Julian Huxley (1957) correct in envisioning a new taxonomic group, the Psychozoa, containing, at this point, only humans? Was H. G. Wells (both himself, 1895, and his alter ego, David Lake, 1981) percipient in envisioning the separate Eloi and Morlocks in our human future? Could *Homo* split?

References

Arnason U, Gullberg A, Schweizer BAS, Janke A, Molecular estimates of primate divergences and new hypotheses for primate dispersal and the origin of modern humans, *Hereditas* **133**: 217–228, 2000.

Ashton EH, Healy MJR, Lipton S, The descriptive use of discriminant functions in physical anthropology, *Proc Roy Soc Lond B* **175**: 73–105, 1957.

Barton RA, Harvey PH, Mosaic evolution of brain structures in mammals, *Nature* **405**: 1055–1058, 2000.

Bond J *et al.*, ASPM is a major determinant of cerebral cortical size, *Nature Genetics* **32**: 316–320, 2002.

Britten RJ, Divergence between samples of chimpanzee and human samples is 5% counting indels, *Proc Nat Acad Sci USA* **99**: 13633–13635, 2002.

Brown WM, Natural selection of mammalian brain components, *Trends in Ecol Evol* **16**: 471–473, 2001.

Brunet M *et al.*, A new hominid from the Upper Miocene of Chad, Central Africa, *Nature* **418**: 145–141, 2002.

Cheng Z *et al.*, A genome-wide comparison of recent chimpanzee and human segmental duplications, *Nature* **437**: 88–93, 2005.

Clark DA, Mitra PP, Wang SS-H, Scalable architecture in mammalian brains, *Nature* **411**: 189–193, 2001.

Cleveland WS, *Visualising Data*, Hobart Press, New Jersey, 1993.

Deacon TW, Rethinking mammalian brain evolution, *Am Zool* **30**: 629–705, 1990.

de Winter W, The bean bag genetic controversy: Towards a synthesis of apposing views of natural selection, *Biol Phil* **12**: 149–184, 1997a.

de Winter W, Perspectives on mammalian brain evolution, Doctoral thesis, University of Western Australia, 1997b.

de Winter W, Convergent evolution in brain 'shape' and locomotion in primates, in Anapol F, German RZ, Jablonski NG (eds.), *Shaping Primate Evolution: Form, Function and Behavior*, Cambridge University Press, Cambridge, 2004.

de Winter W, Oxnard CE, Evolutionary radiations and convergences in the structural organisation of mammalian brains, *Nature* **409**: 710–714, 2001.

Fields RD, The other half of the brain, *Scientific American* (April), 54–61, 2004.

Fields RD, Stevens B, New insights into neuron-glia communication, *Science* **298**: 556–562, 2002.

Finlay BL, Darlington RB, Linked regularities in the development and evolution of mammalian brains, *Science* **268**: 1578–1584, 1995.

Finlay BL, Darlington RB, Nicastro N, Developmental structure in brain evolution, *Behav Brain Sci* **24**: 263–308, 2001.

Groves C, *Primate Taxonomy*, Smithsonian Institution Press, USA, 2001.

Hakeem AY *et al.*, Brain of the African elephant (*Loxodonta Africana*): Neuroanatomy from magnetic resonance images, *Anat Rec* **287A**: 1117–1127, 2005.

Harvey PH, Krebs JR, Comparing brains, *Science* **249**: 140–146, 1990.

Hayakawa T *et al.*, A human-specific gene in microglia, *Science* **309**: 1693, 2005.

Hofman MA, Evolution and complexity of the human brain: Some organising principles, in Roth G, Wullimann MF (eds.), *Brain Evolution and Cognition*, Wiley, New York, 2001, pp. 501–521.

Holloway RL, Brain size, allometry, and reorganization: Towards a synthesis, in Hahn ME, Jensen C, Dudek BC (eds.), *Development and Evolution of Brain Size: Behavioral Implications*, Academic Press, New York, 1979, pp. 59–88.

Holloway RL, Does allometry mask important brain structure residuals relevant to species-specific behavioral evolution? *Behav Brain Sci* **24**: 286–287, 2001.

Huxley A, *Brave New World*, Vanguard Library Edition, 1952, Chatto and Windus, London, 1946.

Huxley J, *New Bottles for New Wine*, Harper, New York, 1957.

Jackson AP *et al.*, Identification of microcephalin, a protein implicated in determing the size of the human brain, *Am J Hum Genet* **71**: 136–142, 2002.

Jerison HJ, *Evolution of the Brain and Intelligence*, New York, Academic Press, 1973.

Jerison HJ, The evolution of neural and behavioral complexity, in Roth G, Wullimann MF (eds.), *Brain Evolution and Cognition*, Wiley, New York, 2001, pp. 523–553.

Lake DJ, *The Man Who Loved Morlocks*, Hyland House, Melbourne, 1981.

Martin RD, Human brain evolution in an ecological context (52nd James Arthur lecture on the evolution of the human brain), *Am Mus Nat His* New York, 1983.

Linardopoulou EV, Human subtelomeres are hot spots of interchromosomal recombination and segmental duplication, *Nature* **437**: 94–100, 2005.

Mekel-Brobov N *et al.*, Ongoing adaptive evolution of *ASPM*, a brain size determinant in *Homo sapiens*, *Science* **309**: 1720–1722, 2005.

Moore DS, *The Dependent Gene: The Fallacy of 'Nature vs. Nurture'*, Freeman, New York, 2002.

Normile D, Gene expression differs in human and chimp brains, *Science* **292**: 44–45, 2001.

Oxnard CE, *The Order of Man*, Hong Kong University Press, Hong Kong, 1983; Yale University Press, Newhaven, 1984.

Oxnard CE, Morphometrics of the primate skeleton and the functional and developmental underpinnings of species diversity, in O'Higgins P, Cohn MJ (eds.), *Development, Growth and Evolution*, Linnean Soc. London, Academic Press, San Diego, 2000.

Oxnard CE, Brain evolution: Mammals, primates, chimpanzees, and humans, *Int J Primatol* **25**: 1127–1158, 2004.

Oxnard CE, Crompton HW, Liberman SS, *Animal Lifestyles and Anatomies*, University of Washington Press, Seattle, 1990.

Oxnard CE, Wessen K, Modelling divergence, interbreeding and migration: Species evolution in a changing world, in Metcalfe I, Smith JMB, Morwood MJ, Davidson I (eds.), *Faunal and Floral Migrations and Evolution in SE Asia–Australasia*, Swets and Zeitlinger, Balkema, Lisse, 2001, pp. 373–385.

Palmer D, *Seven Million Years*, Weidenfeld and Nicholson, London, 2005.

Passingham RE, The brain and intelligence, *Brain Behav Evol* **11**: 1–15, 1975.

Preuss TM, Taking the measure of diversity: Comparative alternatives to the model-animal paradigm in cortical neuroscience, *Brain Behav Evol* **55**: 287–301, 2000.

Preuss TM, Coleman GQ, Human-specific organization of primary visual cortex: Alternating compartments on dense cat-301 and calbindin immunoreactivity in layer 4A, *Cerebral Cortex* **12**: 671–691, 2002.

Reyment RA, Blackith RE, Campbell NA, *Multivariate Morphometrics*, Academic Press, London, 1984.

Ridley M, *Nature Via Nurture*, Fourth Estate, London, 2003.

Roth G, The evolution of consciousness, in Roth G, Wulliman (eds.), *Brain Evolution and Cognition*, Wiley, New York, 2001, Chapter 19, pp. 583–555.

Roth G, Wulliman MF, *Brain Evolution and Cognition*, Wiley, New York, 2001.

Sacher GA, Allometric and factorial analysis of brain structure in insectivores and primates, in Noback CR, Montagna W (eds.), *The Prime Brain: Advences in Primatology*, Vol. 1, Appleton-Century-Crofts, New York, Educational Division, Meredith Corporation, 1970, pp 245–287.

Schoenemann PT, Relationships between prefrontal volume and behavior in normal human females, *Am J Phys Anthrop* (Suppl. 28), 244, 1999.

Schoenemann PT, Brain scaling, behavioral ability and human evolution, *Behav Brain Sci* **24**: 293–295, 2001.

Senut B, Pickford M, Gommery D, Mein P, Cheboi K, Coppens Y, First hominid from the Miocene (Lukeino formation, Kenya), *C. R. Acad. Sci. Paris* **332**: 137–144, 2001.

Sherwood CC *et al.*, Cortical orofacial motor representation on old world monkeys, great apes, and humans, *Brain Behav Evol* **63**: 61–81, 2004.

Sherwood CC *et al.*, Evolution of increased glia-neuron ratios in the human frontal cortex, *Proc Natl Acad Sci* **103**: 13606–13611, 2006.

Stedman HH *et al.*, Myosin gene mutation correlates with anatomical changes in the human lineage, *Nature* **428**: 415–418, 2004.

Stephan H, Andy OJ, Quantitative comparative neuroanatomy of primates: An attempt at a phylogenetic interpretation, *Ann N Y Acad Sci* **167**: 370–387, 1969.

Stephan H, Baron G, Frame H, Comparative brain research in mammals, Vol. 1, *Insectivores*, Springer, New York, 1991.

Tavare S, Marshall CR, Will O, Soligo C, Martin RD, Using the fossil record to estimate the age of the last common ancestor of extant primates, *Nature* **416**: 726–729, 2002.

Waddington CH, *Tools for Thought*, Basic Books, New York, 1977.

Wells HG, *The Time Machine: An Invention.* Heineman, London, 1895; Also in: *The Definitive Time Machine: A Critical Edition of H. G. Wells's Scientific Romance*, Geduld HM (ed.), Indiana University Press, Bloomington, 1987.

Wessen K, Simulating the origin and evolution of ancient and modern humans, Doctoral thesis, University of Western Australia, 2002.

Wessen K, *Simulating Human Origins and Evolution*, Cambridge University Press, Cambridge, 2005.

Chapter 11

The Wonder of Human Evolution

The various chapters in this book are merely the pebbles with which I have played on Newton's beach. They are just the musings of a single, not very well-known, not especially bright and certainly never highly quoted, researcher. However, even these few pebbles have had the capacity to make me wonder about the way human evolution is described by our discipline, and how it is presented by the media and in encyclopaedias to the general public, and in introductory texts to students.

The first pebbles tried to understand the idea of adaptation of structure to function, the relationship between how things are shaped and how they work. At first (in the 1950s) my view of structure was limited to old-fashioned cadaver dissection and bone measuring. My aim was to understand the soft tissue environment (muscles and fasciae) of hard tissues (bones and joints). My view of function started with literature descriptions of how animals move. My aim was to make armchair guesses about how these movements occurred. The only real information I had about how muscles and bones worked were those first electromyographic studies by workers such as Inman *et al.* (1944) during the Second World War. These were solely performed on human subjects. Applications to other primates were simply through anatomical inference.

Over the years, however, these studies were extended. Observation, measurement and simple statistical analysis eventually became multivariate statistics, geometric morphometrics, Fourier Transforms, image analysis and a host of other techniques. They greatly extended what we could learn about anatomical structures. Equally, armchair guesses about function led to techniques such as photoelasticity, strain gauge experiments, finite element analysis and more, to find out what happens during function. Natural history descriptions of animal lifestyles were supplemented by a host of techniques such as cinematography, cine-radiology and kinesiology (though I never did these things myself — younger colleagues did) to discover what really happens in nature. The results of all these studies showed so often that what we expect does not occur, that what we do not expect does occur, and in particular that apparently opposite (paradoxical) situations need explanation. All these provided strong biological information relevant to the assessment of bones in evolutionary investigations. They even gave information relevant to clinical problems.

A particularly fascinating feature in recent years has been how the arrangements of muscles, so clearly related to functional adaptation of bones as the motors of the body, have, as 'ghosts', gradually revealed new information about development and evolution. Of course, the new developmental biology helps to explain some of this anatomical information in the adult. In a surprising reverse manner, however, the new anatomical information in the adult provides ideas, indeed predictions, for testing by developmental biology.

Most fascinating to me, however, has been how, over the years, the complexity of animal form and architecture has yielded to new concepts. Moving from two dimensions to three is not straightforward; nor is moving from anatomical units to whole organisms; nor even, moving from functions to lifestyles. In each case, it seems to be information content that is the key. Two dimensions in bone seem to be the same as two dimensions in mechanics; three dimensions in bone are different from three dimensions in mechanics. Anatomical units are most easily understood in the light of their functions. But the coalescence of anatomical units into whole organisms takes us into a whole new world of information; function (though not lost) slides out of sight; evolution (present though previously hidden) stands revealed; the mechanism seems to be a reflection in adult structure of underlying environmental interactions, developmental processes and hereditary units. This is the reverse of developmental biology which shows how the interaction of hereditary units, developmental processes and environmental interactions produces whole organisms. It all depends upon the questions being asked.

These general ideas on evolution in primates, then led on to 'finding early ancestors', one critical part of the specifically human story. In recent years the number of species recognised as being in the human part of the evolutionary tree has grown so enormously as to provide an evolutionary bush. There are now (depending on how one names them) as many as 20 such species dating back over the past eight or even more millions of years. In contrast, there is only one specimen (and that named only 2 years ago (McBrearty and Jablonski, 2005)) that has been placed in the otherwise totally 'branchless' part of the chimpanzee 'bush'.

Is it not really rather likely that *Homo* as a genus has been around for very much longer than previously guessed? Is it not really rather likely that some of the wide range of fossils that are thought to lie in the evolutionary bush leading to modern humans actually lie in bushes leading to modern apes? Is it not surprising that none of the fossils lie in evolutionary bushes leading to the extinctions of nonhumans and nonapes? What's a bush for the goose is a bush for the gander!

Further, 'recognising recent ancestors' seems to assume yet other improbable events. Is it really the case that modern humans streamed out of Africa and replaced all prior humans elsewhere in the world? Is it not really rather likely that modern *Homo*, no doubt originally from Africa, of course, has emigrated several times into the rest of the world, and has, moreover, become intermixed with other older forms of *Homo* also in the rest of the world?

Finally, 'recognising close relatives' seems to assume yet other improbable factors. So many current views of the relationships between chimpanzees and humans seem concentrated on the degree of similarity between them. Is it not really more likely that the key is to look for differences? Changes in genes, in DNA, are probably the least important; they fit with that 98.6% DNA similarity figure. Changes in other aspects of development distal to the gene, in downstream developmental factors affecting how genes become expressed, in environmental factors likewise affecting those downstream factors, and above all, in changes in timing of expression, are all, surely, far greater keys to differences between us and our closest living relatives. This may be especially true of the human side of the story where these differences can be greatly magnified in a creature that can, uniquely, affect its own foetus, baby, child, youth, family, society, even possibly species (including species extinction), and so on.

When biological aspects of the body are examined, modern humans are indeed incredibly close to modern apes: same flesh, same blood, same bones. Most anatomy and most genes tell us this. There is not much difference between livers and kidneys in chimpanzees and humans; in fact, together, they both differ much more from livers and kidneys in rhesus monkeys. But there are enormous differences between brains in chimpanzees and humans. In brains, in contrast, the similarity is between chimpanzees and rhesus monkeys. It is humans that are different: different brain structure organisations, different brain functions, different brain capabilities, different thoughts (as far as we can tell).

It is true that many of the complexities of human brains, especially their great size and the great sizes of their major components, represent just an extension of a trend common to nonhuman primates. This trend leads from monkeys to apes to humans. In such features, even though somewhat distant from apes, humans are just an extension of the ape condition, and they in their turn just an extension of the monkey condition.

There are, however, a series of other features of human brains, features involving complexities between major brain parts that differ from all nonhuman primate brains, even including ape brains, in new and unsuspected ways. Quantitatively these differences are almost an order of magnitude greater from chimpanzees, than chimpanzees are from rhesus monkeys. In these regards humans seem to be unique. Yet, this finding is based only upon differences of interactions **between** major brain parts. There is no doubt in my mind that if such studies could allow full consideration of interactions **within** major brain parts, especially within the forebrain, the differences will be enormously greater still. Let us be immediately aware that this is not a return to an old, and ill-informed, idea: *pace* the hippocampus minor of a prior generation of scientists.

This brain uniqueness is supported by new 'evidence-based' data. Some of these data are developmental, involving at least four times as much difference in changes of human brain molecules, and possibly a great deal more, as compared with the relative stability of chimpanzee and rhesus monkey brain molecules. Some are anatomical differences in humans that are some 10 times greater between humans and chimpanzees than the differences between chimpanzees and rhesus monkeys.

The extended time and the extended brain allow for new possibilities. Julian Huxley's (1957) view of taxonomy nostalgically wished to return to the old 'grades' of Protozoa and Metazoa, but with a new grade added, the Psychozoa, containing only humans. He wanted to recognise a whole new grade, with whole new modes of evolution essentially unique to humans. The taxonomic argument is an old one and not important. But perhaps recognition of a new place and new mechanisms for humans is newly important.

When read superficially, this is not a popular view. Many workers in human evolution imply that those features of humans that are especially associated with us being human, for example, our aggression, our social structure, our creativity, even our mysticism, (there are many other such categories) are closely related to prior ape characteristics such as ape aggression, ape social structures, ape creativity and precursors to ape mysticism (whatever they might be).

Such human features are, therefore, usually portrayed, to use a geological metaphor, as a

> *very thin veneer, or slice or stratum of human behaviors superimposed upon a very thick multilayered bedrock of complex ape and prior primate, even prior mammal characteristics, and that this is due to very close evolutionary relationships, and a very short period of time.*

This seems to be the viewpoint of many who look to apes and other primates, even other animals, for the primary determinants of human behaviour.

If, however, the new time relationships of the fossils (humans and chimpanzees now possibly three or four times more distantly related than used to be thought), *if* new differences of human brains (human brains now molecularly and size-wise at least three or four times more distant from chimpanzees and, further, at least 10 times more distant in structural complexity) are correct, then new possibilities open up. *If* a brain with a beginning human-like volume (say 900–1100 ccs in *Homo erectus*) has existed for say, 2 or 3 million years rather than just several hundred thousand years, *if* bipedality has existed for possibly even 3 or 4 million years, *if* bipedality of some form or other (not necessarily at all like that of the genus *Homo*) figured on the evolutionary scene for as long as 5 or even 7 million years ago, *if* some kind of fossil and molecular transition from a common ancestor of apes and humans to an ancestor of humans alone really occurred at 8 or even 10 million years (all times permitted by different recent investigators and commentators), or even, *if* at least only some of these things occurred, *then* there will have been so much more time for the neural differentiation of human lineages from ape lineages.

Given that it is likely that brain evolution in humans may have been proceeding at greater rates than the more usual bodily evolution, it may well be that enormous changes have taken place in pre-human and human lineages in the preceding millions of years. It is especially likely, moreover, that the rate of this change, at first rather slow and therefore perhaps hard to discern, has been increasing by greater and greater degrees in shorter and shorter periods of time, becoming, therefore, more and more obvious, like growth by compound interest. This last 2 million years, more, this last 200,000 years, even more, this last 30,000 years, may have left not only our ape-like progenitors far, far behind, but even also our more recent human-like progenitors. It is even possible that this change is still occurring at speeds now so great that we may surprise ourselves.

Given all this, some might think a bit of a tall order, but not at all impossible following the thinking in some of these chapters, then, rather than being a *thin veneer upon a deep bedrock*, we may totally reword the metaphor and think of the cognitive capacities of humans and the brain supporting them as an

> *enormously thick complex neural and cognitive behavioural sludge overlying what is, in comparison, only a very thin sliver of a bedrock of prior ape behaviour.*

As we think of what it is that makes us human this idea does not seem so extreme. Although many investigators have tried to draw close analogies between human behaviour and ape behaviours, we must admit that, as common sense lay individuals, such efforts try our patience to the utmost.

Ape aggression may, indeed, be the initial determinant of human aggression. Let us look, however, at the enormous complexities of human aggression: nastiness of children to one another in the nursery and the playground, bullying in the family and in the workplace, road rage in car parks and on the freeway, family violence — the beating, not only of children, but of spouses and partners, of parents and grandparents, community violence such as robbery, assault and battery, rape, rape with violence, rape even with murder, murder as a crime of passion, murder in the act of committing some other crime, murder in cold blood (as we say), sadistic and sexually oriented murder, socially, politically, ethnically, nationally and internationally determined violence and killing — warfare itself, rape, torture, mayhem and murder as tools of warfare, as a 'normal' way of carrying out some of the business of our 'civilisation', torture and murder to obtain information both in war and 'peace', killing through hand to hand combat in war, indiscriminate killing through terrorism, maiming and killing using child soldiers, killing using suicidal or brain-washed volunteers, killing through long-range weapons, through pilotless machines and push-button techniques, wholesale massacre, attempted and successful genocide, even a vision of the death of the world. There is almost no limit to the aggressive ends of human behaviour; it seems most unlikely that this is a *mere thin veneer lying upon the thick bedrock of ape aggression.*

Likewise we may look upon the development of our societies: our human laws, from the pronouncements of the shamans, the dooms of the kings, the tablets of the priesthood and the customs of the tribes, through the laws of the ancients, the complex forms and fictions of, especially, ancient Greece and Rome, through the court of Torquemada, the star chamber of British constitutional history and through trial by ordeal, punishment by hanging, drawing and quartering, the use of *peine forte et dure* as a method of 'persuading' a person to plead (later not even a punishment but accepted by a debtor for the relief of the family though it lead, or course, inevitably to the death of the debtor), to the extreme complex legalities of modern business, civil, criminal, international, sea and space law, even our new 'laws of ethics'.

It seems highly unlikely that the thin layer of tentative personal interactions seen in ape social structures has much to do with these deep and complex developments in humans.

When, even further, we view the enormous complexities of human creativity, from the making and using of simple tools, through the design of implements that are not only useful but pleasing to the eye and the hand, through delight in our own home-made household articles, the potter's art, architecture, sculpture and painting, through the creations of those artists and scientists who have been among the geniuses of humankind, the art of Michelangelo, the science of Einstein, the performance, more ephemeral but none the less real of a Swan Lake ballet, the many other scientific artistic and humanistic creations of humans, when we look at all these things, it seems very far from a chimpanzee shaping a small twig to eat termites.

On such scores, precursor ape behaviours seem such a low base compared with the enormous heights of human creativity.

Even, finally, we may look at the ethical, moral, religious and mystical creations of the human mind, even some of the perversions of these things such as religious dogma, sectarian heresy, barbaric fundamentalism, torture, murder and terrorism in the name of 'churches', even astrology, Satanism, black magic, the modern cults, and so on, but it seems that we cannot even begin to see

> *in the thin precursor layers in apes, such thick and complex layers of activities, and thoughts and visions as are present in humans.*

An earlier generation of workers saw in these human characteristics developments that were nothing to do with apes and biological evolution, but were purely socially determined. A later generation of scientists saw in them a simple thin extension of the thick bedrock of genetically determined ape behaviours. Today many still emphasize the closeness of humans and apes, some even being willing to place them in the same genus! I agree that this last is a very important development when it is used to emphasize the preservation of species and of life. I support that as strongly as the rest of us.

That humanitarian ('animalitarian') viewpoint, may, however, just get in the way of understanding the science of what has really happened to humans in the recent past, what is still happening to humans today and what in fact may be continuing to happen to us in the future. Perhaps today we should really recognise a very complex situation, the result not just of 'genes' and 'environment', not just even of an interaction between genes and environment, but of the whole new cascades of many interacting molecular and environmental factors and timing. Perhaps we should look to an end, that is to say, of arguments about 'nature versus nurture', even about 'nature-via-nurture' and recognise a much further explication of a complex interactive 'nature vis-à-vis nurture' permitted by new evolutionary mechanisms that do not occur in other primates, that are unique to humans.

As a convinced (but modified) Popperian, all this is predicated upon my genuine desire to use my thoughts and work to test current ideas about human evolution, to attempt to show them wrong, to thereby suggest better ideas for testing in their turn. My efforts are, however, a very far cry from the stupidities of 'intelligent design' and 'creation science'. Pessimistically speaking, I cannot but expect that my continuing desire to test evolutionary ideas will be treated as support, as has been so much of my previous work, for these crazy fundamental beliefs. Optimistically speaking, I hope to see evolutionary studies to be able to progress through adoption of an investigative stance that wishes to test all evolutionary ideas, even those that are most cherished.

References

Inman VT, Saunders JBdeCM, Abbott RC, Observations on the function of the shoulder joint, *J Bone Joint Surg* **26**: 1–30, 1944.

McBrearty S, Jablonski NG, First fossil chimpanzee, *Nature* **437**: 105–108, 2005.

Huxley J, *New bottles for new wine*, Chatto and Windus, London, 1957.

Index